The Climate Crisis
and Other Animals

ANIMAL POLITICS

Danielle Celermajer, Rick De Vos, Katie Woolaston & Chloë Taylor, Series Editors

The Animal Politics series provides a forum for animal studies scholarship that is grounded in and expands political and critical theory. Our understanding of "politics" is expansive, embracing work across disciplines and scales, including but also reaching beyond institutional, cultural, and relational dimensions of politics. We are especially interested in the work of critical animal studies scholars that is intersectional in approach, or that puts considerations of animals as political subjects in conversation with critical race and ethnicity studies, anti-colonialism and Indigenous studies, gender and sexuality studies, feminist and queer theory, critical disability and mad studies, labour, and critical poverty studies.

The Climate Crisis and Other Animals

Richard Twine

SYDNEY UNIVERSITY PRESS

First published by Sydney University Press
© Richard Twine 2024
© Sydney University Press 2024

Sydney University Press
Gadigal Country
Fisher Library F03
University of Sydney NSW 2006
AUSTRALIA
sup.info@sydney.edu.au
sydneyuniversitypress.com.au

A catalogue record for this book is available from the National Library of Australia.

NATIONAL
LIBRARY
OF AUSTRALIA

ISBN 9781743329726 hardback
ISBN 9781743328996 paperback
ISBN 9781743329009 epub
ISBN 9781743329023 pdf

Cover image reproduced by permission of the Australian Broadcasting Corporation – Library Sales. Matt Roberts © 2020 ABC
Cover design: Miguel Yamin

We acknowledge the traditional owners of the lands on which Sydney University Press is located, the Gadigal people of the Eora Nation, and we pay our respects to the knowledge embedded forever within the Aboriginal Custodianship of Country.

Contents

Acknowledgements

I thank the following people who have helped this project in various ways. My colleague and co-director at the Centre for Human Animal Studies, Edge Hill University: Claire Parkinson, as well as former colleague Paula Arcari. Thank you to my present and former PhD researchers Cameron Dunnett, Donelle Gadenne, Abi Masefield and Katherine Jones for their ongoing conversations. Elsewhere in Europe, thanks to Jonna Håkansson and Helena Pedersen (Gothenburg); and to Elisa Aaltola, Saara Kupsala, Salla Tuomivaara and Kuura Irni (Helsinki) for invitations to speak when I could try out ideas that made their way into this book. In a similar vein, thanks to Núria Almiron and colleagues for hosting the European Association for Critical Animal Studies conference in Barcelona in 2019. Thank you to the interview participants from my vegan transition research. In the wider academic community, I must also thank Noel Cass, Amy Fitzgerald, Martin Hultman, Linda Kalof, Richie Nimmo, Paul Pulé, Jeff Sebo, Vasile Stănescu, Neil Stephens and Dinesh Wadiwel. Thanks also to Melissa Boyde, Danielle Celermajer, Rick De Vos and Fiona Probyn-Rapsey as (past and present) editors of the Animal Politics series, and Annabel Adair, Jo Lyons, Penny Mansley, Agata Mrva-Montoya, Susan Murray and Naomi van Groll at Sydney University Press. Finally, thank you to photographer Matt Roberts (and the ABC) for allowing use of the cover image.

This book was a long time in planning. For such a vital topic the balance between accessibility and faithfulness to scholarly debates is hopefully about right. Mitigating the worst effects of the climate crisis will take a monumental effort, but it is a struggle that can reconstruct practices and societies in systemically meaningful, beneficial and just ways. Radically reorganising our anthropocentric cultures is an inescapable part of this struggle.

List of abbreviations

BBSRC	Biotechnology and Biological Sciences Research Council
CAS	critical animal studies
C/AS	critical animal studies and animal studies
CLITRAVI	Liaison Centre for the Meat Processing Industry in the European Union
COP	Conference of the Parties
DEFRA	UK Department for Environment, Food and Rural Affairs
FAO	United Nations Food and Agricultural Organization
GBV	gender-based violence
GLEAM	Global Livestock Environmental Assessment Model
GWP	global warming potential
ILRI	International Livestock Research Institute
IPBES	Intergovernmental Science-Policy Platform on Biodiversity and Ecosystem Services
IPCC	Intergovernmental Panel on Climate Change

LEAP   Livestock Environmental Assessment and Performance
            Partnership

LGBTQI+   Lesbian, Gay, Bisexual, Transgender, Queer or
            Questioning, Intersex, additional gender identities and
            sexual orientations

NFU   National Farmers' Union (England and Wales)

NGO   non-governmental organisation

PPB   parts per billion

PPM   parts per million

UECBV   European Livestock and Meat Trades Union

UNEP   United Nations Environment Programme

Introduction

The aims of this book are to consider human–animal relations in the emergences[1] and effects of the climate crisis and to explore the prospects and strategies for real transformatory change of those relations as an important part of responding to the crisis. The tragedy of climate change has been revealed already in the largely helpless bewilderment of nonhuman animals in the aftermath of extreme weather events, seen starkly in the so-called Black Summer bushfires experienced in Australia in 2019–20 and captured on the cover of this book. Humans increasingly share this perplexed state as the climate crisis makes droughts, heatwaves, wildfires and floods more likely. Humanity can in theory know something about the emergence of the crisis and act to mitigate it. But lived experiences of what, in the worst-case scenario, will turn out to be a great unravelling of the conditions for earthly survival for many species are vital checks on the abstraction of the crisis and how it comes to be understood. The foregrounding of this malaise, a violent imposition of homelessness across species, can act to prevent abstraction from becoming depoliticisation.

---

1   I use the term "emergences" throughout to complicate simplistic narratives of causality and to highlight that sources of greenhouse gases are multiple, overlapping and complex.

Exposing depoliticisation is now a central contribution of the social sciences to the climate crisis, underlining that within its social construction there is much obfuscation and deflection. At first it seems compelling that the crisis is a practical and mathematical problem concerned with reducing the emissions of various gases. Yet to fetishise a single factor within the crisis – for example, "carbon dioxide" – is to risk denial of the inescapably political nature of the climate crisis.[2] At worst this means emissions take the place of understandings of why the climate crisis is happening and the relationships, practices and histories that produce the crisis are obscured. The argument here is not a refutation of data, modelling and statistics, by any means, but serves to underline that this book also aims to promote employment and integration of broad knowledges across the natural–social sciences divide to better understand the climate crisis. Moreover, this aim is contextualised by the recognition that the climate social sciences face an even greater struggle to be heard within policy than climate science narrowly conceived.

Data of all kinds are clearly vital to our understanding of the changing climate, but they must be integrated into sociological and historical understandings of relationships and practices. The following examples are specific and speak to recent decades. I was born in the 1970s, when the concentration of carbon dioxide (CO_2) in the atmosphere was around 328 parts per million (PPM); by June 2023, it was 424 PPM.[3] Another greenhouse gas, methane ($CH4$), rose in concentration from 1,625 parts per billion (PPB) in the early 1980s to 1,910 PPB by the end of 2021.[4] Furthermore, in 2021 global CO_2 levels were 60% higher than they were in 1990,[5] and the majority of CO_2 emissions since 1751 had been produced since 1990.[6] The eight years 2015–22 have globally been the hottest on record.[7] Such statistics speak to a stark failure in policy to mitigate the climate crisis, demonstrating

2   Swyngedouw 2010.
3   2 Degrees Institute 2022.
4   Lan, Thoning and Dlugokencky 2022.
5   Stoddard, Anderson et al. 2021.
6   Stainforth and Brzezinski 2020.
7   Carrington 2022.

that there has been no meaningful attempt to restructure those sectors – energy, transport and agriculture – that generate the majority of emissions. The first major international conference on climate change took place in 1979, yet knowledge of the climate emergency has accompanied its exacerbation.[8] As well as indicating the inadequacy of the response of incumbent institutional leaders, such statistics also emphasise the urgency of the situation, with the amount of "carbon budget" that can still be emitted diminishing all the time before targets to limit the global temperature rise to 1.5 or 2°C are missed. In the absence of ambitious legally binding agreements, 2°C will surely be breached, and the risk of runaway climate change[9] will increase alongside socio-ecological collapse. Indeed, the IPCC *Special Report on the Impacts of Global Warming of 1.5°C*, published by the Intergovernmental Panel on Climate Change (IPCC) in 2018, made clear that rapid, far-reaching and unprecedented systemic changes were required to limit warming to 1.5°C,[10] and the same applies for a 2°C target.

If ever it was time for novel approaches to address the climate crisis, that time is now. Part of this must be the elevation of perspectives from the arts, social sciences, (post)humanities and broader civil society. These are important contributors not only to the analysis of what many have called a classically "wicked problem",[11] a problem which is especially complex and obstinate, but also to alternative social imaginaries of how people and societies may live in such ways that promote the flourishing of the more-than-human world and pull the Earth back from the brink of climate chaos. Stoddard et al. define "social imaginaries" as "collective images of how we might live" and argue that they need to be based upon a radical departure from pre-existing norms and practices.[12] Sociology and the social sciences

---

8   The First World Climate Conference took place in February 1979 in Switzerland and was sponsored by the World Meteorological Organization.
9   "Runaway climate change" is generally understood to mean the crossing of tipping points that positively feeds back into further uncontrollable global heating.
10   IPCC 2018.
11   Mertens 2018.
12   Stoddard, Anderson et al. 2021, 675.

are apt for critical analyses of practices, relations, histories and norms, while philosophy has been reflecting upon environmental and animal ethics for centuries. These are just some examples of critical thinking fields that one generally does not find represented at the annual United Nations Framework Convention on Climate Change's Conference of the Parties (COP).

If the climate crisis on a deeper level is not then primarily, or only, a problem of emissions, but one of relationships, practices and norms, it is also, as these concepts presuppose, one of power. Again, this implies enrolment for social and political analyses. That relations of power are integral to understanding the emergences, obstinacy and effects of the climate crisis is also the position of such framings as "climate justice". However, this framing and others should rightly be seen as open to scrutiny from the aforementioned fields. Nevertheless, there is undoubtedly a consensus between the climate social sciences and a climate justice framing in conceptualising the climate crisis as a problem involving economic, generational, racialised and gendered relations of power. The politics of climate discourse are partly revelatory of struggles (and their marginalisation) to achieve prominence for such thinking on policy agendas.

This apparent symmetry between the climate social sciences and a climate justice perspective could certainly structure a book that explored the alternative social imaginaries made possible by their overlap. However, such a book could ignore how the social sciences and framings like "climate justice" that attain broader popularisation are also capable of producing a confused assessment of the crisis. While this book does not shy away from critiquing the omissions and historical exclusions of the climate social sciences, which can impede their ability to properly capture what exactly the climate crisis is a crisis of, it does aim to demonstrate their importance; there are clear strengths to be teased out. Social scientists are adept at questioning naturalised, taken-for-granted practices and meanings, ways of being hemmed in by tradition, and at catalysing imaginaries for experiments in living differently. They examine processes of social change, detailing how transitions can happen, and underline (contra the reduction of what people do to individualised psychology) the fact that high-carbon practices have emerged from complex socio-historical relations and infrastructure. They are also attuned to the analysis of dominant discourses and the effects of power in which such discourses are bound. This is particularly useful when applied to the climate crisis, enabling critical thinking about longstanding concepts such as sustainability, the Anthropocene or just transition. Critical thinking skills are similarly necessary for identifying when framings which delay meaningful action become dominant. The concept of net zero has been accused of this, for its allowing a largely business-as-usual approach and reliance upon unproven carbon capture technologies.[13] Another framing of relevance to this book is the often-total conflation of the climate crisis with fossil fuels, which may deprive food, agriculture and land-use changes of policy attention. Furthermore, it is important to know about analyses and proposed solutions that are deemed by many to be socially taboo and to understand why this is so[14] – for example, those which question normalised transport or energy- and food-related practices or critique capitalism as a suitable system to address the climate crisis.

This book contends that dominant framings of the climate crisis have been poor at identifying the place of human–animal relations in the emergences, effects and potential ways out of the climate crisis and that the critical knowledges of, for example, climate ethics and the sociology of climate change have not been especially inclusive of nonhuman animals.[15] I also write this book at the point in time when many societies are engaged in a political and cultural struggle over whether transformative changes to instrumental human–animal relations can form part of a response to the climate and biodiversity crises. Such relations of course concern how the development of a global animal-industrial complex dominates the food system and

---

13 Dyke, Watson and Knorr 2021.
14 Almiron 2020a; Gössling and Cohen 2014.
15 There are linguistic politics inherent in writing about human–animal relations due to the embedding of anthropocentrism. Consequently, this book often uses the admittedly imperfect terms "nonhuman animals" and "other animals" to contest a naturalised view of human–animal difference. Terms such as "livestock", which normalise the commodification of animals, are placed within quotation marks to signify a contestation of this normalisation. Homogenising nouns such as "fish", "cattle" and "sheep" are altered to "fishes", "cattle ungulates" and "sheep ungulates".

breeds nonhuman animals on an unprecedented scale.[16] By 2020 this had translated into the annual slaughter of over 73.1 billion farmed land animals.[17] According to 2019 data, of the habitable land on Earth (104 million square kilometres), almost 50% (51 million square kilometres) was used for agriculture. Of this, 78% (40 million square kilometres) was used to farm nonhuman animals, including land used for feed production, despite the fact that this land only contributes 18% to the global human calorie supply and 37% to the global human protein supply.[18] For context, 40 million square kilometres equates to more than four times the size of the United States: this area is devoted globally to animal agriculture. These statistics speak to a human food system, maladaptive for all life on Earth, in which the growth of a sector has been predicated upon capital(isation) rather than human health or the flourishing of other species. The aforementioned struggle to place human–animal relations on the climate agenda unsurprisingly involves the highly politicised debate over emissions; animal agriculture could be responsible for up to one in five of all greenhouse gas emissions. (I cover this debate in Chapter 4.)

Opponents frame this as an example of "trojan horse" politics, in which activists are trying to exploit the climate crisis for an "animal rights agenda".[19] However, this view can only be maintained if one denies the presence of human–animal relations in the emergences of, effects of and responses to the climate crisis. As I cover in Chapters 3 and 4, there is a broad scientific knowledge base pointing to both

---

16 This builds upon my earlier work on the animal-industrial complex (Twine 2012; 2013b). I define it in more detail in Chapter 2 and devote much of Part II of the book to enhancing its theorisation.

17 These data are from the Food and Agriculture Organization of the United Nations (FAO) and are collated in Orzechowski 2022. The figure is an underestimate, because it only includes the major farmed species of chickens, pigs, sheep ungulates and cattle ungulates. It excludes other species like goats, turkeys and rabbits, those that die before slaughter and the trillions of fishes and other aquatic life killed each year. For contrast, a total slaughter number of land animals for 2021 is given as 83.58 billion by *Our World in Data* 2023.

18 These are 2019 data from the FAO and are visualised in Ritchie 2019.

19 This was the position taken in 2021 by conservative politician Lord Deben (John Gummer), chairperson of the United Kingdom's Climate Change Committee (Davies 2021).

systemic climate change impacting other species and vegan eating practices being the lowest emitting. The climate crisis is unavoidably and demonstrably also a question of the ethics of human–animal relations. Denying this can be seen as protective not only of vast profit-making industries but also of an especially elevated human self-image vis-à-vis other species.

For this book, the questioning of this anthropocentrism is pivotal to tackling the climate crisis. Firstly, such questioning assists the understanding of the place of human–animal relations in the enormous remaking of the planet's land mass in recent centuries and the continued influence of the animal-industrial complex upon the climate, biodiversity and other environmental crises. Also, it makes clearer how broader systems of fossil capitalism, in energy and transport, for example, have developed without regard for the value of ecosystems and other species.[20]

To perform a critical analysis of anthropocentrism, this book turns to specific knowledges in the social sciences and humanities of recent decades which can enhance the climate (social) sciences and the frame of climate justice. I refer, in large part, to the so-called "animal turn" and the rise of animal studies, but more especially to that of critical animal studies[21] (CAS), as significant developments that have begun to contest the taken-for-granted, primarily human focus of disciplines such as sociology, literature, history, philosophy, geography, art, cultural studies and politics. The animal turn has involved building upon and innovating a broad range of ethical theory, and in my discipline of sociology, for example, has meant doing work to redress the conflation of society and the social world with the human, underlining the multispecies character of everyday life. Importantly, CAS and animal studies (C/AS) were prefigured by ecofeminism and other ecological theorising that questioned anthropocentrism. I have inhabited this space since the late 1990s, my work located broadly in this turn toward

---

20 The term "fossil capitalism" is used to highlight the historical importance of fossil fuel extraction to the development of capitalism (for example, see Altvater 2007). It is somewhat reductive, because it marginalises the role of agriculture and associated land-use changes in the development of capitalism.

21 For example, Best, Nocella et al. 2007; Taylor and Twine 2014.

animalising the social sciences, casting real doubt upon the sustainability of anthropocentric thinking. More specifically I have sought to theorise meat cultures, the animal-industrial complex and vegan transition and to work with intersectionality approaches interested in the specificities and overlaps between different relations of power, including human–animal relations. Nor are C/AS aloof from the natural sciences, with ethological knowledge (as well as lay knowledge of human–animal relations) important for comprehending the complexity and diversity of nonhuman animal subjectivities, something the long legacy of Cartesianism has held from view. This book is an exploration of what CAS can contribute to our understanding of the climate crisis but simultaneously an attempt to develop the field. CAS is especially suitable for contributing to analyses of the climate crisis, because it is the clearest field to question anthropocentrism, and for the three further reasons outlined below.[22]

Firstly, CAS advocates engaged forms of theory oriented both to lived experience and to radical social change.[23] The influence of ecofeminism on CAS is important, because that body of work when developing in the 1990s created a theoretical framework in which to understand connections between different forms of oppression beyond a humanist focus upon social class, gender and "race".[24] Imperative to this theoretical inclusion of the more-than-human were the critical and historical analyses of dualism in illustrating how discourses of nature, the body, emotionality and animality have operated across gender, class, "race" and species to position nonhuman animals and animalised humans as separate from and inferior to an image of the human predicated upon constructs of masculinity, (self-)control, civility and whiteness.[25] This brought human–animal relations into the orbit of

---

22 For a broader discussion about differences between animal studies and CAS, see Taylor and Twine 2014.
23 Taylor and Twine 2014, 6.
24 This is not to disparage other influences upon CAS, such as anarchist political philosophy, but reflects more my own personal trajectory, which began with a focus in the 1990s on ecofeminism and related bodies of work. An important goal for my book with Nik Taylor (Taylor and Twine 2014) was to recuperate the ecofeminist influence on critical animal work, which we felt had become largely lost in the early formulations of critical animal studies.

theorising class, gender and "race" but also made clear that the animalisation of nonhuman animals was itself based upon ideological constructions and generalisations of the category "animal". Ecofeminism and then CAS are historically significant because they question the claims around (and quests for) humanisation, which, for example, class-based, feminist, disability and civil rights movements have in some cases been based upon.[26] They also counter the way animalisation has worked through such oppressions to argue instead that nonhuman animals themselves should be liberated from human exploitation.

CAS broadens out the social imaginary of critical theory and the social sciences by making the case that nonhuman animals deserve justice for being caught up in a war of endless human capitalisation and for being exploited as a foil, via notions of animality, for a "human" elevated and deemed separate from "nature".[27] CAS also acts as a corrective to the relative silence on other species exhibited within and between the climate social sciences and climate justice perspectives, equipping these approaches to more radically question how discourses of the human naturalise a hierarchical relation over other animals.

There is a further important point to be made here about the uncritical use of justice discourses in ideas of climate justice and just transition. Ecofeminists have long critiqued the abstractness and dualistic aspects of principle-based ethics such as justice frameworks for their inattention to context, relationships, emotions and lived experience. Indeed, such inattention may make the exclusion of nonhumans more likely. In contesting this, ecofeminists have drawn upon the "ethics of care" tradition[28] and, more recently, a notion of

---

25 Plumwood 1993 analysed the gendered and racialised history of culture/ nature, reason/emotion and mind/body dualisms, which hyper-separate these domains and devalue the latter terms.

26 Essentially, they question whether it is liberatory to seek membership of the "human" when that has been historically constituted in dualistic and masculinist ways.

27 For an explicit application of the idea of justice to nonhuman animals, see Garner 2013, and for an elaboration of multispecies justice, see Celermajer, Chatterjee et al. 2020.

28 Donovan and Adams 2007.

"entangled empathy"[29] to better understand how other animals come to matter. This highlights a need, as Gruen argues, to move beyond reason/ emotion dualism, integrating abstract calls for justice with considerations of context, relations, care and empathy.[30]

If tackling the climate crisis means reassessing the human place in nature, overlapping knowledges such as CAS and ecofeminism appropriately question longstanding assumptions around what it means to be human and how such assumptions have acted to maintain a status quo of exploitation. Given their history, CAS and ecofeminism ought to espouse an intersectional understanding of the climate crisis[31] and one that does not repeat the anthropocentric mistake of excluding the more-than-human in its theorisation. This means framing the climate crisis as unfolding historically through a series of complex intersecting relations of power that are hierarchical along lines of social class, gender, species, colonialism and "race", and it is within *these relations* that the rising greenhouse gas emissions of recent centuries can be understood. My own positionality should be noted here, since I am a middle-aged, middle-class white man in the Global North, and it is my demographic that is disproportionately privileged and responsible for blocking change. Wholly mitigating this privilege is beyond the control of the individual, but my approach has been shaped by the empathy of the sociological imagination; by Connell's idea of exit politics, which refers to creative practices of refusing (male) privilege;[32] and by supporting the careers of others from a different positionality. Exiting from meat consumption over 30 years ago, then becoming vegan in 2005, and choosing the bike over the car are examples of how I have contested both anthropocentrism and carbon consumerism.

The second reason that CAS is a suitable perspective from which to contribute to climate analyses is its perspective on social movements. It refuses the societal (and sometimes academic) disparaging of activism. Activists are part of social movements and civil society, and these constitute, in a sociological sense, vital parts of societal reflexivity.

29   Gruen 2015.
30   Gruen 2015, 34–35.
31   See Kaijser and Kronsell 2014.
32   Connell 1995.

While social movements can be objects of critique for social science, they have also unmistakably shaped the history of academic knowledge production. In addition, it is problematic to denigrate the contribution of civil society from the perspective of democratic participation. Climate politics are blighted by forms of denialism and vested economic interests, and this only emphasises the importance of inclusion and open debate.

For CAS, scientists of all hues do better science when they are upfront about their positionality and their own social imaginaries. This is preferable to knowledge construction that either seems unaware of its own framings or tries to conceal its own interestedness. Furthermore, in this way of thinking, the identity of the scholar-activist has more integrity than that of the scholar who fails to change their practices in light of what they know – in this case, about the climate crisis. CAS is a body of knowledge and analyses that are anti-racist, pro-feminist, pro-animal and anti-capitalist, and this is reflected in a community of researchers committed to social change via practices such as veganism and involvement in social movements that reflect these values.

These politics are cross-cutting. For readers unfamiliar with CAS this does entail acclimatisation to a critical framing of animal exploitation as partly shaped by classed, gendered and racialised relations, as this book explores (especially in Chapters 1 and 2). For example, CAS is pro-feminist in order to address gendered societies and oppressive gender-based violence and disadvantage but also because of cultural hegemonies of masculinist values which denigrate emotional attachment to other species. CAS is anti-racist because racism is similarly unequivocally unjust. It understands conceptually that embedded discourses of animalisation have been used to justify white supremacy,[33] that colonialism was and is accompanied by large-scale exploitation of nonhuman animals and that in the era of recent globalisation, "development" has often been understood, in part, via the meatification of diets. The anti-capitalism of CAS centres on the way this economic system has facilitated and intensified the mass exploitation of other animals but is directed toward all those subjected

---

33   See Bennett 2020; Boisseron 2018; Montford and Taylor 2020.

to animalisation, including exploited precarious human workers directly involved in the slaughterhouses of the animal-industrial complex. Expropriation cuts across the species boundary. This take on capitalism has relevance for debates over sustainability and capitalism, getting to the heart of one of the central questions of the climate crisis: is capitalism a system that can meet the challenge of climate change, or does it need to be replaced?

Finally, CAS is a relevant perspective from which to approach the climate crisis because of the work it has done to generate and refine concepts that speak to the relations between humans and other species.[34] Having underlined an aim of foregrounding a critique of anthropocentrism, it is necessary to define this term, beyond a literal human-centredness. CAS philosopher Matthew Calarco has understood the concept in the following way:

> Anthropocentrism is the view that human beings (in opposition to animals and other nonhuman beings) are of supreme importance in ethical, political, legal, and existential matters. … Among the primary characteristics of anthropocentrism are: (1) a narcissistic focus on human exceptionalism; (2) a binary account of human–animal differences; (3) a strong moral hierarchy that ranks human beings over animals and other nonhuman beings; (4) a tendency to de- and subhumanize certain populations; and (5) institutions that aim to protect and give privilege to beings deemed fully human.[35]

This articulates how anthropocentrism incorporates dualism, creating a sharp divide between the human and all other animal species. When philosopher Jacques Derrida critiqued the word "animal" itself, as a form of conceptual violence in which a vast array of difference is homogenised, he had a point.[36] Anthropocentrism then is embedded in human languages and, as Calarco's fifth

---

34 This work has been done by those working across the distinction between animal studies and CAS.

35 Calarco 2021, 18. For an important earlier discussion of the concept, see Plumwood 1996.

characteristic highlights, is institutionalised in politics, the media, law, education and academic knowledge, constituent parts of the animal-industrial complex. Human/animal dualism confers an exaggerated difference between all humans and all other animals and interprets that difference hierarchically. Calarco also alludes to the operation of anthropocentrism and its associated human/animal binary in processes of dehumanisation, which I alluded to earlier. For CAS, the promise of transcending anthropocentrism and the cultural practices which embed it is found in the prefiguration of spaces that no longer exploit nonhuman animals or the myriad humans who are animalised.

Probyn-Rapsey is right to highlight anthropocentrism *within* C/AS which partly contests the view that it is possible to transcend in any straightforward way.[37] For example, in basing the moral considerability of other animals on similarity to the human, the human is centred as a yardstick, and often focus is upon rational and cognitive capacities, a point made by numerous C/AS researchers. Moreover, in making political claims for the value of nonhuman animal lives, humans inevitably speak for these beings and apply human notions such as exploitation, freedom or justice. Yet moral considerability can also be based upon shared emotionality, mortality and indeed difference, with the last emphasis often also being effective in underlining limitations in *human* capacities. The potential risks in speaking for other animals can be balanced by recognition of their capacities for communication and resistance,[38] and I do not think there is any great anthropomorphic mistake being made when it is assumed that other animals have an interest in their own freedom or when practices of animal exploitation are identified.

The Calarco definition also alludes to the related concept of human exceptionalism. Gruen defines this as "the view that we do not have

---

36 Derrida 2008. Consequently, it is an imperfect writing practice to continue to use the word "animal". This book attempts to counter this by underlining the diversity of animals affected by the climate and biodiversity crises (see Chapter 3).
37 Probyn-Rapsey 2018.
38 Hribal 2010; Meijer 2019.

ethical responsibilities to other animals",[39] a position shaped by the dualism of anthropocentrism. In reality, very few people adopt strong human exceptionalism; most have a selective, discriminatory perspective on animal ethics. This takes us to speciesism, a much-used concept in CAS dating back to the work of Richard Ryder, who evoked the idea to "describe the widespread discrimination that is practised by man [sic] against the other species", which he compared to racism in the sense of its denial of similitude and of the interests of others.[40] Subsequent theorists have offered similar understandings, such as Cary Wolfe's definition pointing to "systematic discrimination against an other, based solely on a generic characteristic – in this case, species",[41] which is useful for extending its meaning from an interpersonal prejudice to an ideology that emanates from a broader set of anthropocentric ontology and values. It is also useful for highlighting arbitrary (albeit historical, cultural and economic) inconsistencies around the human attribution of moral considerability, such as valorising dogs over pigs.

This book employs the concepts of anthropocentrism and human exceptionalism but uses speciesism more sparingly, because it is also a limited concept in the sense that it aggregates species and might imply questionable moral equivalences. Since its emergence in the 1970s there has also been debate about what it might imply about the interests of plants.[42] Unfortunately, abstract philosophy, in which much of the debate around speciesism takes place, can fall foul of the same modernist logic of anthropocentrism in failing to see interdependencies, and it is important for CAS not to falsely abstract nonhuman animals from their environments and ecological relationships, which consist of a multitude of animal–plant interactions. In understanding the extinction threat of the climate crisis to nonhuman animals it is also vital to understand these relationships and threats to plant species.

---

39 Gruen 2011, 2.
40 Ryder 1975, 16.
41 Wolfe 2003, 1.
42 For example, see Kagan 2016; P. Singer 2016.

A final useful CAS idea on which this book draws is Annie Potts' notion of meat culture. This refers to the centrality of meat in many cultures but specifically is concerned with the shared cultural representations, discourses, practices and beliefs about the consumption of other animals.[43] Potts draws upon Joy's notion of carnism,[44] which refers to the habituated belief system that naturalises animal consumption and underpins meat cultures. Whether they are called "carnism" or just conceptualised as part of the aforementioned anthropocentrism, the cultures of meat are an integral part of the wider animal-industrial complex, which in turn is a significant part of contemporary global capitalism. One notable impact of the climate crisis has been to fuel the questioning of meat cultures, with many countries presently living through a particular "fleischgeist", a "growing cultural trend of meat consciousness"[45] "fuelled simultaneously by ethical considerations and instrumental logic"[46] and reflected in cultural forms such as literature and film, and a media fascination with meat, but also with veganism and cultured meat. It has attained cultural omnipresence, visible in daily media and regional flare-ups, such as the 2022 contestation of Spanish meat culture.[47] This fleischgeist may be experienced as discomforting, because it brings to the fore violent aspects of the food system which have been partly repressed, treated as what I later refer to as a "cultural secret", a taboo that conflicts with cultural pretensions of civility. The fleischgeist can be read as a complex mixture of meat culture being contested *and* defended, a hegemony being exposed.[48] It is one example of the ways in which the climate crisis is inciting reflexivity to the meanings of the human, again a major concern of CAS, animal studies and ecofeminism, in their overlap with critical posthumanist work that similarly questions dominant discourses of the human.[49] These knowledges favour a

---

43 Potts 2016, 19.
44 Joy 2010.
45 Standen and Wizansky 2007.
46 H. Singer 2016, 184.
47 Burgen 2022.
48 Exposure and visibility are not sufficient for social change; indeed, one way meat culture deflects critique is through brazen visibility (Parry 2010).
49 Twine 2010b; Wolfe 2010.

rejection of fixed essentialisms of the human as a means to reimagine human being and to reject the normative anthropocentrism that has denied human ecological interdependency and partly shaped the emergences of the climate crisis. This book offers the direct undermining of anthropocentrism as an unavoidable and necessary strategy for effectively tackling the climate crisis, albeit one that must be exercised alongside broader, overlapping politics. Indeed, this book emphasises how many of the attempts to understand and tackle the climate crisis end up reproducing anthropocentrism.

Part I is concerned with establishing that human–animal relations deserve their place in analyses and accounts of the emergences, effects and potential ways out of the climate crisis. Chapter 1 connects CAS to broader social science work critical of the concept of the Anthropocene as a means to clarify the nature of the climate crisis. Exploring arguments that the Anthropocene mystifies the climate crisis by being ahistorical and apolitical and by reinforcing rather than questioning anthropocentrism, the chapter argues for a synergy between CAS and Jason Moore's work on the alternative concept, the Capitalocene.[50] This concept properly situates the climate crisis in a historicisation which importantly defines capitalism as extra-economic, delineating its emergence from the 15th century onwards as bound up in colonialism, the oppression of women and the exploitation of nonhuman animals. Moore's world-ecology approach to understanding capitalism is also valuable, as it avoids anthropocentric society/nature dualism by conceptualising capitalism as embroiled in the web of life, as productive of nature. The Capitalocene framework allows us to see the climate crisis not only as a crisis of capitalism but also as a crisis of patriarchy, colonialism and anthropocentrism, relations of power integral to capitalist accumulation strategies which have prefaced the rise in greenhouse gas emissions. The task for Chapter 2 is firstly to improve the Capitalocene framework by detailing the classed, racialised and gendered dimensions of the climate crisis and then, significantly, to begin to historicise the development of the animal-industrial complex and enhance the framework's explanation of human–animal relations as integral to capitalism and the climate crisis. The theoretical affinities

---

50   For example, see Moore 2017; 2018.

identified in the first two chapters contribute to both CAS and the Capitalocene framework but, more importantly, signpost a more accurate and meaningful characterisation of the climate crisis, affording a view of its constituent relations of power and pointing to the transformational change necessary to oppose them.

Chapter 3 demonstrates that climate change is already undermining life for a broad range of species and draws widely on conservation science to outline both the impacts of the Capitalocene and the precarity of life in the face of the warming that has already occurred. It is the necessary documentation of the lived experience of the climate crisis of individual animals across different species and spatial contexts. Attentive to ecological interdependency, the chapter is inclusive, for example, of a wide sample of microorganisms and insects as well as aquatic life, birds, reptiles, amphibians and mammals. The material inevitably incites reflexivity to the operations of the Capitalocene and the extinctions that will multiply without urgent change.

Chapter 4 is key firstly for detailing how the link between animal agriculture and climate change has been omitted or downplayed in such areas as social science, climate ethics, sustainability discourse, the media and non-governmental organisations (NGOs), and secondly for engaging closely with the politicised debate over the quantification of these emissions. Critical of the Food and Agriculture Organization of the United Nations' (FAO) work here and of its role in perpetuating an efficiency framing which prioritises protectionist policies of technical adjustment over larger transformations to the animal-based food system, the chapter presents a compelling evidence base in terms of both peer-reviewed research arguing for reductions to animal consumption and the demonstration of plant-based eating as a diet producing significantly lower emissions. A strong evidence base suggests the view that the climate crisis can be tackled by intentionally protecting the global food system from real change is a fantasy.[51] The chapter ends by

---

51   As a study by Clark, Domingo et al. (2020, 705) found, "even if fossil fuel emissions were eliminated immediately, emissions from the global food system alone would make it impossible to limit warming to 1.5°C and difficult even to realize the 2°C target. Thus, major changes in how food is

describing three scenarios of transformative change: plant-based transition, vegan transition and intersectional veganism, primed for further consideration in the second half of the book.

Part II is organised around the theme of transforming meat cultures and turns once more to a wide range of social science research in an attempt to understand what is involved in such a transformation.[52] Given that anthropocentrism is embedded within many cultures, the second half of the book begins with a focus on childhood in Chapter 5. Here the interest is in how meat cultures are secured through what I call a "generational universalism": the imposition of a set of meanings around animal consumption which construct it as habitual and normative for each new generation. Part of the aforementioned fleischgeist centres around childhood, controversies over meat-free Mondays in schools and vegan children and parenting. This is not surprising, since discourses of child development speak to our assumptions around human being. Following other CAS research that has taken the path into childhood studies, I argue that ideas of childhood innocence act to protect the reproduction of cultural anthropocentrism. Controversies over inadequate climate education imply that the crisis is being read, like the slaughterhouse, as taboo knowledge from which children should be protected, but the inherent issues of generational injustice that it amplifies weaken such taboos. The chapter reflects upon vegan climate activist Greta Thunberg as a killjoy of childhood innocence in this sense and points to CAS research on critical animal pedagogy as indispensable for creating post-anthropocentric cultures. Chapter 6 turns to specific social science theories of transition, many of which

---

produced are needed if we want to meet the goals of the Paris Agreement". Similarly, Hedenus, Wirsenius and Johansson (2014) concluded that reduced meat and dairy consumption is indispensable for limiting warming to a 2°C rise, and Ivanovich, Sun et al. (2023) found that global food consumption alone could add close to 1°C warming by 2100, with meat, egg and dairy production responsible for more than half (+0.5°C) of that.

52 Consequently, this book is more focused on mitigation than adaptation. This is not to downplay the necessity of adaptation policies inclusive of nonhuman animals.

have been used already either to explore meat culture or to theorise vegan transition. Drawing upon my own research employing a practice theory approach to vegan transition,[53] I argue that it affords specific advantages over other approaches. I then consider veganism in terms of what practice theory contends are the three elements of a practice: its associated competences, materialities and meanings.

Chapter 7 continues these themes but begins with a more in-depth examination of the meanings of veganism, returning to the three transition scenarios introduced at the end of Chapter 4, and reflects upon competing definitions of veganism. After summarising understandings of veganism and animal consumption, and possible interventions, the chapter proceeds to engage with key debates in practice theory approaches to transition, examining how they understand power and can work at larger scales. To deliver the co-benefits of transforming and dismantling the animal-industrial complex, transition theories also need to work at large scales. I draw upon work on scale and power to strengthen my own pre-existing conceptual understanding of the animal-industrial complex,[54] pointing toward a practice theory approach which better theorises the animal-industrial complex and is able to generate intervention strategies for its demise.

In Chapter 8, I return to the question of the Capitalocene and capitalism. I begin with a detailed exploration of Chapter 4's scenario 1, noting how plant-based transition, as a social imaginary, is already taking place within pre-existing structures of capitalism. I survey the emergence of this plant-based capitalism, raising doubts over its potential for effectiveness, even as it might take on the veneer of transformative success. I then examine the ways in which capitalist political economy (seen in such practices as lobbying and subsidisation) maintains the dominance of the animal-industrial complex, to question a naivety within plant-based capitalism and to extend the aims of the previous chapter in developing both practice theory and the CAS conceptualisation of the animal-industrial complex. After rejecting scenario 2, I turn to the third scenario,

---

53 Twine 2017; 2018.
54 Twine 2012.

intersectional veganism, also advocated in CAS, as the most promising imaginary, because of its attention to the broader complex of crises that underpin the climate crisis, but also because it has clear affinities with the broader climate justice movement *if* that movement can be convinced to incorporate more clearly counter-politics to anthropocentrism and reflexivity toward the limits of justice frameworks, as noted earlier. It is this sense of veganism that can constitute the most effective opposition to anthropocentrism and that better embodies a transformative ethico-political philosophy.

I turn to the work of Fraser and Jaeggi to locate and develop the CAS opposition to capitalism on functional, moral and ethical grounds and to illustrate how the animal-industrial complex is paradigmatic of capitalism in its prioritisation of capital accumulation as its overriding purpose, irrespective of the commodification of humans and other animals.[55] The remainder of the chapter draws upon a cluster of ideas, including prefiguration, to outline already existing examples of intersectional vegan practice which stress the need to overlap the food and climate justice movements and to advocate for the de-commodification not only of nonhuman animals but of the food system itself. In the short conclusion I summarise the contributions of the book and assess the prospects for tackling the climate emergency.

Other books broadly from within CAS and animal studies are also concerned with nonhuman animals and the climate crisis, including *Animal Crisis: A New Critical Theory*, by Alice Crary and Lori Gruen; *Food, Animals, and the Environment: An Ethical Approach*, by Christopher Schlottmann and Jeff Sebo; and *Saving Animals, Saving Ourselves: Why Animals Matter for Pandemics, Climate Change, and Other Catastrophes*, by Jeff Sebo.[56] I recommend them. They are written by philosophers, following very different formats from this book. This book differs not only in my background in the social sciences but in my diverse and detailed approach to the topic. This translates into a critical attentiveness to dominant narratives which have held sway in climate debates, drawing upon a wide multi-disciplinary range of (social) science research, including my own, contesting many framings

---

55 Fraser and Jaeggi 2018.
56 Crary and Gruen 2022; Schlottmann and Sebo 2019; Sebo 2022.

and opening up much-needed new ways of approaching the problem. While this book will be of interest to those across the environmental social sciences and humanities, it should also be read by climate scientists. Beyond this it is intended to be accessible to students and the general informed reader deeply concerned about the climate crisis. It is with a questioning of what sort of crisis the climate emergency is that Part I of the book begins.

Part I
The climate crisis and human–animal relations

1
Critical animal studies and the Capitalocene

What kind of crisis is the climate crisis?

The naming of a phenomenon as a crisis relates to the necessary processes of contesting dominant narratives, offering alternatives and prompting social action. The contemporary period is beset by major overlapping crises. Although this book asserts that the climate crisis is partly shaped by a crisis in human–animal relations, the first two chapters situate this within a broader context. For many species this crisis is being experienced now in myriad ways from drought to wildfires and loss of food sources as will be outlined in Chapter 3. It is unfortunate that the dominant epochal framing of the Anthropocene has made it harder to discern the broader context which impinges upon so many species. These first two chapters perform the necessary work of articulating a space for a critique of anthropocentrism as pivotal to understanding and tackling the climate crisis.

Nancy Fraser and colleagues refer to a "general crisis" and a "larger crisis complex" made up of the political, economic, social and ecological. One of their concerns is to chart how neoliberalism has maintained its hegemony by superficially co-opting progressive movements such as feminism, LGBTQI+ inclusion and anti-racism, and they cast capitalism in its globalising, neoliberal and financialised form as the underlying cause of this "general crisis".[1] Previously, Naomi Klein highlighted the connection between the neoliberal aversion to regulated, managed economies and the fatal policy inertia on the climate crisis.[2] In other words, Klein suggests that neoliberalism is precisely the wrong economic ideology for the climate crisis. Whether or when "crisis" turns into transformative change is another question entirely, as the history of Marxist thought has shown. Capitalist hegemony is enabled, in part, by a political crisis of democracy, which is culturally reduced to the formal democracy of elections increasingly prone to narrowed choices and sophisticated voter manipulation.[3] Despite corporate power, states still matter, and major governments are led by neoliberal parties consistently unlikely to act meaningfully on the climate crisis, including in key areas of diet and energy production, and especially unlikely to question demand. However, the most effective way to enable capitalist hegemony is to obfuscate the source of the climate crisis – capitalism itself. This chapter frames the Anthropocene concept as an ideological partner in this process of obfuscation.

Karl Marx was keenly interested in how the capitalist mode of production contained contradictions, such as the falling rate of profit, the concentration of capital and the expansion of the working class; he thought these would eventually lead to capitalism's demise.[4] Yet Marx was also aware of the tendency of capitalist production to undermine itself via the degradation of ecology,[5] a point which has been theorised by subsequent *eco*socialist writers.[6] The question posed by this chapter's

---

1   Fraser 2019, 7–8, 36, 37; Arruzza, Bhattacharya and Fraser 2019, 16.
2   Klein 2014.
3   Several writers also link this weakening of democracy to neoliberalism; see Chomsky 2017; Harvey 2005; Sandel 2012.
4   Marx 1981.
5   Marx 1976, 637–38.
6   For example, Foster 2000; 2009; Foster, York and Clark 2010; Moore 2011; 2015; O'Connor 1998; Rudy 2019. This includes reflection on whether the contradiction may be fatal for capitalism (see Harvey 2015, chap. 16). Certainly, capitalism attempts to commodify the climate crisis through, for example, carbon trading, technological innovation, carbon offsetting and the exploitation of "nature-based solutions" by companies (see "Carbon Offsetting is Not Warding off Environmental Collapse – It's Accelerating It"

title finds a partial answer here: the climate crisis is a crisis of capitalism, which is systemically undermining the viability of both its own reproduction and life itself on the planet. More complex has been the necessary endeavour of such ecosocialists to understand the interdependency of this ecological crisis tendency and the classic Marxist terrain of labour exploitation and the concentration of capital. My aims are simpler – namely, to begin to explore how facets of human–animal relations can be theorised within our understanding of the climate crisis. Marx's writings preceded the mass intensification of farmed animal production during the 20th century and the realisation that it contributes significantly to the climate crisis and so to the undermining of capitalist production. The argument here will be that the materialisation of anthropocentrism in the globalised practices of the animal-industrial complex has been partly, though significantly, constitutive of capitalism, its production of nature and the climate crisis. More precisely, this book contends that the climate crisis can be seen as a crisis of the anthropocentrism and human exceptionalism in our dominant human–animal relations – a framing which points toward particular responses explored in more detail in Part II.[7]

There is a danger in crisis-naming that we lose specificity or reproduce unhelpful separations. It is useful to define the ecological crisis as comprising overlapping crises of climate and biodiversity. Talk of climate crisis should be taken to include threats to biodiversity, since a changing climate is one of the leading contributors to species extinction, a topic I explore in Chapter 3. Further, it would be wrong to propose an ecological crisis as separate from the political, social and economic spheres. It is precisely such a tendency that perpetuates dualism, depoliticises climate breakdown and misunderstands the systemic nature of the problem. Fraser's notion of a crisis complex, which I return to below, is useful for bearing in mind the multiple

---

https://tinyurl.com/5n83a45a). Such measures might temporarily protect the status quo, but continuing failure to address emissions will inevitably further destabilise the climate.

7   This book uses "crisis" more broadly than Marx, specifically focused on how the climate crisis can be viewed as a nexus of intersecting crises of capitalist, patriarchal, colonialist and speciesist power relations.

overlaps between these spheres. It is important to note this larger crisis complex even if the scope of this work is inevitably smaller.

In this tricky terrain of crisis proclamations, I contend that overall human–animal relations are in crisis. Dominant norms of human exceptionalism are self-defeating, partly because they are predicated on an assumption of human abstraction from the rest of nature. The planetary scale of animal commodification and destruction intensifies a "war on animals",[8] undermining the very material basis of human existence. To name a crisis in human–animal relations is appropriate because it conveys the extinction stakes which many species face. This recuperates the *experiences* of nonhuman animals into our apprehension of crisis. Thus crisis-naming in terms of for whom a situation may be a crisis also needs to shed its anthropocentric baggage.

The first element of a critical animal studies (CAS) response to the climate crisis is necessarily a rebuke of the dominance of human exceptionalism, underlining it as a pivotal reason for climate breakdown. One path would be to try to clarify how this crisis in human–animal relations relates to the crisis complex inclusive of the broader political and social landscape. I do not fully go in this direction, because such a large project would take discussion away from a focus on climate. However, I will inevitably overlap with this since the economic sphere is best understood as extra-economic. Understanding how the climate crisis involves human–animal relations does necessitate a broader focus than might be initially imagined. From a CAS perspective, human–animal relations cannot be disentangled from their intersections in capitalist, gendered and colonial discourses and practices. To ignore these dimensions would be ahistorical.

The overarching aim of Part I of this book is to clarify my argument that a crisis in human–animal relations is integral to the climate crisis. This is potentially significant, not least because, as I argue later, it suggests specific pathways out of climate breakdown. The process of clarification can be achieved by interrogating the idea of the Anthropocene. The notion of a novel human-instigated geological epoch has come to dominate climate discourse in both academic and popular contexts. Within the social sciences and humanities this idea

---

8   Wadiwel 2015.

has received important criticisms that have not necessarily become mainstream. One of the drawbacks of the idea of the Anthropocene is that it artificially abstracts the ecological crisis from a broader constellation of crises. Drawing upon key criticisms of the Anthropocene allows this chapter to: firstly, clarify debates over the origins of the climate crisis in general; secondly, to consider how a strong candidate for an alternative concept – namely, the Capitalocene – avoids the key criticisms of the Anthropocene; thirdly, to show how this alternative better opens a space for including human–animal relations in climate discourse; and finally, to contend that CAS can both strengthen the Capitalocene concept and benefit from its theorisation of capitalism.

The Capitalocene, a concept most clearly identified with Jason Moore,[9] emerged in the 2010s as an important critical alternative to the Anthropocene by embedding the climate crisis in a longer history of capitalist, colonial and patriarchal development in the West. The Capitalocene avoids the narrower historical and political domestication of the climate crisis that is encouraged by the Anthropocene. It promises a more accurate framework for understanding the crises that underpin climate breakdown and signposts those relations of power which must be contested to address it. Although CAS researchers are closely interested in the climate crisis,[10] they have yet to engage significantly with the Capitalocene concept. It could, however, be a mutually beneficial engagement, because the Capitalocene strengthens the CAS understanding of capitalism, and CAS can extend the Capitalocene interest in a multispecies ontology. Indeed, it is noteworthy that both are influenced by ecofeminist thought ontologically and politically.

The most significant connection underlining this fruitful encounter is a theoretical overlap in Jason Moore's and Nancy Fraser's work which stresses the extra-economic dimensions of capitalism. As I outline later, this affords a broader (and more accurate, I contend) understanding of the history of both capitalism and the climate crisis,

---

9 For example, Moore 2016.
10 For example, Almiron and Zoppeddu 2015; Best 2009; Stănescu 2010; Twine 2010a.

which opens a space for better comprehending the role of human–animal relations in the latter. To date, Moore has appeared more interested than Fraser in acknowledging multispecies relations, but the analyses of both can be strengthened by its fuller inclusion. This can inform a richer understanding of the climate crisis wherein it emerges not just as a result of economic contradictions. While the ecological crisis, as Fraser contends, is connected to a larger crisis complex, it is further shaped, this chapter argues, by longstanding gendered, racialised *and* species hierarchies. In this CAS perspective, the climate crisis is a class issue, a gender issue, a "race" issue *and* a human–animal relations issue.

CAS is in broad agreement with the left's claim that the climate crisis must be addressed by breaking the inequality of capitalist class relations. However, it intends to highlight how hegemonic masculinity, white supremacy and human exceptionalism are further intersecting forms of power which must be rejected if the climate crisis is to be avoided or minimised. The effect of this claim is to redefine the climate crisis and hand leading roles to the social movements of feminism, anti-racism and animal advocacy as part of broad anti-capitalist politics in the struggle for climate justice. This constitutes a contemporary difference for the left in this advocated political formation: for the first time the left must become inclusive of politics of human–animal relations. Whereas in the past, movements for social justice concerned with gender, "race" and class have politically struggled for the right to be included in the realm of the "human", this must be conjoined to a reflexive reconceptualisation of the human.[11] Habituated conformity to a business-as-usual understanding of the human premised upon human exceptionalism does not address human alienation from the more-than-human and will do little to contest normalised human–animal relations, which are at the heart, directly and indirectly, of the climate crisis. In late 2018, UK journalist George Monbiot suggested changes to the dominant discourse about climate change which conformed well to the aforementioned academic turn to talking of crisis. Some of his suggestions concerned nonhuman animals (for example, saying "fish populations" rather than "fish stocks"), but the

---

11   Plumwood 1992.

most widely reported recommendation was his call to start referring to "climate change" as "climate breakdown".[12] This was underlined by a separate opinion piece in *The Guardian* newspaper in 2019 calling for the retiring of "climate change" in favour of "climate crisis".[13] The same year, *The Guardian* updated its style guide accordingly, to reflect a reconstruction of the problem as both serious and urgent.[14] In the United Kingdom there have been concerted attempts to force institutions, local councils and political parties to declare a "climate emergency". Indeed, over 300 district, unitary and metropolitan borough councils had done so by 2022.[15] A concern is that councils may not fully comprehend the scale and speed of change required to achieve meaningful mitigation and adaptation or lack the resources to do so. The danger then is that declarations become performative gestures. Moreover, action taken will inevitably be shaped by understandings of how climate breakdown has emerged. Such developments may indicate a growing consensus of a crisis, but again, the nature of that crisis is not exactly clear.

The Anthropocene

The Anthropocene is an inaccurate concept for the contemporary period; however, discourse surrounding it helps to clarify the nature, emergences and causes of the breaking climate. For example, the aforementioned ecosocialist writers view climate breakdown as a fundamental crisis of capitalism – a narrative the Anthropocene concept struggles to accommodate. It is highly challenging to consider how the climate could be stabilised without addressing the inequalities, consumerism and growth imperatives of global capitalism. However, in this and the subsequent chapter, I go further than this and argue from a CAS perspective that climate breakdown is the result of *multiple*

---

12   *Deeply Good Magazine* 2018.
13   Redlener, Jenkins and Redlener 2019.
14   Carrington 2019.
15   https://tinyurl.com/bde4xu86.

intersecting crises and contradictions, including a crisis in human–animal relations.

To begin, it is necessary to go back to the problematic roots of the Anthropocene concept. My argument implies counterintuitively that the lofty status of climate scientists (within the environmental movement, that is – their low status in right-wing media notwithstanding) is itself an ideological hierarchy. This helps to background a whole swathe of historical and social scientific knowledge pertinent to the endeavour of gaining meaningful explanation of the climate crisis. This epistemological tension plays out exactly in the original naming of an "Anthropocene" and its subsequent critique from the social sciences and humanities. Concealing this critique from wider public dissemination in the cultural normalisation of the Anthropocene concept encourages a misunderstood framing of the climate crisis as a mathematical problem about various emissions. To be clear, most work by climate scientists is essential. This might read as if I am reproducing a quite traditional distinction between the natural sciences and the social sciences and humanities; this is not the intention. Rather, I am advocating a somewhat academically mainstream view that climate breakdown can only be understood from an interdisciplinary perspective which acknowledges degrees of disciplinary difference. The empirical measuring of climate changes and attempts at future forecasting are indispensable and become part of the political and social movement narratives calling for action. However, the prioritising of specific forms of knowledge takes place within a context of cultural scientism which takes the authority of the natural sciences for granted.

Contesting the dominant reading of climate breakdown as a mathematical problem about various emissions and instead arguing that it is a problem that has *unfolded historically through a series of complex, intersecting relations of power* makes it clear that climate scientists need help from other disciplines when it comes to understanding and tackling the crisis. Also, once these other perspectives are included, our understanding of what the crisis is, and what the Anthropocene might really be, inevitably shifts. This alternative framing has the added advantage of helping climate (social) scientists anticipate some of the inequitable consequences and effects of a breaking climate. It is unsurprising sociologically that the instability of the climate is increasing within the context of a decidedly unlevel social playing field. Outcomes are already shaped by the social divisions of class, gender, "race" and species. The obligation to understand these further necessitates the inclusion of the social sciences within climate policy.[16]

The Anthropocene originated as a natural science concept, but it was not authored by climate scientists. It has become such an influential discourse that it inevitably shapes research and provokes responses from all disciplines engaging with the climate crisis. Work on understanding and naming different epochs has involved not only geological knowledge; the two men who are commonly associated with the Anthropocene concept and epoch, Paul Crutzen and Eugene Stoermer, are an atmospheric chemist and a biologist respectively.[17] Crutzen argued, "It seems appropriate to assign the term 'Anthropocene' to the present, in many ways human-dominated, geological epoch, supplementing the Holocene – the warm period of the past 10–12 millennia".[18] It is important to note that the concept has also come in for substantial criticism from within the natural scientific community, which contests this "appropriateness". Geological epochs occur over large timescales, whereas human civilisation capable of leaving a meaningful imprint on the geological record has been in existence for a far shorter period. Arguing that the Anthropocene is better seen as an event rather than an epoch, Brannen points out, "[T]he longest-lived radioisotope from radioactive fallout, iodine-129, has a half-life of less than 16 million years. If there were a nuclear holocaust in the Triassic [252–201 million years ago], among warring prosauropods, we wouldn't know about it".[19] This implies that the Anthropocene concept may be caught up within overinflating human capacities and so, anthropocentrism, a theme developed further below. This could be because the concept itself was arguably not the sudden

---

16 Recent work by the Intergovernmental Panel on Climate Change has been markedly more inclusive of social science; see IPCC 2022.
17 Crutzen 2002; Crutzen and Stoermer 2000.
18 Crutzen 2002, 23.
19 Brannen 2019. See also Haraway 2015, 160.

historical rupture that is assumed. Several writers have pointed to conceptual antecedents which could provide theoretical context to the Anthropocene.[20] Simpson argues that these antecedents (such as the Anthropozoic era, and the Noösphere) firmly place the Anthropocene concept within an Enlightenment and colonial discursive lineage,[21] a point which I return to later.

Brannen stresses that the field of geology, like that of cosmology, is markedly decentring of, and typically disinterested in, the human.[22] It is odd to find a concept that comes close to celebrating the power of the human within that field. It may be more accurate to focus on the ways in which "humanity" has produced the biosphere, rather than only impacting geology; this is a more accurate representation of how past epochs have been decided upon, by regarding which species have been present during different epochs. Brannen concurs and argues that the first wave of human-driven extinctions of megafauna, which began tens of thousands of years ago, may be a contender for an Anthropocene start date.[23] A reading of the Anthropocene as foregrounding human impacts on the planet's biosphere[24] is bolstered when linking climate breakdown to the extinction crisis and genomic technologies. In this vein Bennett et al. have argued that "broiler chickens vividly symbolize the transformation of the biosphere to fit evolving human consumption patterns and show clear potential to be a bio-stratigraphic marker species of the Anthropocene".[25] The ability of climate breakdown to redraw maps and to contribute to mass extinction is of more import than whether our species will leave long-term traces in the geological record, even though the presence of chicken bones would be a particularly telling marker of capitalism and its animal-industrial complex.

---

20 Schulz 2017; Simpson 2020; Steffen, Grinevald et al. 2011.
21 Simpson 2020.
22 Brannen 2019.
23 Brannen 2019.
24 This dualistic language is criticised below in the switch to a Capitalocene framework.
25 Bennett, Thomas et al. 2018, 9.

Crutzen himself recognises "strongly increasing species extinction"[26] and the significance of land-use changes[27] in his understanding of the Anthropocene. His work asserts an Anthropocene start date of the late 18th century, because that is when "analyses of air trapped in polar ice showed the beginning of growing global concentrations of carbon dioxide and methane". In addition, he mentions James Watt's design of the steam engine in 1784.[28] Interestingly, Crutzen, when discussing rising emissions, recognises that these have been caused "by only 25% of the world population",[29] thus offering some sense of differentiation in the "Anthropos" which has not always been apparent in how the concept has travelled and been used since and is not implied by the word itself. Indeed, this possibility of offering sociological and political nuance is absent from his later co-written work.[30]

Such discussions signpost three interconnected points of critique from which the Anthropocene has been subject to scepticism or outright dismissal: homogenisation, naturalisation and anthropocentrism. Explaining these further opens up key debates on the Anthropocene in the social sciences and humanities and also, I contend, brings us closer to an understanding of climate breakdown which avoids the abstraction of the problem in terms of emissions calculations, instead providing historical and socio-political context. This examination affords the opportunity to better clarify what type of crisis is constituted by a climate in breakdown. After discussing these three critiques in brief, this chapter then turns to the alternative idea of the Capitalocene.[31] This centres capitalist political-economic relations in understanding the climate crisis. I argue that it is in affinity with a CAS approach to the climate crisis, understanding it as arising via multiple intersecting crises. I contend that the Capitalocene can be brought into conversation with CAS and similar feminist approaches to understanding climate breakdown,[32] to constitute an even stronger alternative narrative.

---

26 Crutzen 2002, 23.
27 Crutzen and Stoermer 2000.
28 Crutzen 2002, 23.
29 Crutzen 2002.
30 Steffen, Crutzen and McNeill 2007.
31 For example, Moore 2016.
32 Fraser and Jaeggi 2018; Kaijser and Kronsell 2014.

Homogenisation and the problem of the generalised Anthropos

A key point which has characterised social science and other critical responses to the Anthropocene concept has been to critique an undifferentiated and implied equal responsibility for the climate crisis. There has been a tendency for the Anthropocene to be used in this generalised way in natural science[33] and popular[34] contexts. In such instances "we" are all amalgamated together in a "human enterprise". It is in such moments that the concept of the Anthropocene appears most obviously as an ideological concept expunging social and political conflict from history. The historical subjects, institutions and states most embedded in, and most responsible for, the climate crisis become lost within a generalised Anthropos. Similarly, those social groups less responsible, perhaps through lower energy consumption or through traditions of vegetarianism, are hidden. This critique of homogenisation has been eloquently made by several writers allied to ecosocialist positions. Moore accuses the Anthropocene of reducing the "mosaic of human activity in the web of life to an abstract, homogenous humanity".[35] Elsewhere, he mocks the idea: "It's a trick as old as modernity – the rich and powerful create problems for all of us, then tell us we're all to blame".[36] In their critique of the Anthropocene, Malm and Hornborg argue that the climate crisis has

> arisen as a result of temporally fluid social relations as they materialise through the rest of nature, and once this ontological insight – implicit in the science of climate change – is truly taken onboard, one can no longer treat humankind as merely a species-being determined by its biological evolution. Nor can one write off divisions between human beings as immaterial to the broader picture, for such divisions have been an integral part of fossil fuel combustions in the first place.[37]

---

33 For example, Steffen, Crutzen and McNeill 2007.
34 For example, *The Times* 2019.
35 Moore 2016b, 82.
36 Moore 2017, 599.
37 Malm and Hornborg 2014, 66.

Although the conflation of emissions (only) with "fossil fuel combustions" is simplistic, their critique is a call for the important retention of critical social sciences analyses of culture and power and a reminder of the relevance of inequalities to the trajectory of, and resistances to, the crisis.[38] The inadequacy of a generalised Anthropos is recognised not only in academic writing but in much critical journalism on the climate crisis and in well-known concepts such as climate debt, concerned with inequitable national histories of emissions, and climate justice, which further recognises how the climate crisis is playing out intersectionally according to intra-human relations of (geo)power.

Naturalising the Anthropocene

A second critique of the Anthropocene relates to tendencies for proponents to advance a historicisation which explains the new epoch as the inevitable result of human development. This shares the inattention to historical and socio-political detail of the generalised Anthropos and is accompanied by an impoverished understanding of the social context of technological development. The assumed inevitability of the Anthropocene is typically located in a notion of human nature or key technical innovation. Pointing to particular technological landmarks in an overall narrative of human development and progress might provide an easy and attractive way of doing history, but it is woefully inadequate. As Hartley has argued, "Technological determinism is always tempting, and much easier to communicate than the messy processes of class struggle".[39] In other words, to do history better, one must locate change within the complex milieu of social relations, class struggle against inequality being just one example of this. This is not to downplay the importance of new technologies but an argument against their naturalistic and depoliticised presentation. Andreas Malm makes this point in his 2016 book, *Fossil Capital*. While Malm can be seen to have something in common with Anthropocene proponents in affording significance to the invention of the steam

---

38   Di Chiro 2017, 498–99.
39   Hartley 2016, 156.

engine in England and the subsequent increases in the burning of coal,[40] he places that technological development within the context of the class relations of the time. Malm produces a more nuanced social and economic history in explaining the eventual ascendency of coal-fired steam power over water mills. For example, one of his important findings is that the "immobility of direct-drive waterpower" acted as a spatial constraint upon access to more profitable sources of labour power.[41] Consequently, the transition to coal-fired steam power was not some benign and joyous evolution to a better technology[42] but was embedded within class conflict between workers and factory owners, and the latter's desire to have a malleable and controllable natural "resource" in the shape of both coal and human labour. Ultimately the transition happened because it afforded more profitability. This analysis reiterates the importance of historical and sociological knowledge in understanding the climate crisis. It also checks the prevailing representation of the crisis in "post-political" terms, which replace ideological contestation with techno-managerial planning and construct the crisis as a mathematical, emissions-based problem.[43]

The Anthropocene as anthropocentric

A third, overlapping point of critique has been to detect in some Anthropocene arguments a celebratory and overly human-centred account of change. The heralding of the Anthropocene risks sounding like a coronation rather than a much-needed moment of social and political reflexivity and reconstruction. That the Anthropocene is unintentional should undermine any temptation to use it to celebrate the power of the human. The climate crisis contests the logic of the Western modernist project based around the mastery of an externalised

---

40   Crutzen 2002; Steffen, Crutzen and McNeill 2007; Steffen, Grinevald et al. 2011.

41   Malm 2016, 162.

42   See Hartley's (2016, 157) critique of Steffen, Grinevald et al. 2011 for uncritically assuming a "Whig view of history", in which history is seen as an "endless story of human progress and enlightenment". See also Royle 2016, 71.

43   Swyngedouw 2013.

nature and as such is an intimation of *post*humanism, the realisation of the limits of human rationality. It is indeed an understanding of rationality that has instrumentalised the more-than-human that has contributed to the crisis and, as climate change has become undeniable, an orchestrated effort in corporate and state "dither and denial",[44] especially since 1990. Haraway even humorously suggests "the Dithering" as an alternative epochal name for the Anthropocene.[45]

Furthermore, the climate crisis makes clear the non-separability of "humans" and "nature", since the changes with which it is concerned have arisen via a combination of the agential properties of varied species, elements and ecosystems. Although human activities in agriculture, energy and transport are key, these have never been examples of an external "human" acting on "nature" but processes of enmeshment and entanglement. Any Anthropocene that is caught in such abstracted ontology perpetuates the notion of the human as centred and as separate from other species.[46] Eileen Crist has been a significant critic of the self-celebratory aspect of the Anthropocene, inviting us to

consider the shadowy repercussions of naming an epoch after ourselves: to consider that this name is neither a useful conceptual move nor an empirical no-brainer, but instead a reflection and reinforcement of the anthropocentric actionable worldview that generated "the Anthropocene" – with all its looming emergencies – in the first place.[47]

Crist critiques a dominant, tacit worldview which

esteems the human as a distinguished entity that is superior to all other life forms and is entitled to use them and the places they live. The belief system of superiority and entitlement – or human supremacy – manifests in a range of anthropocentric

---

44   Watts, Blight et al. 2019.
45   Haraway 2015, 161.
46   Royle 2016, 72.
47   Crist 2013, 129.

commonplace assumptions, linguistic constructs, institutional regimes, and everyday actions of individual, group, nation-state, and corporate actors.[48]

Crist's analysis is in close affinity with that of CAS, which has long identified human exceptionalism[49] and the animal-industrial complex[50] as arbitrary dispositions and practices enabling and enacting violent relations with other species. This constitutes a human will to power, the exercise of a naturalised human sovereignty over other species,[51] which prefigured the climate crisis and threatens the sustainability of future generations of humans and other animals. Simpson, mentioned above, writing on conceptual antecedents of the Anthropocene, focuses on the 19th-century idea of the "Anthropozoic" era[52] and the later notion of the "Noösphere".[53] He discerns in such ideas familiar modernist tropes which present (white, European) humans as the highest level of civilisation, symbolised in mastery over an externalised idea of nature.[54] While I return to the relevance of this research for understanding a colonial Capitalocene later, it is useful to note here the human exceptionalist thinking at play in these antecedents.

Human exceptionalism can be viewed as giving rise to a *crisis* in human–animal relations, because as an ideology it is self-defeating and unsustainable. This concept, like anthropocentrism, is imperfect. Like the Anthropocene, it has an in-built generalisation. From one CAS perspective anthropocentrism does not confer benefit on humans because it takes us to a place of depleted ecology, endangering human flourishing. This is encapsulated well in the contradiction between the pressing need to decarbonise the global economy and the unabated expansion of the animal-industrial complex. Anthropocentrism as an ideology appears just as open to the charge of falsely generalising the

---

48 Crist 2018, 1, 242.
49 Gruen 2011.
50 Noske 1989; Twine 2012.
51 Wadiwel 2015.
52 Stoppani 1873.
53 Le Roy 1928.
54 Simpson 2020.

Anthropos as the Anthropocene, since it is hardly of benefit to humans in a general sense. This is partly the argument of Kidner, who favours the rejection of the critical use of "anthropocentrism". He writes:

> The accusation of "anthropocentrism" … serves an ideological purpose by diverting attention from the concealed origins of environmental destruction to the human behavioural characteristics that are themselves symptomatic of these origins. While it is true that humans are the visible agents of environmental destruction, to refer to our behaviour as "anthropocentric" is to ignore the roots of this behaviour within the industrial system we are socialised into.[55]

Kidner does not discount the term as a useful descriptive shorthand but says its use as a critical analytic fails to explain "forces and influences that have no concern for human well-being, and are in fact highly damaging to human welfare".[56] While anthropocentrism certainly informs social norms to which most people consent, the main (short-term) benefactors are those who profit from industries that exploit nonhuman animals. There may be a degree of short-term general benefit to all, but overall it is highly uneven, because the structural organisation of the exploitation of animal lives intersects with class, gender and racialised relations. The suspicion of the likes of Crist is that the Anthropocene concept as articulated by its founders lacks a critical distance from anthropocentric ways of thinking.

So, rather than as the pinnacle of human sovereignty, the Anthropocene (inclusive of the climate crisis) is better understood as the outcome of particularly powerful human agents (in tandem with the agential capacities of the nonhuman) taking the life-supporting capacities of the planet to the brink of collapse, which constitutes a phenomenon of unprecedented criminality. The tension points of an anthropocentric Anthropocene are visible in how the climate crisis is discussed. Is it to be addressed to safeguard human survival? Or do other species matter too? Are human–animal relations (most clearly

---

55   Kidner 2014, 476.
56   Kidner 2014, 474.

agricultural) included when historicising the crisis? Does the food system make it onto the agendas of the United Nations Framework Convention on Climate Change's Conferences of the Parties? And, perhaps most fundamentally, in our visions of necessary transformatory change, are we inclusive of a radical revision of human–animal relations? If our answers to these questions are negative, we are stuck within an anthropocentric framing of the climate crisis and are less likely to adequately address it.

Taken together, these three points of critique – homogenisation, naturalisation and anthropocentrism – provide important correctives to the concept of the Anthropocene. They resist a post-political reading of the climate crisis and direct us toward more accurate explanations and solutions more attentive to the economic, social and historical relations which have raised levels of greenhouse gases and have begun to kill humans and nonhuman animals in their millions. Considering these critiques, the aim is to re-present the Anthropocene as constituted by multiple crises but ultimately to reject and move beyond it. The most useful place to start is by considering the leading alternative narrative, that of the Capitalocene.[57]

A different story: The Capitalocene

Several thinkers with overlapping ideas participated in the early use of the term "Capitalocene". One of these, Donna Haraway, indicates via another, Jason Moore, that it was Andreas Malm who originally coined the term, in 2009.[58] Since then, it is undoubtedly Moore who has put most work into outlining the concept. On the face of it, the Capitalocene seems like a straightforward corrective to the problem of the generalised Anthropos of the Anthropocene and a recasting of the climate crisis as decisively emerging from global capitalist relations, with key responsibility residing with major Global North states and the corporations they have facilitated. However, this would be a simplistic

---

57   Haraway 2015; Moore 2014; 2016; 2017; 2018.
58   Haraway 2015, 163.

reading that would do a disservice to the theorising of the concept. As I shall make clear, the Capitalocene, as an alternative narrative, avoids easy criticisms of economic reductionism, because it does not draw upon an understanding of capitalism as a hermetically sealed concept of economics and class conflict.[59] This is vital, because in eschewing economism it admits into the theoretical conversation patriarchal and colonialist relations and, as I will demonstrate, also opens the door to consideration of human–animal relations.

To appreciate Moore's Capitalocene concept it is necessary to also understand dimensions of his earlier development of a "world-ecology" framing of capitalism.[60] An important aspect of relevance here is that Moore is critical of the way the economy and society have been understood as external to nature. He allies himself to the anti-dualist tradition of ecofeminist theory and much social science since the 1980s which has contested culture/nature dualism.[61] His world-ecology framework is intended to show that capitalism should be viewed as thoroughly embroiled in the web of life, as a form of human organisation and historical force which has become hegemonic, producing nature and being shaped by it. For example, it is instructive

---

59 Moore 2016b, 81.
60 Moore 2011; 2015. There are heated arguments between those supportive of Moore's position and those who follow other ecosocialist positions reflected in the work of Foster, York and Clark (2010). I come to these works more from the traditions of ecofeminism and CAS – and aware that both these fields have, with important exceptions, lacked conceptual analysis of capitalism. There is, then, a need for synthesis, and Moore's framework is more open to the sort of "intersectional" framework favoured by ecofeminism and CAS; he cites feminists (Silvia Federici) and ecofeminists (Val Plumwood) in his work. This makes Moore's research a better prospect for a broader liberatory framework and one that, I argue, is more empirically accurate. At the same time, I do not intend to gloss over disagreements about Marxist theory in this ecosocialism debate or to negate the value of, for example, Foster's body of work. As much as ecofeminism and CAS have lacked conceptual sophistication around capitalism, ecosocialism has tended to theorise capitalism aloof from an awareness of how the commodification of animals contributes to both capitalism and the climate crisis. Ecosocialist exceptions to this criticism typically come in the form of CAS interventions (see Dickstein, Dutkiewicz et al. 2022; Sanbonmatsu 2005).
61 Twine 2002.

to think of the radical land-use changes and myriad forms of extractivism of the last 500 years in particular. Moore underlines capitalism as an environment-making project that refashions natures; as he phrases it: "we make environments and the environments make us" acting through, not upon, the web of life.[62] Externalising capitalism from nature has been a way of perpetuating the view of humanity as "society" and as separate from "nature". Moore locates this tendency in other ecosocialist works[63] in which he otherwise finds much of value.[64] On first reading this may seem like a minor ontological quibble. However, the history of this entrenched Western dualism has also been hierarchical. This has been one of the central arguments made by ecofeminist theory,[65] which Moore cites.[66] Dualism has not simply been a statement of difference but has been the rendering inferior of a sphere named "nature" and, as Moore highlights, following similar arguments in both ecofeminism and CAS, the ideological marking and exclusion of women, the racialised, the poor and nonhuman animals from culture.[67] In other words, it has been and remains an ontology of control over "nature" and those associated with this constructed realm.

For Moore, this dualist ontology is entirely what the notion of the Anthropocene uncritically reproduces, and he adds a vital fourth element – the critique of being dualistic – to the three points outlined above. The Anthropocene does acknowledge interrelations of "humans" and "nature" and the ability of the human to shape the climate, but it maintains the dualism, in part through an impoverished historical understanding. Moore makes clearer the connection between the notion of the generalised Anthropos and this dualism. The generalised Anthropos occludes analyses of power and inequality, and "cleansed of such differences Humanity appears as a kind of Cartesian virgin birth. Nature appears, in this same imaginary, as 'out there', somehow pristine and untouched".[68] This compromises the Anthropocene concept's

---

62 Moore 2017, 599.
63 Moore 2017.
64 For example, Foster, York and Clark 2010.
65 Plumwood 1993.
66 Moore 2017.
67 Moore 2018, 242.
68 Moore 2017, 597.

ability to properly embed the human within nature or to understand how capitalism has reproduced "nature" successively. An explanatory concept that is "captive to the very thought-structures that created the present crisis" will inevitably be found lacking not only in its historicisation of climate breakdown but in its advocated transformatory narrative.[69] The Anthropocene is caught within a post-political malaise and is ill-equipped to understand or analyse the capitalist emergences of the crisis, to understand how gender relations or colonialism or geopolitics are implicated. It could be better positioned to grasp *how* a crisis of human–animal relations is inherent to climate breakdown and extinction (a relic of the dualistic division of labour, which placed nonhuman animals with the natural rather than the social sciences), but it is unlikely to understand the philosophical, sociological and historical reasons *why* this is so.

Moore's Capitalocene is a more accurate explanatory framework of the climate crisis and a catalyst for a decisive transformatory narrative because it understands capitalism not just as a way of organising nature but as extra-economic and has centred the importance of "cheap" or "unpaid work/energy". This is a further dimension of Moore's world-ecology framing of capitalism, which is central to his understanding of the climate crisis. He writes:

> This [thriving of capitalism through the appropriation of Cheap Natures] entails a reconstruction of capitalism's value-relations to encompass exploitation (surplus value) within more expansive movements of appropriation: the extra-economic mobilization of unpaid work/energy in service to capital accumulation. In this approach, unpaid work comprises work, energy and life reproduced largely outside the cash nexus, yet indispensable to capital accumulation. I speak of work/energy rather than simply work because we are dealing with work in a broadly biophysical sense, comprising the activity and potential energy of rivers and soils, of oil and coal deposits, of human-centered production and reproduction.[70]

---

69 Moore 2017, 604.
70 Moore 2018, 242.

Moore further makes clear that his understanding of "work/energy" includes the appropriation of the colonised, women and nonhuman animals. He benefits from his engagement with feminist perspectives that for decades have criticised a masculinist Marxist theory[71] for insufficiently incorporating the unpaid work that historically and disproportionately has been done by women, also referred to as the "social reproduction of labour power" to denote the gendered domestic servicing of paid work.[72] This emphasis on the capitalist appropriation of "unpaid work/energy" enables the Capitalocene concept to include gender relations, colonialism and geopolitics in both the history of capitalism and the historicisation of the climate crisis. However, it is further inclusive of the more-than-human. Specifically, it offers the potential to theorise the appropriation of species including nonhuman animals as sources of unpaid work. This applies across many examples of the appropriation of animals for profit beyond the obvious example of agriculture, which is captured by the broader concept of the animal-industrial complex.[73]

Moore's focus on unpaid work is tied to his critique of Cartesian dualism, because it is precisely those groups (women, the poor, the racialised and nonhuman species) that have been historically associated with "nature" rather than "society" and have been appropriated as "free", unpaid labour alongside the exploitation of the cheap labour of the impoverished working class. As he outlines:

> That so many humans could be reassigned to the domain of the not-human (or not-quite human) allowed capitals and empires to treat them cheaply – even as this cheapening was fiercely resisted.
>
> This Cheapening is twofold. One is a price moment: to reduce the costs of working for capital, directly and indirectly. Another is ethico-political: to cheapen in the English language sense of the word, to treat as unworthy of dignity and respect. These moments

---

71 Moore's (2018) main point of reference is Federici (2004), but there is a long and rich tradition of socialist feminism associated with, and predating, second-wave feminism.
72 Arruzza, Bhattacharya and Fraser 2019.
73 Noske 1989; Twine 2012.

of Cheapening work together, rendering the work of many humans – but also of animals, soils, forests and all manner of extra-human nature – invisible or nearly so. These movements of Cheapening register practically in low- and non-wage labor and dramatic forms of violence and oppression.[74]

For Moore, a "new world-praxis" of "Cheap nature"[75] cutting across humanity/nature dualism, exploiting human and nonhuman bodies, has been integral to the profit accumulation of the Capitalocene. With more specificity Moore refers to labour, food, energy and raw materials as the "Four Cheaps", which have been fundamental to the development and perpetuation of capitalism. If these can all be maintained as cheap, then the accumulative goals of capitalism can be too. Preserving cheapness has necessitated moments of "primitive accumulation", understood not just as the early capitalist forced seizure of property and processes of proletarianisation but, from the outset of capitalism, a continual relation of economic, gendered and colonial power reliant upon new scientific knowledges.[76] These cheaps are interrelated. If food can be kept cheap, the reproduction of labour is cheap. Imperatives to keep food cheap have had an enormous exploitative consequence for nonhuman life, and the oppression of nonhuman animals has been inextricably linked to the perpetuation of cheap human labour. Moreover, if energy is cheap, the extraction of more raw materials becomes viable. The Four Cheaps aspect of Moore's theory is useful for understanding the historical link between capitalist and ecological crises. Capitalism toils when it struggles to maintain (or appropriate new) cheapness. Currently the climate crisis and the crisis of capitalism seem to converge upon a failure, most clearly, to maintain cheap energy, in which the externalisations of emissions from the costs of production, ignored for so long, have become both critical and menacing.

In Moore's terminology, "appropriation" (of unpaid labour) and "exploitation" (of paid labour) are codependent within capitalism. This is crucial because it ties together an ecofeminist analysis of dualism

---

74   Moore 2017, 600.
75   Moore 2017, 600.
76   Moore 2017, 606.

(read as a significant ideological narrative in Western history which has been useful for capitalism) with an ecosocialist reading of capitalism, consolidating both in the process. Since Moore is aware that both a dualistic worldview and capitalism as a nature-transforming system predate the late 18th century it is not surprising that he favours an earlier understanding of the emergence of capitalism and thus an alternative historicisation of the climate crisis. While most Anthropocene theorists may claim as significant the period that witnessed the intensification of coal production and the invention of the steam engine, at the end of the 18th century,[77] in Moore's view, the Capitalocene – *and the roots of the climate crisis* – emerged at least 300 years earlier.[78] Not only does this periodisation include significant developments in class relations and conflicts; it allows Moore to attend to the role of colonialism and patriarchy in the early formation of capitalism as a world-ecology-making system and to consolidate his view of capitalism as extra-economic. He is, not surprisingly, critical of the Anthropocene historical narrative of modernity and capitalism, because it misses the "remarkable remaking of land and labour" which took place after 1450.[79] His critique is evidenced by sharp rises in labour productivity during the period in multiple sectors, such as printing, sugar, iron, shipbuilding, mining and water mills, which extended across nations and was enabled by colonialist exploitation.[80] This broadens out the origins of capitalism beyond the narrow confines of England to include the significant emergences of other countries, such as the Netherlands, Spain and Portugal, as colonial powers. As Moore says:

---

77 For an exception, see Lewis and Maslin 2015. This is an important paper which can be read as being broadly in sympathy with the Capitalocene framing. It argues for an Anthropocene start date of 1610 which is associated with the Orbis spike, in which carbon dioxide levels dipped. Following the decimation of the Indigenous peoples of the Americas and the resultant temporary decline in farming, afforestation occurred and acted as a carbon sink. Defining epochal change in terms of capitalist-colonialist development places the analysis in line with Capitalocene theory.
78 Moore 2016b, 81.
79 Moore 2016b, 94.
80 Moore 2017, 612–13.

> Between 1450 and 1750, a new era of human relations in the web
> of life begins: the Age of Capital. Its epicentres were the seats of
> imperial power and financial might. Its tentacles wrapped around
> ecosystems – humans included! – from the Baltic to Brazil, from
> Scandinavia to Southeast Asia.[81]

This Age of Capital has been systemically violent, inextricably made
possible by the colonialist appropriation of free slave labour from West
Africa and the broader genocides against Indigenous peoples notably
in what are now referred to as North and Latin America as well as
British and other European colonialism in Africa, Asia and Oceania.
British economic power is based in part on colonial expropriation, here
meaning "the dispossession of bodies and property", which was always
also an expropriation of the nonhuman species of occupied countries
(sometimes to the point of extinction) or the working of the land by
slaves. Slave labour is written into the urban geography of many British
cities and contributed to the British industrial revolution. As Moore
succinctly puts it, "For every Amsterdam there is a Vistula Basin; for
every Manchester, a Mississippi Delta".[82] Colonialist expansion was
associated with deforestation, mining, agriculture and the plundering
of animals through hunting, trapping and the establishment of animal
breeding. Vergès points out that expropriation was accompanied by a
colonising flow in the other direction: a transfer of plants, animals,
diseases and goods *from* Europe.[83] It is also noteworthy that Haraway,
writing with anthropologist colleagues, muses over the idea of the
"Plantationocene" as an alternative to the Anthropocene to underline
the forced assemblages of plants, animals, microbes and slave labour in
the "pre-industrial" colonial period.[84]

In aspects of Moore's critique, it is possible to discern his unease
with the historical human exceptionalism of the social sciences, which
has denied human enmeshment with the rest of nature,[85] the efforts

---

81 Moore 2017, 610.
82 Moore 2018, 266.
83 Vergès 2017.
84 Haraway, Ishikawa et al. 2016, 557.
85 Moore 2017, 596.

of animal and environmental social scientists notwithstanding. Abandoning this exclusionary approach in the historical analysis of capitalism means foregrounding human–animal relations and again underlining the *hierarchical* dimension of the dualistic separation. Furthermore, this entails the recognition that capitalism has simultaneously been enacted as a narrative of human supremacy over other (animal) species.[86] If capitalism is to be understood as a world-ecological system that reorganises nature, then historical and contemporary analyses must incorporate nonhuman animals, not just as endless victims of human violence but also as agential beings within capitalist development at times capable of resistance[87] or of co-shaping economic development and the production of ecology.[88] Moore makes clear that "[t]he Capitalocene argument posits capitalism as a situated and multispecies world-ecology of capital, power, and re/production".[89] He later asks, "[I]s there any reasonable way to think through capitalism abstracted from its relationship with non-human animals …?"[90]

This echoes Sarah Franklin's useful history of the word "stock" and its relationship to the emergence of animal agriculture and its use in human slavery.[91] The etymology of "cattle" derives both from the mid-13th-century Anglo-French "*catel*", meaning "property", and from the medieval Latin "*capitale*", meaning "property" and "stock",[92] underlining the prominence of animal agriculture to proto-capitalistic economic activity. From the 15th century onwards, animal "husbandry" became embedded in English capitalistic practices of "improvement",[93] adding credence to my own insistence that sociologists of capitalism (and the climate) cannot credibly exclude analyses of the animal-industrial complex.[94] This confirms not only that historically the commodification of nonhuman animals has been

---

86   Nibert 2017a; 2017b.
87   Hribal 2010.
88   Cronon 1991, 220.
89   Moore 2016b, 94.
90   Moore 2017, 599.
91   Franklin 2007, 51–53. See also Gunderson 2013; Schwabe 1994.
92   See https://tinyurl.com/5jkak6c7.
93   Grau-Sologestoa and Albarella 2019.
94   Twine 2023.

closely tied to the development of capitalism but that a narrow understanding of capitalism based only on exploitative wage-labour relations contains a humanist error which has obscured a truer understanding of how capitalism works in practice, along racialised, gendered and speciesist lines. Furthermore, in later work, with Raj Patel, Moore uses the example of the intensive farming of broiler chickens in the animal-industrial complex as paradigmatic of the way the Capitalocene has operated via intersecting "cheapenings" of, for example, "nature, money, work, care, food, energy and lives".[95] This overlaps, but is in significant conceptual contrast to, the suggestion quoted above that the broiler chicken may be the best candidate for a bio-stratigraphic marker of the Anthropocene.[96]

Moore's alternative historicisation of capitalism and his theories of unpaid labour and Cheap Nature have a distinctive relevance for how the coevolution of gender and capitalist relations is understood. His main influences here are Silvia Federici's *Caliban and the Witch: Women, the Body and Primitive Accumulation* and Maria Mies' Patriarchy *and Accumulation on a World Scale: Women in the International Division of Labour*, which I read as part of the ecofeminist tradition.[97] These related texts (Federici wrote the foreword to the 2014 reissue of the Mies text) posit the emergence of the witch hunts of Europe and North America, which overlapped this period of such interest to Moore (1450–1750), as highly contextualised by the development of capitalism. The witch hunts and their associated misogynistic representations of women contributed to the general confinement of women to the private sphere and a primarily reproductive role, while labour power, production and emergent norms of property ownership were masculinised. Most women were captured in a backgrounded, unpaid role consisting of reproductive labour and the servicing of male waged labour.[98] This began to change in a gradual, faltering manner in the United Kingdom during the 19th century and later by periods of higher employment for women as a reserve army

---

95   Patel and Moore 2018, 3.
96   Bennett, Thomas et al. 2018, 9.
97   Federici 2004; Mies 1986.
98   Federici 2004, 181–82; Hartley 2016, 161–62.

of labour,[99] though not significantly until after World War II. These intersecting developments of a patriarchal capitalism undoubtedly reinforced now familiar dualistic associations of men with mind and rationality (also a classed and racist ideology) and women with body and emotionality, which remain influential today despite feminist contestation.

As noted earlier, Moore's work has affinity with that of certain contemporary socialist-feminist theorists of capitalism, particularly that of Nancy Fraser,[100] which has made important links between the climate crisis and the capitalist crisis of social reproduction. Indeed, in 2016 Fraser and Moore shared a speaking panel at which Fraser described the experience of reading Moore as like discovering a doppelgänger.[101] They have similarly followed the path of underlining extra-economic dimensions of capitalism beyond the classic narrative of the exploitation of labour. Labour exploitation certainly involves far larger numbers of women and people of colour currently than it did in the past, contradicting representations of the working class as white and male. But processes of social reproduction on which capitalism relies (childcare, elder care, foodwork, housework and education) remain gendered. This means many, if not all, women are subjected to the dual burden of expropriated unpaid housework and exploited wage labour. Indeed, the marketisation of education and care in the neoliberal era constitutes audacious attempts to commodify the very processes which facilitate labour power. Fraser has tended to use the Marxist concept of expropriation, meaning "dispossession" or, for Marx, "appropriation without exchange",[102] in preference to Moore's distinction of "exploitation" and "appropriation". While both Moore and Fraser are referring to the same capitalist relations, "expropriation" seems a more accurate term to describe what occurs in the capitalist relations of

---

99   Beechey 1977.
100 Arruzza, Bhattacharya and Fraser 2019; Fraser and Jaeggi 2018, 94.
101 This was a panel discussion that took place at the New School University, New York, in February 2016: "Jason W. Moore, Nancy Fraser, and Eli Zaretsky, Capitalism in the Web of Life: A Conversation" and is available here: https://www.youtube.com/watch?v=ZhQmjqTcdlA.
102 As discussed in Foster and Clark 2018; see also Marx 1976.

slavery, free domestic work and, indeed, direct dispossession of the more-than-human.

This Moore–Fraser affinity returns us to the discussion of interlocking crises with which I began this chapter. In their book *Feminism for the 99%: A Manifesto*, Arruzza, Bhattacharya and Fraser characterise contemporary capitalism as beset by three interrelated contradictions: ecological, political and social-reproductive. They write:

> Behind capitalism's official institutions – wage labour, production, exchange, and finance – stand their necessary supports and enabling conditions: families, communities, nature; territorial states, political organizations, and civil societies; and not least of all, massive amounts and multiple forms of unwaged and expropriated labour, including much of the work of social reproduction, still performed largely by women and often uncompensated. These, too, are constitutive elements of capitalist society – and sites of struggle within it.[103]

In common with Moore and previous socialist feminists they expand the definition of capitalism beyond the narrow focus on labour and production. This is crucial, because it reveals a series of denied dependencies which capitalism requires to exist and illustrates, following Marx, how, as an economic system, capitalism pervades everyday life. For Arruzza, Bhattacharya and Fraser there is an "ecological contradiction" that means capitalism is "structurally inclined" to degrade habitats and ecosystems (and harm nonhuman animal life, though they omit explicit mention of this).[104] The "political contradiction" concerns the marketisation of public institutions and the erosion of democracy, and the "social-reproductive contradiction" concerns the capitalistic harnessing of social-reproductive labour without economic recompense, long a concern of feminist critique. They discuss crises in care sectors as evidence of the social-reproductive contradiction. Indeed, they later frame the neoliberal capitalist assault on social democracies as advancing through

---

103 Arruzza, Bhattacharya and Fraser 2019, 64.
104 Arruzza, Bhattacharya and Fraser 2019, 65.

the cannibalisation of social reproduction, as seen in policies of austerity pursued in many Western countries after the global economic crash of 2008.[105] I note their work not just because of its theoretical sympathies to Moore. Both afford a broad conceptualisation of capitalism that underlines its inherent tendency to produce, exploit and exhaust "nature", and both advocate oppositional ecosocialist politics. Also, while *Feminism for the 99%* has some shortcomings,[106] its analysis of social reproduction has specific relevance for climate justice politics and the need to grasp interconnections between the climate crisis, gender, "race", capitalism and poverty.

These similar ways of understanding capitalism are important for Anthropocene critiques, because not only do they – and Moore's work specifically – historicise capitalism differently, but they also portray it more broadly, as embedded in the reproduction of nature, gender and "race". This allows for an attentiveness to power, which has been largely absent from the Anthropocene concept. Anthropocene adherents might criticise Moore's longer historicisation of capitalism on the grounds that actual carbon-related emissions are not seen to have risen significantly until the latter half of the 18th century. However, this would be to miss the point of the Capitalocene thesis: that the social and economic relations which ultimately play out with exponential increases in greenhouse gases have a far longer history and are inseparable from class, gender, colonial and species relations. In other words, "the scale, speed, and scope of landscape transformation across the expanse of early capitalism"[107] set capitalism on a normalised trajectory of exploitation, appropriation and capital accumulation. "Nature" (inclusive of human labour) was not suddenly misused from

---

105 Arruzza, Bhattacharya and Fraser 2019, 77.
106 Arruzza, Bhattacharya and Fraser (2019, 46) state clearly in their ninth thesis: "Fighting to reverse capital's destruction of the earth, feminism for the 99 percent is eco-socialist". As stated, one of the three co-authors, Nancy Fraser, is aware of and influenced by Jason Moore's work (see Fraser and Jaeggi 2018, 94). What is harder to explain is the absence of ecofeminism in the work. Perhaps relatedly, it is silent on nonhuman animals. While it is a strong work of joined-up and coalitional thinking it would be improved by attention to these omissions.
107 Moore 2016b, 96.

the 18th century; the appropriation and its ideological framing had been in place considerably earlier. To illustrate this Moore uses coal, undoubtedly a key fossil fuel for any explanation of the climate crisis. English coal production had *already* markedly increased during the 16th century. This significant establishment of coal mining facilitated further increases, in the late 18th century, which, as noted earlier, Malm ties to specific social class relations in England at that time.[108]

The reading of capitalism presented here brings into relief the deficiencies of the Anthropocene concept while providing explanations for the climate crisis and its classed, gendered, racialised and speciesist characteristics. Taking stock, I have identified the four key deficiencies of the Anthropocene as homogenisation, naturalisation, anthropocentrism and dualism. Homogenisation thwarts attempts to identify the power relations of the crisis. Naturalisation operates similarly, via an impoverished understanding of history, and an anthropocentric Anthropocene perpetuates the very same abstracted hierarchical and dualistic ontology that has facilitated the climate crisis. Dualism is found in the impoverished historical approach of the Anthropocene concept, falsely separating humanity and the economy from nature. The Capitalocene framework adeptly deals with all four of these deficiencies. It sets the stage, I contend, for a mutually beneficial conceptual and political alliance with CAS. For CAS, Moore and Fraser provide a more sophisticated conceptualisation and historicisation of capitalism. Their work has affinity with the CAS opposition to economism and dualism and stresses how the expropriation of slave labour, gendered domestic labour and nonhuman animal bodies has enabled capitalism's development. In return CAS can bolster both the understanding of capitalism as a multispecies world-ecology and the partial critique of human exceptionalism offered in Capitalocene discourse.

In the next chapter it is necessary to develop this affinity further – specifically, to examine how a CAS perspective contends that a crisis in anthropocentric human–animal relations is also constitutive of the climate crisis, and how the Capitalocene critique of the generalised Anthropos can be sharpened by linking it more strongly with other critical work on the climate crisis. This reinforces the argument made

---

108 Malm 2016.

above, takes the Anthropocene to its moment of retirement in the work of both critical theory and activist politics, and proceeds toward the adoption of this alternative concept.

2
Detailing the Capitalocene

Class, "race", gender, species

Much critical social science on the climate crisis has close affinity with the Capitalocene framework. Indeed, outside of the Capitalocene theory narrowly conceived, other work has made a clear point of linking the climate crisis with capitalism.[1] I begin by covering quite mainstream arguments that position the climate crisis as "classed" in various ways, before considering similar and overlapping work that has more explicit focus on "race" and gender. However, I do so by considering their intersection with human–animal relations. To reiterate, these inherently inequitable dimensions of the Capitalocene are products of the philosophical heritage of capitalism that has apportioned value according to humanity/nature dualism, constructing multiple (human) beings as "less than human", as cheapened, and more amenable to exploitation.[2] This is a significant point of overlap between Capitalocene theory, ecofeminism and critical animal studies (CAS). Importantly, the chapter then proceeds to more fully embed human–animal relations within the Capitalocene framework.

---

1   Klein 2014; Koch 2012.
2   Moore 2017, 600; Plumwood 1993; Twine 2010a, 10.

The climate crisis and social class inequalities

An important accompanying point to Moore's longer historicisation of capitalism (and the Capitalocene) is to underline the more recent acceleration in greenhouse gas emissions in the post–World War II period. Indeed, both are important for revealing antecedents and historical inequalities between nations in their emissions and their disproportionate responsibility for the climate crisis. These comprise fatal blows to the myth of the generalised Anthropos implied by the Anthropocene. Data for annual global carbon dioxide (CO_2) emissions show that between 1850 and 1950 most emissions emanated from Europe and the United States. Significant acceleration occurs from 1950 with corresponding regional increases in Europe and the United States. Emissions from China started to increase markedly by the 1990s.[3] Between 1950 and 2007, Europe (excluding Russia) and the United States were responsible for 52% of the total share of CO_2 emissions from fossil fuels and cement.[4] Taking a longer time slice of between 1751 and 2017, and using the broader North America, Europe (this time including Russia) has emitted 33% of global cumulative CO_2 emissions and North America 29%, giving a total of 62% for the two continents. Breaking this down, the United States is by far the largest national emitter on 25% of global cumulative emissions, with China on 12.7%, approximately half the emissions of the United States.[5] Although these statistics refer only to CO_2, they give a good indication that it is the economic activity of specific regions and countries that bears greater responsibility for the climate crisis. This historical, national and geopolitical inequity ought to call into question the practices, institutions and infrastructures instigated by these countries especially as they pertain to key sectors such as energy, construction, transport and agriculture. It implies one answer to the climate crisis as the radical transformation of economically privileged ways of living. However, this inevitably must involve very significant infrastructural changes

---

3   Ritchie, Roser and Rosado 2020.
4   Dow and Downing 2011, 46–47.
5   Ritchie, Roser and Rosado 2020.

across societies that tackle class inequalities in consumption.[6] For the Capitalocene perspective, such a "radical transformation" must address the systemic nature of the crisis within capitalism. To recall, "Capitalocene names capitalism as a system of power, profit and re/production in the web of life".[7] Profit accumulation is divorced from both morality and need and much of the damage capitalism perpetrates against people and other species stems from its incessant imperative to keep the Four Cheaps of food, labour, energy and raw materials low-cost.[8] If the costs of the gendered social reproduction of labour can also be reduced or externalised, then "all the better". For the Capitalocene perspective, these logics of capitalism are incompatible with resolving the climate crisis because attempts to increase profit rates are inextricably linked to emissions rises. One drawback of such emissions statistics is that they gloss over economic inequalities and different ways of living *within* (both richer and poorer) countries. Consequently, *overall* income levels across national boundaries can be a better indicator of the relationship between class privilege and greater complicity in the climate crisis. A 2020 study examined exactly this, looking at 86 countries including both rich and poor, finding that the wealthiest 10% of people in a country consume approximately 20 times more energy overall than the bottom 10%, wherever they live. The study also found that wealthier people consume disproportionately more energy in transport and domestic use and that even the poorest 20% of the UK population consumes over five times as much energy per person as the bottom billion in India.[9] Research has suggested emissions may rise in some poorer regions as a result of combating poverty, putting an even greater onus on rich, high-emitting countries to substantially reduce their emissions.[10]

A further important way that the crisis is classed is in its effects. There are longstanding debates over the relationship between class-specific and environmental risk,[11] and the longer-term worst-case

---

6   Oswald, Owen and Steinberger 2020.
7   Moore 2017, 606.
8   Moore 2017.
9   Oswald, Owen and Steinberger 2020.
10   Bruckner, Hubacek et al. 2022.

scenario of climate breakdown could see the erosion of the insulation abilities of the class-privileged. It is a significant climate injustice that in the shorter term those with greater wealth (and the countries with greater complicity in the crisis) also have a greater ability to adapt and avoid the worst effects of climate-related events. This is not true across the board as shown by bushfire events across much of Australia and in California, but it is true in general. Certainly, it is likely that some of the adaptive responses of the wealthy (such as increased use of air conditioning) will hasten climate breakdown. Further exacerbation of climate injustice is found within projected modelling of temperature changes. In a far-reaching analysis of the human mortality impacts of temperature increases, Carleton et al. found hotter poorer countries less able to adapt will be hardest hit.[12] Bathiany et al. have modelled future temperature variability and shown "marked predicted increases in temperature variability in tropical land regions including many of the world's poorer countries, with the Amazon being a particular hotspot of concern. These hotspots result mainly from soil drying in the Southern Hemisphere and from increased atmospheric variability in the subtropics of the Northern Hemisphere".[13] Sociologically it is also important to highlight the mediation of heatwave impacts by age. The *Lancet* Countdown report found that "from 2000 to 2018, heat-related mortality in people older than 65 years increased by 53.7% and, in 2018, reached 296,000 deaths, the majority of which occurred in Japan, eastern China, northern India, and central Europe".[14] It is also true that sea-level rises, a further decisive aspect of the ensuing climate crisis, may ultimately have radical effects on wealthy nations such as the United Kingdom and the United States, but low-lying nations and island-states face virtual or complete annihilation. It is for this reason that many poorer countries have been at the forefront of arguing for capped global emissions to reduce overall temperature rises to 1.5°C

---

11   Beck 1992.
12   Carleton, Jina et al. 2020.
13   Bathiany, Dakos et al. 2018, 8.
14   Watts, Amann et al. 2021, 135. Chapter 5 returns to human age as an important dimension of the climate crisis with a focus on childhood.

above "pre-industrial" levels. Such a goal is realistically now impossible despite political rhetoric to the contrary.

These points around the classed nature of the climate crisis bolster the Capitalocene thesis. They also intersect with and illuminate further differentiations of the falsely homogenised Anthropos. For example, it is also the case that animal consumption is a classed practice. Moreover, although animal consumption is transcultural, the historical establishment of the animal-industrial complex emanated from the industrialising West, which I return to later. Meat consumption is markedly higher in more wealthy regions of the world, noticeably Western Europe, North America and Australia. Feeding people in poorer countries with animal-sourced foods remains a dominant discourse within Western humanitarian, agricultural economics and food security research institutes as an appropriate response to food poverty.[15] The well-known phenomenon of the "nutrition transition"[16] outlines when poorer countries see rising income levels their average diet tends to undergo changes with decreases in the consumption of grains and fibre and increases in the consumption of sugar, fat and animal-sourced foods. I have previously criticised the naturalising tendencies of framing the nutrition transition as only demand driven,[17] and it would be naïve to think that it has not involved the export of meat/dairy from rich nations and the import of Western agricultural practices. The assumption of the nutrition transition centres and valorises Western eating practices and rests upon a Eurocentric developmentalist logic, now being undermined by the climate crisis which exposes fossil capitalism, and so Western modernity, as a form of *mal*development. The nutrition transition subjects poorer countries to a double burden of both communicable and non-communicable disease,[18] instead of working with local diets, traditions and plant-based transitions as an alternative approach to addressing food insecurity issues.

---

15 Headey, Hirvonen and Hoddinott 2018.
16 Popkin 1998.
17 Twine 2010a, 129. See also Rivera-Ferre 2009.
18 Boutayeb 2006.

A classed and racialised climate crisis

Such themes constitute indications of intersections between class, species and "race" in the injustices of the climate crisis. It was noted above in the work of Moore that the Capitalocene has accorded prominence to colonialism in both the development of capitalism and the climate crisis, indicating a general failure in Anthropocene discourse to attend to practices of racialisation and white supremacism in the present-day articulation of climate breakdown. It is possible to strengthen this attention given by the Capitalocene with further research which, although it has not always allied itself to the Capitalocene concept, has certainly articulated the Anthropocene as racialised.

Firstly, the above points about historical carbon debt and differential impacts of the crisis are an articulation of racial injustice as much as class injustice. It is hard to think of a clearer colonialist expression of domination than the literal disappearance of low-lying island states. As Amitav Ghosh puts it, "The Anthropocene has reversed the temporal order of modernity: those at the margins are now the first to experience the future that awaits all of us".[19] Secondly, climate effects and extreme Capitalocene weather events exacerbated by a warming planet also can bring into relief pre-existing, overlapping class and racial injustices. The effects of Hurricane Katrina conjoined with the inequalities of US society rendered such a scenario upon the city of New Orleans in 2005, killing over 1,300 people in Louisiana alone. The racialised and classed segregation of the city had already meant that poorer African Americans were more likely to live in flood-prone areas.[20] Similarly, the island of Puerto Rico, an unincorporated territory of the United States, already experiencing debt and budget cuts, lost around 3,000 people to the impacts of Hurricanes Irma and Maria in September 2017. Wealthier citizens were able to move to mainland US cities just before the hurricanes struck. This underlines the classed and racialised politics of (im)mobility which are already playing out with economic migrants and will increasingly be at the fore with climate

---

19   Ghosh 2016, 62–63.
20   Lavelle and Feagin 2006.

refugees, especially, though not only, in low-lying Pacific island-states.[21] A 2020 report by the Institute for Economics and Peace found that there could be 1.2 billion climate refugees by 2050 with the majority comprising vulnerable populations in Africa and Asia.[22] In the cases of Katrina, Irma and Maria, the US government was strongly criticised for a slow and inadequate response that itself may have been racialised. Such events compound poverty, racism, disability, homelessness and trauma for the already vulnerable. Moreover, an important point in moving from an Anthropocene to a Capitalocene reading of such events is to reject a dualistic anthropocentrism and to produce accounts of interspecies relations that acknowledge shared vulnerabilities and kinship. For example, in the case of Katrina it was reported that 600,000 animals were killed or stranded, and that seven years later New Orleans was still affected by a stray dog problem, worsened by the hurricane.[23]

Other writers take the analysis of the racialised Anthropocene further, enhancing the Capitalocene narrative. Firstly, in returning to the work of Simpson[24] on Anthropocene antecedents, he argues convincingly that their discursive elements persisted and influenced the initial theorising of the Anthropocene. He names these elements as the following three tropes: i) the notion of the gradual progression of human cultures advancing through stages, ii) within this process the idea that humans transcend from a state of nature to one of civilisation, and iii) this transition is seen as having a pseudo-religious teleology whereby the human attains its all-powerful destiny of mastery.[25] This has resonance with a white racialised transhumanism and recalls my earlier points about the Anthropocene containing a hubristic celebratory sub-text. Significantly, Simpson identifies all three tropes in the work of Steffen et al.[26] and is concerned how they perpetuate and re-centre classic narratives of Western colonial modernity which tie progress to a Eurocentric mastery of externalised nature.

---

21 Suliman, Farbotko et al. 2019.
22 Henley 2020.
23 Jonassen 2012.
24 Simpson 2020.
25 Simpson 2020, 61–63.
26 Steffen, Crutzen and McNeill 2007.

Secondly, it would be remiss not to embed critical analysis in the longer history of environmental justice and environmental racism research. Françoise Vergès in her essay "Racial Capitalocene – is the Anthropocene racial?"[27] outlines that for several decades research has highlighted that US toxic waste dumps are more likely to be located beside Hispanic and African American communities,[28] a finding explainable not only in terms of social class. This is now accompanied by novel variations such as the racialised exposure to heat stress.[29] Vergès is critical of the Anthropocene concept for ignoring these racialised histories and perpetuating an illusion of a shared fate across difference. Furthermore, environmental racism has not been confined to the United States; consider, for example, the histories of "nuclear colonialism", testing by countries such as the United Kingdom, France and the United States with their vicious impacts upon Indigenous people and nonhuman animals around the world.[30] Environmental justice concerns are also pertinent in attempts to prolong fossil capitalism in such cases as fracking (e.g., Australia), tar sands (Canada) and the controversy over the Dakota Access pipeline. A CAS perspective on environmental racism corrects the omission of ecological repercussions for other species, underlining the fact that these injustices extend beyond species boundaries. Anthropocene writers should have been aware of environmental racism research and primed for understanding differentiated social effects as climate breakdown unfolds.

Thirdly, if Anthropocene naming centres geology, it would be surprising to exclude the very imbrication of geological knowledge production in the intertwined histories of colonialism and extraction. Addressed by Kathryn Yusoff in *A Billion Black Anthropocenes or None*,[31] this underlines a tendency toward scientific illiteracy among some natural scientists. I specifically mean an illiteracy toward the social, historical and political contexts of scientific knowledge production that

---

27 Vergès 2017. See also Karera 2019.
28 Bullard 1990; Higgins 1994.
29 Hsu, Sheriff et al. 2021.
30 Yusoff 2018, 44–48.
31 Yusoff 2018.

has been the mainstay of social studies of science. Anthropocene writers[32] have lacked this science studies reflexivity. For example, fields such as genetics and biology emerged as racialised discursive formations[33] and Yusoff presents a similar history for the geological. Contesting the tendency to see geology as innocent and without a subject, Yusoff makes clear its links to colonialism, extractivism, and the expropriation and commodification of land and slave labour. She writes:

> The histories of the Anthropocene unfold a brutal experience for much of the world's racialized poor and without due attention to the historicity of those events (and their eventfulness); the Anthropocene simply consolidates power via this innocence in the present to effect decisions that are made about the future and its modes of survival. The sleight of hand of the Janus-faced discipline of geology (as extractive economy and deep-time palaeontology of life forms) is to naturalize (and thus neutralize) the theft of extraction through its grammars of extraction. Recast as "development", the colonial and settler-colonial dispossession of the relation to land and geography was never something chosen without coercion.[34]

This has affinity with arguments above noting how the Anthropocene glosses over historical and present-day differentiations of violence. What becomes clear from Yusoff is that geological knowledge has been inseparable from the development of capitalism, itself inseparable, as noted, from the cheap or free expropriations of colonialism. Yusoff's discussion proceeds to cover the role of geology in furnishing a white exclusionary model of the human and is critical of Anthropocene start dates which cleave off early capitalism's colonial enmeshments. Such points are in synergy with Moore above, but Yusoff appears to not know his work, or the Capitalocene's interest in colonial history, instead offering a premature dismissal of the concept and mistakenly associating it with viewing the 19th century as the crucial birth date

---

32 Steffen, Crutzen and McNeill 2007.
33 Gould 1981; Kevles 1995.
34 Yusoff 2018, 11–12.

of change.[35] Despite this confusion, the analysis weaves a closer understanding of geological imaginaries into the Capitalocene framework. Had Anthropocene advocates offered a command of this history (in some cases the history of their own discipline) then, once more, they would have been less likely to falsely homogenise the human, instead better comprehending the Anthropocene as inherently racialised and colonial.

A gendered climate crisis

The Anthropocene narrative has also been noticeably silent on the gendered dimension of its own theorising and the gendering of the climate crisis. The Capitalocene era has evidently also been a patriarchal era wherein an evolving social construction of hegemonic masculinity has been inflected by tropes of white supremacy and nature mastery. The complex dynamic pathways of hegemonic gender constructs have furthermore, as the Capitalocene concept acknowledges, been partly shaped by the historical male domination of work and public life. Moore's work on the Capitalocene, as was noted, also afforded an important role to gender in the historical formation of capitalism involving the exclusion of women from wage labour, a position in accordance with Nancy Fraser, and in affinity with other ecofeminist work focused on class and capitalism.[36] Furthermore, as also noted, Moore's understanding of capitalist development has also been influenced by the ecofeminist critique of dualism.[37] This all primes the Capitalocene to have significant advantages over the Anthropocene in being able to understand the gendered emergences and effects of the climate crisis. The generalised Anthropos of the Anthropocene is an effective erasure of gender/nature connections, concealing the patriarchal history of the climate crisis. In contrast, the Capitalocene concept signposts mitigative feminist politics critical of hegemonic masculinity, a point returned to shortly. What the

---

35   Yusoff 2018, 39.
36   Mies 1986; Oksala 2018; Salleh 2010.
37   Plumwood 1993.

Capitalocene lacks, however, beyond this significant priming, is a direct engagement with contemporary feminist scholarship on the climate crisis.

To speak of crisis in relation to masculinity is a fraught process given the history and longevity of both academic and media pronouncements of masculinity "in crisis".[38] Some of this discursive terrain speaks to an inability of many men to respond in just ways to feminist critique, and of the injurious effects of socially constructed gender norms played out in poorer health and empathic atrophy. In the 1990s Connell[39] famously conceived "hegemonic masculinity" as a dominant, shifting cultural construct of masculinity which people orientate themselves to and which continues to guarantee male dominance, despite a partly delegitimised patriarchy in certain countries. While misogynist, structurally violent practices toward women and the absurd self-injurious nature of masculine gender norms are sufficient in themselves to engender a profound questioning of masculinity and patriarchy, I intend to argue that the climate crisis is also, in part, a crisis of gender, a crisis of hegemonic masculinity. This is a neglected claim in the public discourse on the climate crisis. The critical examination of the trans-cultural valorisation of similar social constructions of masculinity that take for granted a dispassionate human exceptionalism and marginalise more ecological masculinities[40] and perspectives deserves its place at the table in trying to both historicise and address the climate crisis. Longstanding and recent deliberations of ecofeminists[41] on this matter illustrate, I think, that there is no need to lapse into gender essentialisms in this endeavour, such as portraying "special affinities" between "woman" and "nature" or viewing "men" as hopelessly fated to dominate the more-than-human. Instead, ecofeminism emphasises the historically entrenched ways in which human/nature dualism has been a gendered hierarchical separation, feminising and devaluing the very connective knowledge and compassionate wisdom that enable the human to comprehend a

---

38   Horrocks 1994; Levant 1997; Rogers 2008.
39   Connell 1995.
40   Hultman and Pulé 2018; Hunnicutt 2020.
41   Adams 1990; Gaard 1993; Hunnicutt 2020; Plumwood 1993.

shared ecological embeddedness with other species. As with the notion of hegemonic masculinity, these deliberations require us to conceive of gender in an institutional, or "structural", sense,[42] to appreciate how gender is built into practices, institutions and geopolitical decision making, and how gendered subjects are constructed vis-à-vis nature (non) relationality. Necessarily this can generate ideas for how critical feminist and ecopedagogy can construct subjects, practices and institutions differently.

While it can be challenging to pinpoint exactly how people's embodiment of hegemonic masculinity curtails relationality and ethical considerability to the more-than-human, the disavowal of care and empathy typical of it lends itself to a routinised instrumentalism of other species. In male-dominated societies which valorise a dispassionate reason and feminise the qualities of care, compassion and empathy, meaningful climate change policy is already, structurally, at a disadvantage, because a rationale of economic utility will dominate, and prevent notions of justice, respect or care from being applied to the more-than-human. Given that political power itself around the world also remains patriarchal, male dominance, to the extent that it embodies hegemonically masculine social norms, contributes to political dithering on climate action.[43] Research on ecomasculinities has also theorised ways in which masculinity has been bound up in the ecological crisis. This identifies two forms of masculinity which have been enabling for environmental exploitation, namely industrial/ breadwinner masculinities and ecomodern masculinities.[44] Industrial masculinities are associated with entrenched norms of masculinity which have positioned white men as chief instigators and benefactors of extractivism. "Industrial" is used

> to refer primarily to those individuals who possess and manage the means of production and support service corporations who are handsomely rewarded by wealth-creating practices. Notably,

---

42   Acker 1992; Risman 2004. In Chapter 6 I draw upon a practice theory approach which aims to go beyond the agency/structure dichotomy.
43   See MacGregor 2009, 129.
44   Hultman 2017; Hultman and Pulé 2019.

the term industrial is used here to also note the ways that broader social and environmental implications of industrialisation are backgrounded for the sake of capital growth. Some examples of these kinds of masculinities manifest in fossil fuel and mining executives, financial managers and bankers, corporate middle and senior level managers and administrators – the vast majority being Western, white and male.[45]

Ecomodern masculinities, on the other hand, are performed by men who take on a façade of environmentalism. Informed by the notion of ecological modernisation, which tries to reconcile capitalism and the environment but arguably ends up protecting the status quo, ecomodern masculinities are the equivalent of "new men" who purport to take on feminist "points of concern". As Hultman and Pulé outline:

> Ecomodern masculinities represents [sic] the embodiments of masculine identities that valorize sustainability through ecomodern notions of reform; they collapse social and ecological crises with economic concerns as a supposed balance point between competing forces. While distinct from industrial/ breadwinner masculinities in their willingness to recognize glocal challenges, ecomodern masculinities have emerged paradoxically; aligning global responsibility and determination with increased care for the glocal commons but held strictly within the market forces of industrial and corporate capitalism.[46]

This conception may be an appropriate way of thinking about eco-entrepreneurialism which sees the climate crisis as an opportunity to accumulate capital without subjecting, for example, capitalism and human exceptionalism to critical scrutiny. Hultman and Pulé's distinctions also signpost intersectional understandings of class, "race" and gender privilege in producing climate breakdown.

The masculinisation of animal consumption, long highlighted within ecofeminist and CAS research,[47] and in the broader social

---

45 Hultman and Pulé 2019, 478.
46 Hultman and Pulé 2018, 47.

science of food practices,[48] provides a specific example of how social scripts of gender mediate relations with other species. This tends to also translate into more women than men adopting vegetarianism and veganism,[49] although certainly most women continue to consume meat and dairy. This is not surprising since women's practices are also shaped by investments in and distancing from hegemonic meanings of masculinity. The legacy of this research is that a *degendering* of diet is necessary for both the ethical transformation of human–animal relations and sustainable food transitions. Social science research on animal consumption is examined more closely in Part II of this book but here provides one example of a gendered high-carbon practice. Western cultural icons of masculinity such as red meat and the automobile[50] not only symbolise nature mastery; they highlight that certain key high-carbon practices of capitalism also overlap with a specifically gendered (non) relationality to the more-than-human and imply that men generally contribute more to carbon emissions.[51] Relatedly, forms of ecological masculinities are likely to be found in counter practices such as cycling and veganism.[52] McCright found greater levels of both concern about climate change and assessed levels of climate science knowledge among women.[53] Later research found higher levels of climate denialism among conservative white men in the United States,[54] a finding later repeated for Norway.[55] Whether the new narrative of ecological masculinities can be politically effective and attractive to men remains to be seen. It strongly implies that such an identification needs also to be pro-feminist to best avoid reproducing the usual patterns of male dominance commonly found in "progressive" left politics. Furthermore, some men prefer to avoid the

---

47   Adams 1990; Potts and Parry 2010; Rogers 2008.
48   De Backer, Erreygers et al. 2020; Ruby and Heine 2011; Sobal 2005; Sumpter 2015.
49   Ipsos 2016.
50   Polk 2009.
51   Räty and Carlsson-Kanyama 2010.
52   Twine 2021a.
53   McCright 2010.
54   McCright and Dunlap 2011.
55   Krange, Kaltenborn and Hultman 2019.

language of masculinity completely, informed by queer discourse critical of simplistic identifications between "sex" and "gender". Noteworthy is also the specific emergence of research on gender and climate change[56] which has overlaps with the aforementioned longer-standing work on ecofeminism[57] and "gender and the environment" research.[58]

These bodies of work have long highlighted the gendered nature of environmental impacts such as the toxic effects of pollution upon embodied reproductive abilities and intersections between gendered divisions of labour and socially constructed vulnerabilities in the face of environmental breakdown. MacGregor, for example, contends that women cross-culturally are more likely to be providers and carers (i.e., the gendered social reproduction of labour highlighted earlier), and the everyday practices that comprise this work, such as related to subsistence, may be disrupted by climate breakdown. Furthermore, reinforcing the class-related points above, and acknowledging the global feminisation of poverty where women disproportionately comprise the poor, cannot sustain a generalised narrative of climate impacts,[59] instead necessitating an intersectional understanding of complex effects.[60] Thurston et al. have also highlighted that in the aftermath of disasters women and girls are more likely to be displaced and become victims of violence.[61] Yet this risk of gendered violence takes place not only with the impacts of climate-related disasters but in the very high-carbon industries that are fuelling climate breakdown. For example, the animal slaughter industry is low paid, highly male dominated and populated disproportionately in many locations by poor and precarious immigrant labour.[62] These are forms of violence on the workers themselves as well as on the disproportionately female animals they kill. Moreover, in some instances, cases of sexual assault

---

56 Alaimo 2009; Dankelman 2002; 2010; Denton 2002; Grusin 2017; MacGregor 2009; Pearse 2017; Zylinska 2018.
57 See Gaard 2015.
58 See MacGregor 2017.
59 MacGregor 2009, 130–31.
60 Kaijser and Kronsell 2014.
61 Thurston, Stöckl and Ranganathan 2021.
62 McSweeney and Young 2021.

against women have been found to be higher in communities with slaughterhouses.[63] Similarly, in their analysis of the Bakken oilfields of North America fuelled by the boom in fracking, Parson and Ray highlight increased violence toward women in temporary labour towns and the sexualised ad campaigns produced by energy companies wherein resource extraction is mediated through a predictable penetrative heteronormative symbolic lens.[64] Further evidence linking environmental degradation and gender-based violence (GBV) can be found in the wide-ranging International Union for Conservation of Nature 2020 report *Gender-based Violence and Environment Linkages: The Violence of Inequality*.[65] As it outlines:

> This analysis reveals the complex and interlinking nature of GBV across three main contexts explored in this paper: access to and control of natural resources; environmental pressure and threats; and environmental action to defend and conserve ecosystems and resources. Gender inequality is pervasive across all these contexts. National and customary laws, societal gender norms and traditional gender roles dictate who can access and control natural resources, often resulting in the marginalisation of women compared to men.[66]

The report goes into depth on such areas as gender-based violence as a form of control over land and resources, links between GBV and illegal environmental exploitation, and the impacts of agribusiness and the extractive industries on GBV.

Additional feminist work has directly and critically engaged with the framing of the Anthropocene.[67] These are critical of the generalised Anthropos notion as falsely degendered, non-racial and classless. Di Chiro also sees the Anthropocene as bound up in anthropocentrism and Zylinska discerns in the cultural imaginary of the Anthropocene two

---

63 Fitzgerald, Kalof and Dietz 2009.
64 Parson and Ray 2020.
65 Castañeda Camey, Sabater et al. 2020.
66 Castañeda Camey, Sabater et al. 2020, xiii.
67 Di Chiro 2017; Zylinska 2018.

dominant masculinist tropes which re-centre the human as supremely agential. She names these as Project Man 2.0 and World 2.0, both forms of technological escapism from the Anthropocene.[68] The former refers to the realm of transhumanist fantasy[69] whereby "man" might "upgrade" and escape from ecological crisis via human techno-enhancement and the remaking of the more-than-human is framed as an achievement that is somehow under control. World 2.0, on the other hand, names the familiar trope of giving up on Planet Earth and colonising other planets or moons, encapsulated in the contemporary projects of Elon Musk and Jeff Bezos. As well as featuring in many post-apocalyptic cultural texts, both tropes conform to what has been argued as a tendency[70] in the dominant masculinist imaginary to pursue particular technological fixes as a response to climate crisis. Similarly, Gaard names "the linked rhetorics advocating population control, anti-immigration sentiment, and increased militarism"[71] as a further example of the masculinisation of climate debates that serves to protect the status quo, or act parasitically upon it, as in the disaster capitalist[72] rubric of exploiting crises to consolidate inequality. Project Man 2.0 and World 2.0 fail to address the underlying "causality" of the climate crisis and are also adept at leaving intact relations of power rather than subjecting them to both critical scrutiny and redress.

This broader coverage of gender and climate analyses highlights important lessons both for enhancing the Capitalocene conceptual analysis and, more broadly, for political approaches to climate breakdown. The emergences and effects of the climate crisis are gendered, as are denialist protections of business-as-usual. Significantly, these points imply that part of the problem of the climate crisis can also be found in the social re/construction of gender norms which inform everyday practices at multiple social sites and scales. For example, this highlights the importance of the prefiguration of counter-hegemonic masculinities as a strategy for meaningful climate

---

68   Zylinska 2018, 31.
69   Twine 2010b.
70   Denton 2002; MacGregor 2009.
71   Gaard 2015, 24.
72   Klein 2007.

policy, underlining the relevance and urgency of the analysis and practice of ecomasculinities.[73] This must include contesting masculinity and male dominance at all levels of policy. The lessons are also for feminism as a social movement that is no longer sustainable as only a humanist project. Indeed, the posthumanist[74] and ecofeminist[75] strands of feminism have made clear for some time now that the emancipation of women is inseparable not only from struggles against capitalism, racism and heterosexism but also from the intersection of gender in human exceptionalism and vice versa. As mentioned earlier, *Feminism for the 99%: A Manifesto*[76] concurs that feminism must have an ecological dimension.

The Capitalocene and other animals

I end this chapter by further exploring its key assertion in the broader context of this book: that Capitalocene theory and critical animal studies (CAS) are mutually enriching. Due to its anti-hierarchical anarchist and intersectional ecofeminist heritage, CAS is inherently suspicious toward the implicit assumptions and post-political naiveties of the Anthropocene. In these respects, the Capitalocene framing is an understandable ally for CAS and, as alluded to above, they share some theoretical heritage such as the critique of Western dualism. If Capitalocene theory had, with regard to other species, only been concerned to contest the ontological separations and dualistic conceptions of nature/society, mind/body, human/animal, then there would still be much to do to get the two approaches in the same ballpark. However, Jason Moore's work goes noticeably further than that. It was already noted in Chapter 1 that Moore deemed it unreasonable to theorise capitalism apart from its relationship with

---

73   Allister 2004; Cenamor and Brandt 2019; Hultman and Pulé 2018; MacGregor and Seymour 2017; Pease 2019; Pulé and Hultman 2021; Twine 1997; 2021a.
74   Braidotti 2022.
75   Plumwood 1993.
76   Arruzza, Bhattacharya and Fraser 2019.

nonhuman animals.[77] This is a central tenet of CAS and represented in some key work.[78] Yet Moore's Capitalocene theory goes further still; for example, he accuses the Anthropocene of denying the "multi-species violence and inequality of capitalism"[79] and elsewhere calls for "multi-species politics of emancipation".[80] Thus, Moore's analysis at times overlaps with CAS although he is not engaging with the field to a significant extent. These are interesting connections but let us consider what a broader engagement with CAS could bring to Capitalocene theory.

The counterargument to the anthropocentrism of the Anthropocene is to underline the importance of human–animal relations to the climate crisis. Furthermore, in trying to clarify the exact character of the crisis I contend that human–animal relations are themselves in crisis and that this needs to be addressed to fully understand and remedy climate breakdown. This means several important correctives for including nonhuman animals in our understandings of how and why the climate is imperilled. These correctives are broadly historical, scientific, ethical and political; and they are to be found in the re-examination of the emergences of the climate crisis, its impacts, and how the systemic transformation of human–animal relations needs to be at the centre of reversing climate breakdown.

I have already noted an important disagreement in Moore's work in terms of the Anthropocene's historical assumptions. His historicisation and broader spatialisation of capitalism together with an understanding of its colonialist, patriarchal and human exceptionalist politics, already underway in the 15th century, are constitutive of the alternative narrative of the Capitalocene. The next step is to more fully understand how these developments, relations and practices gradually come to prefigure substantial rises in greenhouse gases. Malm performs important work regarding understanding the political economy of the scaling up of coal.[81] Environmental histories of oil such as Frehner's

---

77   Moore 2017, 599.
78   Best 2009; Nibert 2017a; 2017b; Wadiwel 2015.
79   Moore 2018, 239.
80   Moore 2017, 599.

Finding Oil: The Nature of Petroleum Geology, 1859–1920 also add to our understanding of the development of a fossil fuel that became so important for complexes of (transport) practice.[82] Something similar is required for properly understanding the histories of animal breeding and their eventual transformation into large-scale world-ecological making practices substantially contributing to the ecological crisis.

Previously I have developed Barbara Noske's concept of the animal-industrial complex[83] to give conceptual weight to the multiple scales and sites in which capitalism expropriates nonhuman animal life. Specifically, I offered a definition of the animal-industrial complex as

> a partly opaque and multiple set of networks and relationships between the corporate sector, governments, and public and private science. With economic, cultural, social, and affective dimensions it encompasses an extensive range of practices, technologies, images, identities, and markets.[84]

However, this begs the question of its historical development – how can the emergences and intensifications of animal exploitation in the animal-industrial complex be better situated within the development of capitalism itself? Such research is overall underdeveloped, but examples from historical animal studies and from environmental history are useful. Firstly, animal historian Harriet Ritvo's examination of the early professionalisation of animal breeding in 18th-century England[85] illustrates how it emerged from forms of instrumentalised experimentation aimed at coaxing new profitability from farmed animal bodies. In particular, Ritvo emphasises the normalisation of a concept of genetic property and the heightened control over, and manipulation of, farmed animal reproduction. Such professionalisation eventually afforded more manipulation of economically valued traits of meat and milk, further profitability, and expansion, prefiguring

---

81   Malm 2016.
82   Frehner 2011.
83   Noske 1989.
84   Twine 2012, 23.
85   Ritvo 1995.

20th-century developments in quantitative genetics and then, more recently, farmed animal genomics[86] and so-called gene-editing technology. It is not possible to theorise the expansion of the animal-industrial complex without including these technological developments in how farmed animals came to be bred and controlled. Neither are these developments separable from the human labour consequences of the intensification of the animal-industrial complex. As Neo and Emel underline, the animal sciences have contributed to the shift toward large-scale farms with lower production costs, larger numbers of animals, and fewer overall farms and human workers.[87]

Secondly, environmental historian William Cronon's *Nature's Metropolis: Chicago and the Great West*,[88] a work cited by both Jason Moore[89] and Fraser and Jaeggi,[90] is another example of necessary historical research. To be clear, this is not just a case of doing better environmental and animal history or animalising the history of capitalism; this enables a better account of the emergences of the climate crisis itself. Fraser offers the following summary of the significance of Cronon's text:

> In *Nature's Metropolis*, Cronon shows how the small, early nineteenth century, mixed Euro-Amerindian settlement of Chicago was transformed in a few short decades into the single most important US entrepôt for trade in lumber, grain, and livestock; how the city reconstructed the entire landscape to the West as its hinterland as it went about supplying the East; how the pull of its markets transformed biodiverse prairies into monocultural farm and grazing lands; how their produce became standardized and subject to abstraction in the buying and selling of futures; how fortunes were made and lost; how town/country relations, political power, class relations, and regional ecosystems were transformed together in an integrated symbiosis. In

---

86   Twine 2010a, 100.
87   Neo and Emel 2017, 46.
88   Cronon 1991.
89   Moore 2018.
90   Fraser and Jaeggi 2018.

Crononâ€™s account, "nature" and "society" are inextricably entangled with one another. Yet neither is reduced to the other. What we get, rather, is a precise, dialectical, and thoroughly historical account of the socioecology of US capitalism in the nineteenth century. To my mind, this is the right ontological starting point for an eco-critical theory of capitalist society.[91]

Like Moore, Fraser is methodically attentive to the classed, racialised and gendered dimensions of capitalist development, though unlike Moore, and despite this extract, she offers no specific multispecies dimension to her analysis. Nevertheless, what Cronon affords us is a way to understand the emergence of the animal-industrial complex in a specific time and space, which highlights the centrality of the human exploitation of other animals as "livestock" to capitalist development, and ultimately the climate crisis. Moreover, it bears stating that the lumber, grain and "livestock" industries closely overlap not just in this site of the animal-industrial complex (a very significant one given the subsequent development of the Chicago stockyards and the US role in the meatification of the global diet) but in many others, leading to great scales of change to land globally, and ultimately raised greenhouse gas emissions especially during the 20th century and since. It is telling that Cronon names his chapter on meat "Annihilating Space".[92] His economic history of the Chicago meat industry conveys its spatial influence on the United States during its rapid development in the 19th century. Much land was expropriated for animal production and key moments such as the development of the railroad, and refrigeration, enabled the broad infrastructure of the US animal-industrial complex to expand. Although Chicago would lose its entrepôt status in the 20th century as the expansion of the animal-industrial complex became more diffuse and less tied to place,[93] the ability of the expanding, global meat, dairy and egg industry to radically remake world ecology was established. For Cronon, "The packers' triumph was to further the commodification of meat. To alienate still more its ties to the lives

---

91   Fraser and Jaeggi 2018, 93–94.
92   Cronon 1991, 207.
93   Cronon 1991, 259.

and ecosystems that had ultimately created it".[94] This instigated a new set of human–animal relations able to radically expand the capitalist expropriation of nonhuman animal bodies and the exploitation of human workers in the service of capital accumulation. Similar to Cronon's work is Specht's *Red Meat Republic*, a history of the US beef industry.[95] This analysis highlights how ranching was not only responsible for the destruction of bison but heavily embroiled within wars against American Indians, their dispossession from their land and the creation of the reservation system.

Such histories of the animal-industrial complex are important because, in contrast to the Anthropocene, they reinstate a history of multispecies violence as inherent to both capitalism and the Capitalocene. Moreover, the violence of the animal-industrial complex has been and continues to be a violence not just against nonhuman animals but against ecologies and those humans disproportionately impacted by being part of its productive infrastructure, or by being targeted for its mass consumption. In line with Moore's Capitalocene theory *and* CAS, this reinstated history of violence is inseparable from, internal to and constitutive of the histories of capitalism, colonialism and patriarchy. Though the influence of some ecofeminist work on Moore was noted earlier, it is fair to say that a broader recognition of ecofeminist and CAS research underlining nonhuman animal subjectivities would enhance Capitalocene theory further.[96]

The historical developments of the animal-industrial complex have undoubtedly served to normalise the contemporary and staggering degree of violence toward nonhuman animals today. The animal-industrial complex developed from a capitalist context in which human exceptionalism was already normalised, but it then took that commodification further and made its violence mundane and routine. Indeed, the word "violence" has largely been removed from the ritualised mass slaughter of animals. It has taken the CAS field to remind academics of their largely unacknowledged immersion in social norms. Consider Erika Cudworth's important critique of the sociology

---

94   Cronon 1991, 256.
95   Specht 2019.
96   Thanks to Darren Chang for underlining this point to me.

of violence for example,[97] which forms part of the broader sociological CAS endeavour to convince the discipline that "social" should not be conflated with "human". Other animals are not only involved in pre-existing areas of interest for the sociology of violence (domestic violence, war); Cudworth also outlines how "violence towards domesticated animals is routinized, systemic and legitimated. It is embedded in structures of authority, such as the nation state, and in formations of social domination".[98] Cudworth's argument makes a mockery of the exclusion of nonhuman animals from the sociology of violence, and in doing so, highlights the ideological masking of what is done to other animals as denied the word "violence". This assumption of human exceptionalism and its marginalisation of disturbing human–animal relations may also offer a part explanation for why so many historians of capitalism have managed to avoid meaningful inclusions of human–animal relations, and why such relations continue to be ignored from discourses related to the emergences, impacts of and potential transformative solutions for the climate crisis.

Capitalism as a war on animals and animality

Dinesh Wadiwel's work may explain the disqualification of violence from being applied to normalised human–animal relations and provide insights into thinking through the history of the ideology of human exceptionalism.[99] Wadiwel makes the argument that our systems of violence perpetrated against nonhuman animals can be conceptualised as a war against animals. Drawing upon Foucault's work on sovereign power,[100] and Agamben's work[101] on the politically constitutive role of human/animal dualism, Wadiwel affords an important place to the role of human sovereignty in shaping human–animal relations. The assumption of a right to domination, a human entitlement to animal

97　Cudworth 2015.
98　Cudworth 2015, 14.
99　Wadiwel 2015.
100　Foucault 1990.
101　Agamben 1998.

bodies,[102] exercised by sovereign power over the more-than-human has been an integral part of expansionist capitalism and the Western humanist self-conception. This sense of sovereignty has informed the direct expropriation of other species in the unfolding story of the Capitalocene. Wadiwel argues that human sovereignty precedes ethics, which is problematic because "ethics constructed after sovereignty works only to regulate or mitigate the violent effect of that sovereignty, while leaving the basic structure of that domination intact".[103] This normalisation of sovereign power already wrong-foots attempts to construct animal ethics narratives, rendering them unintelligible and unimaginable, and reducing them to the more palatable framing of welfare.[104] Before animal ethics can be practised and effective, human sovereignty must be contested. For example, although voices in support of the abolition of animal agriculture come increasingly to the fore, assisted by its enmeshment in multiple crises, for a long time this goal was politically unthinkable and unspeakable. This is also why Wadiwel's pronouncement of a *war* on animals might appear hyperbolic.[105] However, as he explains, it is exactly part of the epistemological violence of this war to disguise and routinise itself, to distract from its *own* excess so that "at its most diabolical, [it is] a knowledge system that denies that this violence is occurring".[106] At this point Wadiwel is influenced by, and reminiscent of, the philosopher Jacques Derrida

---

102 Arcari 2020, 225.

103 Wadiwel 2015, 22.

104 Wadiwel 2015, 22.

105 Notions of a human war against the more-than-human are not restricted to CAS. In December 2020 the United Nations secretary general António Guterres gave a speech in which he said, "Humanity is waging war on nature. This is suicidal. Nature always strikes back – and it is already doing so with growing force and fury. Biodiversity is collapsing. One million species are at risk of extinction. Ecosystems are disappearing before our eyes … Human activities are at the root of our descent toward chaos. But that means human action can help to solve it." In the same speech he was further quoted as saying, "We must declare a permanent ceasefire and reconcile with nature" (Rowlatt 2020). Further evidence in support of Wadiwel includes the finding by Darimont, Cooke et al. (2023) that one third of all nonhuman vertebrate species have been exploited.

106 Wadiwel 2015, 27.

who also spoke of a tradition of war against animals shaped by
Judeo-Christian-Islamic theology and, later, Cartesianism.[107] In a
detailed philosophical exposition of what the CAS field would now
call the "animal-industrial complex", Derrida was also concerned about
cultural denials of violence to other animals. He writes:

> Such a subjection ... can be called violence in the most morally
> neutral sense of the term ... Neither can one seriously deny the
> disavowal that this involves. No one can deny seriously, or for very
> long, that men [sic] do all they can in order to dissimulate this
> cruelty or to hide it from themselves, in order to organize on a
> global scale, the forgetting or misunderstanding of this violence
> ...[108]

This concealment and disavowal are undoubtedly bound up in the
scale of such violence being culturally and morally incongruous to
definitions of civility. The aforementioned sociology of violence can be
seen paradigmatically as a (sub)field that forgets or refuses an inclusive
multispecies narrative of violence and in doing so perpetuates this
epistemic violence and human(ist) sovereignty. I have previously
critiqued the field of bioethics for excluding animals[109] and I now
augment those analyses in terms of underlining the uncritical
normalisation of human sovereignty historically at play in the bioethics
field. Other animal omissions in, for example, climate ethics or the
sociology of climate change can also be read similarly, and these are the
focus of the first part of Chapter 4.

Wadiwel's framing adds to our understanding of the
animal-industrial complex and points to the importance of
undermining human sovereignty if any progress is to be made on
improving the circumstances of presently commodified animals and
those whose habitats are threatened. Indeed, McBrien has argued that
accumulation and death are inseparable, positing the Capitalocene as
also a Necrocene,[110] a time of traumatic extinction. In thinking through

---

107 Derrida 2008, 101.
108 Derrida 2008, 25–26.
109 Twine 2005; 2007; 2010a.

the maintenance of the war against animals, Wadiwel argues that inter-subjective, institutional and epistemic violence all act to secure the continued domination and plunder of nonhuman animals. To these levels of theorising violence can be added Nixon's influential and complementary notion of "slow violence".[111] This he defines as "a violence that occurs gradually and out of sight, a violence of delayed destruction that is dispersed across time and space, an attritional violence that is typically not viewed as violence at all".[112] As Nixon and others[113] have noted, this conception of violence is particularly appropriate for theorising resource extraction, deforestation, ocean acidification and the multiple delayed effects of the climate crisis. It captures well the emissions lag whereby the warming experienced today is the result of cumulative emissions from some decades ago. This "incremental and accretive"[114] understanding of violence is further critical for theorising multispecies violence wherein the climate crisis enacts slow violence across class, gender, "race" and species. Although the ecocidal onslaught of climate breakdown does increasingly manifest itself in tangible spectacular events such as floods, fires, hurricanes and cyclones, it is also the slow creep of rising emissions and, for example, the delayed effects of trauma, acidification and sea-level rise which cut across species differences.

Wadiwel's war against animals is also partially a war against animality and this point is pivotal for appreciating the CAS relevance to Capitalocene theory. This pertains to a key argument of the CAS field, namely that the ideology of human exceptionalism and the animal-industrial complex also maintain systems of *human* domination[115] in a two-step move that a) constructs and denigrates animality, and b) projects animality onto humans marked out[116] by class, "race", gender, sexuality and disability. This would suggest that the colossal scale of violence achieved by the animal-industrial complex

---

110 McBrien 2016.
111 Nixon 2011. See also Arcari 2023.
112 Nixon 2011, 2.
113 Malm 2016, 8–9; McBrien 2016, 116; Parson and Ray 2020, 262.
114 Nixon 2011.
115 Wadiwel 2015, 9.
116 Twine 2001.

is not simply reducible to the expropriated, material "spoils of war" as capital accumulation, but, furthermore, that its performative reiteration consolidates entrenched symbolic meanings which enable and perpetuate hierarchical constructions of difference *within* the human. I have previously referred to this as an internally torn understanding of the human[117] that works as a significant generative dynamic in constructing human difference to underline how the dualistic polarisation of the human from "nature, animals and animality" has been not only the basis of the human exceptionalism that has led us to the climate crisis but also one of the main ways to ideologically justify the oppression of humans and other animals alike. Significantly, the climate crisis can be understood, from a CAS perspective, as a crisis of human exceptionalism and the unravelling of that way of thinking. One potential of CAS, in contesting devaluations of nonhuman animals and this construct of animality, is that it may be possible to destabilise the rhetorics of animalisation that have been a consistent part of intersecting forms of exploitation.

Furthermore, this proposes a fundamental point of reflection. If we are to be serious about addressing the climate crisis there is no alternative but to grasp at its discursive roots, which means contesting human sovereign power, human exceptionalism and anthropocentrism, and fashioning radically new ways of being in-relation with other species. This chimes with Wadiwel's position on the need for recognition of the war against animals and for relational spaces of "truce", meaning "new forms of connection, friendship, topography, love and living-together that have been previously unimaginable, and, as a result, lead to reconstruction of the human/animal binary in ways which might recognize multiple non hierarchized difference".[118]

This CAS perspective of a war on animals and animality, born in part from its historical overlap with ecofeminist theory, also already connects with Capitalocene theory. To reiterate, when Moore talks about the emergence of a new world-praxis of Cheap Nature in the history of capitalism which assigns value through a nature/humanity dualism[119] and accordingly justifies the exploitation and expropriation

---

117 Twine 2010a, 10.
118 Wadiwel 2015, 8–9. Examples are included in Chapter 8.

of humans, animals and ecologies alike, he is in partial synergy with both CAS and ecofeminism. What I have attempted to do in this chapter is to enhance aspects of Capitalocene theory in quite simple ways, either by extending it further into feminist, anti-racist and CAS literatures or to strengthen it via further conceptual perspectives. In doing so, I have argued that via a closer relation with CAS, Capitalocene theory can better address the anthropocentric roots of the climate crisis.

The intention has been to clarify from a CAS perspective that the Capitalocene constitutes assuredly a crisis of capitalism, but within that proclamation, the interrelated projects of patriarchy, colonialism and anthropocentrism similarly appear as bound up in the historical trajectory toward climate breakdown. As Moore would agree, gender, "race" and species hierarchies have always been internal to, and constitutive of, capitalism. This necessitates seeing the climate crisis through a lens of intersectionality,[120] a lens not smeared by human exceptionalism.[121] A prerequisite for intersectional thinking is the ability to empathise from the perspective of others. In clarifying what sort of crisis the climate crisis is, there is no alternative but to prioritise as many other perspectives as possible since experience and perception of crisis will be shaped by multiple vantage points. Already substantial numbers of nonhuman animals have experienced the direct and slow violence of the climate crisis. Chapter 3 explores this in more detail.

An intersectional framing which contests human sovereignty ensures the visibility of that violence. Kaijser and Kronsell, who take a multi-levelled and more-than-human approach to intersectionality, argue that "an intersectional analysis of climate change illuminates how different individuals and groups relate differently to climate change, due to their situatedness in power structures based on context-specific and dynamic social categorisations".[122] Merging such an approach with Capitalocene theory is one answer to the highly limited generalised Anthropos of the Anthropocene. By leaving behind a narrow scientism

---

119 Moore 2017, 600.
120 Kaijser and Kronsell 2014.
121 Twine 2010c.
122 Kaijser and Kronsell 2014, 417.

and instead valuing social scientific and humanities perspectives on the climate crisis, critical social scientists, other researchers and activists are likely to produce a more inclusive narrative of climate justice[123] and not read the climate crisis as an ahistorical present but to better understand its real historical emergences and the broad political challenges ahead.

In this and the previous chapter I have noted how the anthropocentrism of dominant Western cultures has shaped responses to the climate crisis in the form of the Anthropocene narrative. This has meant that several cultural and academic responses to the climate crisis have been proffered with their anthropocentrism intact, including startling omissions of human–animal relations that contradict climate science. The cultural ambivalence and dissonance outlined above in relation to the violence of the animal-industrial complex seeps into theorising about, and policy on, the climate crisis precisely because the crisis amplifies the fleischgeist. This, I contend, translates into a discomforting reluctance to confront the animal-industrial complex not only because of the ambivalence its violence engenders but also because it is so normalised, and most people are complicit with it. This turns it into a tainted and even shameful evidence base and so it becomes easier to prioritise a focus on other climate-relevant areas such as energy and transport. Even though there is now substantial scientific weight and consensus behind the view that animal agriculture makes a significant contribution to the climate crisis (see Chapter 4), it still struggles to make it onto climate policy agendas, including the United Nations Framework Convention on Climate Change's Conference of the Parties in 2021 (COP26).

Such exclusions are conceptually related to more general omissions noted in this chapter such as those of human–animal relations from the historicising of capitalism, or from the sociology of violence. Chapter 4 details further animal *omissions*: examples of academic and cultural knowledge production that have managed, despite the evidence, to conceptualise (the emergences of) the climate crisis as having little to do with human–animal relations. It will also focus on the issue of animal *emissions*: the question of the contribution of animal agriculture

---

123 Ellis, Maslin et al. 2016; Gaard 2015, 27–28.

to the climate crisis and the compelling evidence base that illustrates the necessity of foregrounding and contesting human–animal relations in any coherent response to the climate crisis and the inadequacy of their omission. However, firstly, in the next chapter, I examine the way the climate crisis is contributing to an unprecedented undermining of life.

3
How climate breakdown is undermining animal life

If the focus of tackling the climate crisis mainly attends to how it plays out for humanity, this perpetuates the very anthropocentric framing which has contributed to it. Moreover, it prolongs the damaging notion that human flourishing is not interdependent with other species. Already in extreme weather events such as bushfires, hurricanes, heatwaves and droughts, it is becoming clear that humans and other animals are dying as a consequence of climate change in increasing numbers.

This chapter aims to rectify an anthropocentric framing of the climate crisis and does this by outlining the myriad ways in which other animal species have already been affected by a climate in breakdown. Substantial space is devoted to outlining case studies of impacts of the climate crisis on specific species. This is intended as an effective approach for "animalising the climate crisis" and underlining how different environmental changes are impacting individual animals, specific species, or networks of interrelated species, and how they are, in turn, responding. Climate breakdown is also ecological breakdown as the viability of life becomes increasingly undermined and begins to unravel.

Knowledge here pertains to a broad field of scientific research and so a secondary aim of the chapter is to provide a useful guide for how to continue this research. For example, ecologists and conservation biologists have been diligently producing evidence for several decades into how a changing climate is impinging upon the sustainability of life, making them an important part of the climate science field *as well as* of animal studies. Some of this work has been covered by social and conventional media[1] engendering awareness of the ongoing severe biodiversity crisis and sixth mass extinction event.[2] Periodic reports are further significant, none more so than the global assessment report on biodiversity and ecosystem services of the Intergovernmental Science-Policy Platform on Biodiversity and Ecosystem Services,[3] which asserted that "[n]ature and its vital contributions to people, which together embody biodiversity and ecosystem functions and services, are deteriorating worldwide" and that 1 million species face extinction in the next few decades. Although the framing of "ecological services" is now a familiar and limited anthropocentric valuing of the more-than-human, it at least underlines interdependencies between the human and other species.

In a series of studies Ceballos et al. have analysed biodiversity loss, extinction and defaunation in detail.[4] Their 2015 analysis, which they characterise as methodologically conservative, found that the average rate of vertebrate species loss over the last century was up to 100 times higher than the background rate. The background rate refers to levels of extinction that might be expected based on fossil evidence. This actual figure may be scientifically contested; for example, other research has estimated that current extinction rates are 1,000 times higher than the background rate.[5] The 2017 analysis by Ceballos et al. corrected the potential exclusions of only focusing on species loss because that does not tend to expose significant degrees of population loss or "defaunation" that may also be occurring. This corrective, they argue, suggests that the sixth mass extinction is more severe than indicated by the data on species loss alone. Within their sample, comprising

1   Vaughan 2015.
2   Kolbert 2014.
3   IPBES 2019, 2–3.
4   Ceballos, Ehrlich and Dirzo 2017; Ceballos, Ehrlich and Raven 2020; Ceballos, Ehrlich et al. 2015.
5   De Vos, Joppa et al. 2015.

nearly half of known vertebrate species, 32% are decreasing, in both population size and range. Mindful of species interdependencies, they highlight the cascading consequences of such losses for ecosystem viability. Indeed, they use the phrase "biological annihilation" in their paper title. The later (2020) paper examined 29,400 species of terrestrial vertebrates and determined which are on the brink of extinction because they have fewer than 1,000 individuals. It found that 515 species are on the brink (1.7% of the evaluated vertebrates). Concluding, it argued that the influential criteria of the International Union for Conservation of Nature for defining when a species is "critically endangered" should be increased to include all species thought to have under 5,000 mature individuals remaining. Moreover, its view was that "the conservation of endangered species should be elevated to a national and global emergency for governments and institutions, equal to climate disruption".[6] This research alludes to how the different categories of the International Union for Conservation of Nature's Red List – ranging from least concern, to near threatened, to vulnerable, to endangered, to critically endangered, to extinct in the wild, and to extinct – can be definitionally contested. A species may also be defined as data deficient, as the union's task to assess more species is ongoing. As of October 2022, it defined over 41,000 species as threatened with extinction, including 41% of amphibians and 27% of mammals.[7] To be placed in a category such as endangered or critically endangered has significant consequences for conservation policy,[8] underlining the importance of the above research by Ceballos et al. Work by the International Union for Conservation of Nature, the Intergovernmental Science-Policy Platform on Biodiversity and Ecosystem Services and Ceballos and his team paints a vivid and perilous picture of the precarity of life for an increasing number of nonhuman animals.

An initial problem in confronting how the climate crisis is affecting nonhuman animals may be an apparent need to disentangle actual climate changes as causes of biodiversity loss from a wider

---

6   Ceballos, Ehrlich and Raven 2020, 6.
7   https://www.iucnredlist.org/.
8   Braverman 2015, 191.

range of other explanations. While the climate crisis and biodiversity crisis at first appear distinct yet overlapping, there are reasons why so many species are at risk of extinction which seem to have little to do with a changing climate. Habitat loss and degradation is being exacerbated by climate change but is also caused by economic development, such as losses from roadbuilding.[9] Other modes of capitalisation on the lives of other animals and sources of biodiversity loss include hunting,[10] poaching,[11] fishing,[12] chemical pollution,[13] wildlife trafficking and trade,[14] human recreation,[15] the consumption of bushmeat[16] and the further issue of invasive species.[17] All these contribute to population declines, or defaunation. An example where climate change does not seem to yet be a major reason for defaunation is found in a 2019 analysis of the decline of freshwater megafauna[18] which is explained more by hunting, pollution, dam construction and the use of freshwater for crops. At the same time, the climate and biodiversity crises *are* caused by some of the same economic developments, most notably in mining, deforestation and animal agriculture. The growth of animal agriculture (often inclusive of deforestation) has been a significant contributor to greenhouse gas emissions and climate breakdown. This growth has similarly contributed to the biodiversity crisis. Machovina et al. argue that the key to biodiversity conservation is reducing meat consumption, pointing out that "livestock production is the single largest driver of habitat loss, and both livestock and feedstock production are increasing in developing tropical countries where the majority of

---

9   Quintana, Cifuentes et al. 2022.
10 Benítez-López, Alkemade et al. 2017. Importantly, much hunting is state sanctioned and intended to protect animal agriculture. For example, the US Department of Agriculture's Wildlife Services disclosed that they killed 1.75 million animals in 2021 (Milman 2022).
11 Gore, Ratsimbazafy et al. 2016.
12 Rijnsdorp, Bolam et al. 2018.
13 Bernhardt, Rosi and Gessner 2017.
14 Sollund 2022; Symes, Edwards et al. 2018.
15 Larson, Reed et al. 2016.
16 Ripple, Abernethy et al. 2016.
17 Gallardo and Aldridge 2013.
18 He, Zarfl et al. 2019.

biological diversity resides".[19] Similarly, the World Wildlife Fund for Nature report *Appetite for Destruction*[20] highlighted the impact of animal feed crop production on biodiversity loss. As well as underscoring the reliance of UK agriculture on animal feed crops harvested overseas, the report highlighted key areas around the world where animal feed crop production is being intensified, areas which are vulnerable and important for biodiversity. These include the Northern Great Plains of North America, the Atlantic Forest, Amazon and Cerrado of Latin America, the Congo Basin and Great Rift Lakes of Africa, and the Eastern Himalayas, Central Deccan Plateau, Amur-Heilong and Yangtze River Basin of Asia. As the next chapter shows (Table 4.4), the World Wildlife Fund for Nature has joined the chorus for reducing meat consumption.

It is relevant now to recall the Capitalocene concept from Chapter 1. In its understanding, the colonialist, patriarchal and anthropocentric character of capitalist development becomes embroiled in and reproduces the web of life to foster extensive projects of capital accumulation. Consequently, it makes less and less sense to exact a clear distinction between the climate and biodiversity crisis because both are underpinned by the same capitalist development which assumes human sovereignty and the perpetuation of a war against nonhuman animals.[21] So, while, for example, species face varied threats from climate change, hunting and the trade in wildlife, these dangers are all underpinned by more or less direct attempts to capitalise nonhuman animals. This is important to bear in mind even as this chapter aims to make clear in a scientific sense specific effects of climate changes on nonhuman species. The overlapping climate and biodiversity crises merge under the broader sixth mass extinction event.[22]

As noted in Chapter 1, the classed, racialised and gendered Capitalocene confers complex intersecting vulnerabilities as climate breakdown worsens. The anthropocentric character of the Capitalocene which plays out in assumed human sovereignty and exceptionalism

---

as well as a general cross-cultural social unintelligibility of ethical frameworks for valuing other animal species arguably bestows starker vulnerabilities for more-than-human species. If other species are not seen as ethically significant, and as "beautiful, fascinating, and culturally important",[23] then they are likely to disappear via direct expropriation or as the "collateral damage" of climate breakdown. While a broad understanding of the ideas of animal resistance[24] and agency *could* include the very adaptive responses of species to climate breakdown discussed below, the risks to nonhuman animals are further exacerbated through their general reliance upon human advocacy in the context of largely unsympathetic anthropocentric cultures. A stark example underlying the different scales at which deaths are occurring from climate breakdown was the 2019–20 Australian bushfire season, since known as "Black Summer",[25] killing 34 humans directly and a further 445 from smoke inhalation[26] but thought to have killed or harmed over 3 billion nonhuman animals.[27] Although media coverage tended to focus on iconic Australian mammals such as the koala, the range of species affected was very broad. Australia's Natural History Museum Directors put the figure in the trillions "when considering the total of insects, spiders, birds, mammals, frogs, reptiles, invertebrates and even sea life impacted over such a vast area".[28] Such numbers are difficult to comprehend but they translate into individual animal suffering on a colossal scale.

This quote also hints at one simple way of approaching a review of climate impacts on species, by conducting a sampled examination according to the sub-phylum vertebrata which encompasses fishes, mammals, birds, reptiles and amphibians, as well as major invertebrate groupings including insects, arachnids, molluscs, crustaceans and corals. It is worth bearing in mind that there are only approximately 70,000

---

23   Ceballos, Ehrlich et al. 2015, 3.
24   Coppin 2003.
25   Hitch 2020a.
26   Hitch 2020b.
27   BBC 2020.
28   Australia's Natural History Museum Directors (2020). Additionally, the Brazilian wildfires of 2020 (partly caused by climate change) directly killed 17 million nonhuman animal vertebrates (Tomas, Berlinck et al. 2021).

known vertebrate species, and 1.3 million invertebrates, with most of these insects and arachnids. Despite this imbalance, researchers have typically granted more attention to vertebrate species, which has been referred to as "institutional vertebratism"[29] or "insect speciesism".[30] For reasons specified shortly, this approach of assessing climate impacts via categories of vertebrates and invertebrates, while relatively straightforward, is not quite sufficient for grasping species interdependencies and how they may be undermined by climate breakdown. A second approach is to proceed with a review using the dominant social constructions of other animals. Indeed, it could be assumed that any such review is focused only on "wild" animals since when discourses of biodiversity and conservation are present this is often taken for granted. However, this would be a limited review given current populations. To exclude impacts on, for example, captive animals, companion animals and animals used within agriculture would give a skewed representation especially when underlining how the biomass proportions of mammals have shifted in the recent history of the Capitalocene.

For example, the biomass of "livestock" animals now dwarfs that of "wild" animals (in these analyses focused on mammals) and is also considerably larger than the human biomass. Shockingly, the entire mammal biomass is 60% farmed animals, 36% humans and only 4% "wild" animals.[31] The total "livestock" biomass is far greater than "wild" animals (mammals) yet with its biodiversity miniscule in comparison.[32] A 2023 paper underlined that the biomass of farmed pigs is almost double that of all wild land mammals.[33] Not surprisingly, given the economic value of the animal-industrial complex, research has been underway for some time into climate impacts upon farmed species, some of which I attend to later. Farmed animals, already experiencing the suffering related to their commodification, now face the compounded suffering of climate-related events such as flooding, fires and heatwaves. A third approach involves augmenting a focus on species

---

29 Leather 2013.
30 Gunderman and White 2020.
31 Bar-On, Phillips and Milo 2018; Smil 2011.
32 Zeller, Starik and Göttert 2017, 122.
33 Greenspoon, Krieger et al. 2023.

with geographical regions or biomes such as marine, tropics, grasslands, temperate, Mediterranean, polar, freshwater, mountainous, boreal forests and so on. This allows for the documentation of large-scale ecological changes such as desertification, ice melt, precipitation and temperature changes, and the identification of geographical biodiversity hotspots.[34] Although with a limitation in coverage, the approach here is to combine elements of all three approaches by presenting cases from the literature which cut across species classifications, geographical spaces and cultural constructions. Before proceeding, a further issue of atomisation versus interdependency requires attention.

When engaging with even the basics of environmental thought, let alone more detailed analyses of the impacts of climate breakdown, a point of contention is the historically entrenched framing of the human as above and separate from the "rest of nature". This failure to understand and take seriously human interdependencies with other species has been a hallmark of the Capitalocene. However, as one delves into the wealth of ecological knowledge accrued around the emergent impacts of climate breakdown upon other species, the required understanding of species interdependencies becomes increasingly salient. This can also be read as a reminder to (critical) animal studies (C/AS) to be reflexive to its own possible exclusions. Specifically, if the same atomistic thinking that has historically been applied by the social sciences to studying humans is reproduced then it is possible to end up in a similar anti-ecological ontological mess. It is then conceivable that ecologically obvious and fundamental animal–*plant* interactions become de-emphasised because plants are somehow not seen as the subject of animal studies (just as the social sciences have traditionally excluded both nonhuman animals *and* plants). To counter this, it makes sense to learn from ecologists who are accustomed to analysing "ecological assemblages",[35] multiple interrelated species present in diverse communities. The interrelation can take many forms such as plants providing food and habitat, and animals playing a role in pollination and seed dispersal. It also makes sense to re-orient our thinking in this way because plant species are also becoming extinct

---

34   Pacifici, Visconti and Rondinini 2018; Manes, Costello et al. 2021.
35   Anderson and Thompson 2004.

for the same reasons animal species are. For example, Humphreys et al. report that close to 600 plants have become extinct since the industrial revolution and that this itself is likely to be an underestimate.[36] A 2020 report by the United Kingdom's Royal Botanical Gardens viewed 40% of the world's plants at risk of extinction.[37]

A further potential impediment to grasping ecological interdependencies has been an over focus on charismatic megafauna in (critical) animal studies (C/AS), be they animals of agriculture, companionship or conservation. At times this resembles a conflation of "animal" with vertebrate, to the exclusion, strictly speaking, of most animal species. This bias of scale risks excluding many species, which would both broaden the scope of C/AS and improve its understanding of climate change and ecological interdependency.[38] For example, ectothermic animals, whose regulation of body temperature relies upon external sources, such as fishes, reptiles, amphibians and invertebrates, are particularly vulnerable to minor changes in temperature. Indeed, such species and those which have shorter reproductive cycles may be especially important because they can act as early warning bioindicators of climate change.[39]

The public representation of conservation biology in relation to climate breakdown also often latches onto charismatic megafauna perhaps most noticeably in the case of the image of the polar bear,[40] and it may be fair to conclude that such aestheticisation of climate breakdown has not been particularly effective in promoting social change. What becomes clear when reviewing the scientific literature on how a changing climate is impacting species is the importance of smaller scale animals such as insects and microorganisms. Small changes to climate such as temperature rises can have considerable effects upon other species and processes which rely upon them. With increasing awareness of insect decline and the important ecological

---

36 Humphreys, Govaerts et al. 2019.
37 Antonelli, Fry et al. 2020.
38 Concurrently there is a need for framings of conservation and biodiversity to transcend their own anthropocentrism (see Treich 2022).
39 De Groot, Ketner and Ovaa 1995.
40 Slocum 2004; Yusoff 2010.

role of some insects in pollination, it may be that prior exclusions of invertebrates are starting to be addressed within C/AS. One challenge, especially for critical animal studies, is to reflect and convey how invertebrates matter as animals.[41] For in this discussion of climate impacts one is inevitably constrained into talking of such species in terms of their "ecological role", an issue which animates historical tensions between animal studies and ethics versus environmentalism and environmental philosophy.[42] In addition to pollination, insects (and microorganisms) are also involved as detritivores and decomposers, making a valuable contribution to ecosystems and to soil health. Moreover, the threat of insects acting increasingly as disease vectors in a warming climate has brought certain species more into the policy, media and scientific limelight. It is noteworthy that although all these ways in which insects matter in relation to climate change emphasise human–insect interdependencies, they also perpetuate an anthropocentric framing of their mattering in terms of their contribution to agriculture, as novel foodstuffs themselves, as disease threats to the human, or as a risk to the bio-capitalisation of "livestock".

Before commencing a review of climate changes together with ecological and physiological responses in a broad range of species, I suggest that climate change can be partly viewed as a systemic form of slow violence against animal life. The idea of slow violence, introduced in Chapter 2, is applicable to a gradual and less visible violence associated with environmental exploitation.[43] Although the original formulation is not applied specifically to climate impacts on nonhuman animals, it is equally applicable. The slow, gradual changes of temperature increases, ocean acidification and deoxygenation, ice loss and shifting seasons undermine the very sustainability of life for a multitude of species in a systemic effect of Capitalocene violence. As indicated above, its applicability must be set alongside the more dramatic and representable aspects of climate breakdown increasingly

---

41 For example, Gunderman and White (2020) are critical of attempts to mass produce insects for Global North consumers as a strategy to tackle unsustainable food systems.
42 Fitzgerald 2018.
43 Nixon 2011.

seen in extreme weather events. Furthermore, much climate science research has looked at the possibility of abrupt climate change related to potential tipping points[44] speeding up the further breakdown of the climate in the future. Abrupt biodiversity loss could even occur without abrupt climate breakdown. In their analysis of annual projections (from 1850 to 2100) of temperature and precipitation across the ranges of more than 30,000 marine and terrestrial species to estimate the timing of their exposure to potentially dangerous climate conditions, Trisos et al. "project that future disruption of ecological assemblages as a result of climate change will be abrupt, because within any given ecological assemblage the exposure of most species to climate conditions beyond their realized niche limits occurs almost simultaneously".[45] Given ongoing rates of defaunation over what is an abnormal timespan, argued by the World Wildlife Fund for Nature's *Living Planet Report* to be an overall decline of 60% in vertebrate population sizes between 1970 and 2014,[46] this should not be a surprise. In 2022 the *Living Planet Report* updated this to say that on average, global populations of mammals, birds, fishes, amphibians and reptiles had declined by 69% since 1970.[47] Such a decline over a relatively short period of time is unprecedented. While these dramatic rates of defaunation are not being produced by climate change alone, it would be a mistake to overplay the notion of slow violence since it may imply a steady and predictable unfolding of climate change, at the cost of non-predictability and changes happening at different rates around the world. With this qualification in mind, it can be noted how nonhuman animals are already subject to the extreme weather and slow violence dimensions of a novel climate reality, alongside the direct and indirect violence of the animal-industrial complex and the other modes of capitalisation including hunting and the wildlife trade. It is consequently predictable that the ultimate consequence for species may be extirpation or extinction as the 21st century unfolds. The science, as I aim to show, ultimately backs up discursive framings of a sixth mass extinction event.

---

44   Bathiany, Dijkstra et al. 2016; Jansen, Christensen et al. 2020.
45   Trisos, Merow and Pigot 2020, 496.
46   WWF 2018.
47   WWF 2022b.

Reviewing environmental and species impacts of climate breakdown

Given the aforementioned valorisation of megafauna, it seems appropriate to begin this main section of the chapter with insects. Thinking about insects has not been intuitive to the social sciences or (critical) animal studies. However, there are recent exceptions to this which have emerged from sociologists working in animal studies. For example, Moore and Kosut examine bees and beekeeping, underlining human dependencies on bees due to their role in pollination.[48] As Nimmo further points out, "Cultural imaginings of bees and 'bee society' have been entangled with historically changing – and contested – political inflections of collectivism, individualism, cooperation, subservience, altruism, and self-interest".[49] Environmental sociologists have also examined the controversy over insecticides and colony collapse in bees.[50] Insects have crawled onto the sociological radar, figured in some policy circles as more "sustainable" sources of protein for human consumption.[51] Insects are interesting for their capacities, differences, ecological importance, social symbolism, the common association of many varieties with human emotions of fear and disgust, and their assumed lowly moral status vis-à-vis more charismatic megafauna. Bees, butterflies and ladybirds, for example, appear to be exceptions to such devaluation. Contrastingly, with mosquitoes and ticks, the very term "vector" reduces their status to disease carriers and transmitters.

The impact that climate change is having on insect populations and will have in the future currently constitutes an area of considerable research focus.[52] Insects are very important for biodiversity generally, for example, as food sources for birds and many vertebrate animals. Changing temperature patterns, varying precipitation and altered food sources for insects are likely to exert considerable pressure upon

---

48 Moore and Kosut 2013.
49 Nimmo 2018, 31.
50 Kleinman and Suryanarayanan 2013.
51 Wilkie 2018.
52 Johnson and Jones 2017.

ecosystems as they currently exist. This can be explored further with some examples of ecological assemblages involving interrelationships between trees, caterpillars and bird species. Ecologists use the term phenological mismatch to refer to a lack of synchronicity between predator and prey species which may occur either temporally or spatially. This mismatch emerges as one of the most significant temporal disruptions of climate breakdown. At its simplest, if at the time a species gives birth its usual food supply is absent its survival is at risk. This may include animal–animal interactions, plant–animal interactions, and complex variations thereof.[53] In their analysis of the oak-caterpillar-bird system in the United Kingdom, Burgess et al. found that asynchrony between peak caterpillar numbers and peak nestling demand of blue tits, great tits and pied flycatchers increases in earlier, warmer springs.[54] Assuming climate change confers continued spring warming, they predict that temperate forest birds will become increasingly mismatched with peak caterpillar numbers. The leafing oak trees in this case provide the habitat for caterpillars and are themselves subject to shifting growth times. A similar study, but looking at Dutch populations of pied flycatchers, found "pied flycatchers have declined by about 90% in areas with the earliest food peaks, but have only declined by about 10% in areas with the latest food peaks".[55] This example encapsulates insect–plant interactions as part of the assemblage, but the case of pine beetles in North America underlines this dimension further. Different species of pine beetle such as the mountain pine beetle in the north-west of North America and the southern pine beetle in the south-eastern states of the United States have exhibited substantial range expansion shifting north in both cases to reflect warming temperatures.[56]

These insects are usually constructed as pests for the economic and ecological damage they cause to forests, in some areas (such as California) compounding the climate breakdown impacts of wildfires. Although the example of the shifting ranges of pine beetles may

53 Donnelly, Caffarra and O'Neill 2011.
54 Burgess, Smith et al. 2018.
55 Both, Bouwhuis et al. 2006, 81.
56 Lesk, Coffel et al. 2017; Raffa, Aukema et al. 2008.

emphasise how climate changes favour some insect species, the broader research indicates an overall crisis of insect decline. Again, this picture of defaunation is not reducible to climate change but it is one of the main drivers. In their review of worldwide decline, Sánchez-Bayo and Wyckhuys forecast the potential extinction of over 40% of insect species worldwide over the next few decades.[57] Moreover, they argue that the four main drivers of decline, in order of importance, are habitat loss and conversion to intensive agriculture and urbanisation; pollution, mainly that by synthetic pesticides and fertilisers; biological factors, including pathogens and introduced species; and climate change. The co-extinction of species on which insect species are reliant is a further factor.[58] Localised and longitudinal studies have already sounded the alarm, notably Hallmann et al., who found that since 1990, in 63 nature protection areas in Germany, there was an estimated seasonal decline of 76% in flying insect biomass.[59] The seriousness of overall declines prompted a group of leading entomologists to publish a warning to humanity on insect extinctions in which they argued for the importance of insects on intrinsic, ecological and economic grounds.[60]

Apprehending how climate change will specifically affect the insect vectors of human pathogens is of paramount importance to health provision around the world. Specifically, climate change is extending the range at which insect "vectors" such as ticks and mosquitoes travel.[61] Milder winters also assist in the breeding cycle of certain insects. Such impacts as they pertain to novel human vulnerabilities to disease infection underline a clear interconnection between climate breakdown, insects and human health. Belying dualistic assumptions, climate impacts on "animals" are often also impacts on "humans". For example, as Chaves underlines, vector-borne diseases account for 17% of the total burden of infectious diseases affecting humans,[62] with over 1 billion people becoming infected and over 1 million dying each year.[63]

---

57 Sánchez-Bayo and Wyckhuys 2019. For a critique of this paper, see Thomas, Jones and Hartley 2019.
58 Cardoso, Barton et al. 2020.
59 Hallmann, Sorg et al. 2017.
60 Cardoso, Barton et al. 2020.
61 Altizer, Ostfeld et al. 2013.
62 Chaves 2017, 127.

This consolidates socio-economic and racialised health inequalities. Climate change has the capacity to introduce this risk to countries lacking the experience and infrastructure to deal with the problem. In the case of malaria, spread by certain types of mosquito, it has been decreasing globally but has remained stable or increased in specific locations. For example, it has returned to Greece and modelling predicts it could return to the United Kingdom later this century.[64] Larger scale modelling work suggests that climate change is expected to make "malaria burdens higher than they would otherwise have been".[65]

Chaves is clear that vector-borne disease transmission is an ecosocial phenomenon, "where many factors, beyond the biology of vectors and parasites under changing environments, play key roles on emerging patterns of disease transmission that have occurred along with global warming".[66] Similarly, Parham et al. argue that it is "important to view climate-driven disease systems as complex socio-ecological dynamical systems".[67] Thus, although increased temperatures, up to a certain point, encourage vector spread, human practices can increase the likelihood of disease transmission. This includes the degree of urbanisation and population density and extends to practices recognised to constitute part of adaptive or mitigative responses to climate change. For example, both flood protection through provision of new wetlands and increased urban greenspace to minimise an urban heat island effect can affect the risk of vector-borne disease.[68] Several species of mosquito have colonised parts of Europe since 1990, including *Aedes albopictus*. This species has been "implicated in the last ten years in the transmission of chikungunya (>200 human cases in Italy) and dengue (isolated cases in France and Croatia), after these non-native mosquitoes acquired infection by blood-feeding from infected travellers".[69]

---

63   Campbell-Lendrum, Manga et al. 2015, 1.
64   Medlock and Leach 2015.
65   Campbell-Lendrum, Manga et al. 2015, 3.
66   Chaves 2017, 127.
67   Parham, Waldock et al. 2015, 2.
68   Medlock and Leach 2015, 1.
69   Medlock and Leach 2015, 3.

The United Kingdom is likely to face similar threats this century. The *Ixodes ricinus* species of tick is now found throughout the United Kingdom. It is a vector for Lyme disease and tick-borne encephalitis virus – both potentially serious conditions in humans and dogs. It is expected to increasingly affect mainland Europe.[70] Writing in the North American context of this tick, Ostfeld and Brunner point out the further importance of factoring in the climate change impacts on hosts.[71] An important host in this context is the white-footed mouse and continued northward range expansion could extend Lyme disease into new parts of Canada.[72] In the United Kingdom, the incidence of Lyme disease has increased in recent years and there is some evidence that total numbers of cases have been underestimated.[73] Such shifts have consequences for advising preventative behaviour and the training of health professionals in many societal contexts, constituting just one aspect of projected human health impacts of climate change.[74]

The shifting ranges of insect vectors also has consequences for both companion and farmed animals. In fact, this constitutes a significant amount of the academic literature on the climate change impacts upon companion animals. Abdullah et al. have conducted several surveys of cats and dogs in the United Kingdom to better understand which infections are transmitted via tick bites and as a surveillance exercise regarding knowledge of changing distributions.[75] Companion animals also face other threats of extreme weather events, especially heat stress from heatwaves and hurricanes, as noted in Chapter 2. Similar threats face animals held in zoos with zookeepers more often having to think of ways to keep them cool as the frequency and severity of heatwaves increases. While some research has looked at the effects of climate change on the occurrence and distribution of diseases (including vector-borne disease) which affect "livestock" animals,[76] the impact of climate breakdown on the breeding of animals for human consumption

---

70   Medlock and Leach 2015, 6.
71   Ostfeld and Brunner 2015.
72   Roy-Dufresne, Logan et al. 2013.
73   Cairns, Wallenhorst et al. 2019.
74   McMichael, Woodruff and Hales 2006.
75   Abdullah, Helps et al. 2016; 2018; 2019.
76   Bett, Kiunga et al. 2017.

is likely to be far broader. This is of note because the sector tends to self-project considerable future increases in meat and dairy production to feed a growing human population. Again, extreme weather events such as heatwaves, floods and wildfires are already killing species such as sheep or cattle ungulates. Rojas-Downing et al. provide a useful overview of impacts in terms of the effects of rising levels of carbon dioxide, rising temperatures and variation in rainfall.[77] For example, increases in temperatures are likely to lead to increases in the need for water consumption in an already water intensive mode of agriculture. Temperature increases could also impact the availability of crops for animal feeds, and have negative impacts on reproduction, health and the "productivity" of animals. For example, drawing on previous research they suggest that increased temperatures may reduce body size, carcass weight and fat thickness in ruminants, and in pig production, larger pigs will see reduction in growth, carcass weight and feed intake.[78] Furthermore, the Intergovernmental Panel on Climate Change's report *Climate Change and Land* expressed high confidence that "in drylands climate change and desertification are projected to cause reductions in crop and livestock productivity".[79] Unsurprisingly, animal scientists have turned to farm animal genomics in the hope of breeding animals which are more robust, heat tolerant or less methane emitting, under a general rubric that I have previously referred to as the molecularisation of sustainability.[80] Alongside feed interventions, such technologies can be seen as protectionist of the profitability status quo of the animal-industrial complex. Yet broader climate mitigation policies could also have considerable impacts on the scale of "livestock" production, for example, as conflicting views over land use may argue for a decrease in the land afforded to animal agriculture and an increase in afforestation.

As illustrated, the impacts of insects intersect with a broad range of social categories of nonhuman animals ("wild", "companion", "livestock") as well as producing new risks for human health. Another

77   Rojas-Downing, Nejadhashemi et al. 2017.
78   Rojas-Downing, Nejadhashemi et al. 2017, 149.
79   IPCC 2019, 16.
80   Twine 2010a, 137–38.

set of animals which similarly could be easily omitted from a review of climate impacts are microorganisms. They have already been implicated in the discussion of vector-borne disease above, but they are significant to this review in broader ways. As far away in terms of scale and visibility as is possible from charismatic megafauna, microorganisms are nevertheless a crucial component in reviewing the impact and interconnections of climate breakdown with the biosphere. The 2019 publication of a Consensus Statement by a large group of microbiologists entitled "Scientists' warning to humanity: microorganisms and climate change"[81] aimed to bring to wider attention how microorganisms are affected by and, in turn, affect climate change. As they point out, the ecological role of microorganisms in both terrestrial and marine biomes is vast. Soil microorganisms "regulate the amount of organic carbon stored in soil and released back to the atmosphere, and indirectly influence carbon storage in plants and soils through provision of macronutrients that regulate productivity (nitrogen and phosphorus)".[82]

Climate change disruptions to these roles can enable increases in greenhouse gas emissions as with the well-known example of methanogens releasing methane as permafrost melts. In marine biomes, microorganisms like phytoplankton constitute the base of several food webs. Rising temperatures and other effects of climate change can interrupt their distribution and have further cascading ecological impacts. As Cavicchioli et al. point out, "Increases in solar radiation, temperature and freshwater inputs to surface waters strengthen ocean stratification and consequently reduce transport of nutrients from deep water to surface waters, which reduces primary productivity".[83] Marine phytoplankton also play an important role in carbon dioxide sequestration, which could be disrupted by climate change.

Such statements are reminders of the intricate multiple-scale ecological interdependencies at risk as the climate breaks down. Acknowledging the significance of marine microorganisms is also a

---

81   Cavicchioli, Ripple et al. 2019.
82   Cavicchioli, Ripple et al. 2019, 574.
83   Cavicchioli, Ripple et al. 2019, 571.

reminder for (social) scientists of climate change not to neglect the oceans and the varied marine biomes which support such a diversity of life and play crucial roles in the planetary carbon cycle. Sociologically, oceans have been neglected as social spaces[84] with their ecological importance not given due attention.[85] Payne et al. identify threats of oceanic mass extinction finding that those animals of larger body size are most threatened. This sets up a particular risk to marine ecosystems as they outline:

> The preferential threat to large-bodied marine animals poses a danger to ecosystems disproportionate to the percentage of threatened species. Large-bodied animals are critical to ecosystem function because of their preferential position at the top of food webs and importance to nutrient cycling and bioturbation of sediments. Removal of large-bodied predators can also trigger trophic cascades affecting many other species.[86]

Concern has also been raised recently about ocean deoxygenation and the formation of "dead zones" (hypoxic areas with reduced levels of oxygen) which have insufficient oxygen to sustain life. In September 2018 hundreds of scientists signed the *Kiel Declaration on Ocean Deoxygenation*, urgently calling for more marine and climate protection. The two main causes of deoxygenation are climate change and intensive farming, and the number of such dead zones has increased in recent decades. Run-off from farmed animal waste and fertiliser use creates an excessive nutrient load, largely of nitrogen and phosphorous, in coastal waters in a process known as eutrophication, causing algal blooms and deoxygenation. In 2017, for example, the media reported upon the contribution of the US meat industry to the creation of a large dead zone of nearly 21,238 square kilometres in the Gulf of Mexico.[87] One of the processes by which climate change contributes to ocean deoxygenation is that rising sea temperatures

---

84 Cusack 2014.
85 Longobardi 2014.
86 Payne, Bush et al. 2016, 1286.
87 Milman 2017.

decrease the solubility of oxygen in water. A 2019 International Union for Conservation of Nature report on ocean deoxygenation points out that between 1960 and 2010 oceans worldwide have lost on average 2% of their oxygen content and outlines the risks to oceanic life, from microorganisms to megafauna.[88]

Another environmental change capable of systemically undermining marine life is ocean acidification. Oceans absorb about 30% of carbon dioxide emissions and as these have increased over the last few hundred years they have combined with water to form carbonic acid, increasing the acidity of the oceans. Much aquatic life with shells and skeletons (and coral) are particularly vulnerable to a more acidic ocean environment. Coral bleaching cases are primarily caused by rising sea temperatures rather than acidification.[89] Already by 2012 research on an Antarctic sea snail showed severe levels of shell dissolution caused by acidification.[90] Fishes are thought to be able to compensate for small increases in acidity but past a certain point exhibit a negative physiological response.[91] Other studies have modelled likely end-of-century Ph levels and found deleterious effects on Atlantic cod larvae.[92] Larger fish species such as sharks were thought to be more robust, but research has found that they may be more susceptible than previously assumed.[93]

Environmental changes associated with climate change are seen in various physiological responses from fish species. Changing sea temperatures result in shifting ranges, changes to the phenological timing of life cycle events, and effects on body size. For example, in reviewing changes to the waters around the United Kingdom, Wright et al. point out that recent warmer waters have led to increases in populations of sea bass and anchovy.[94] A further example here is reported by Tanaka et al. in their study of great white sharks.[95] They found that juvenile great whites have moved 600 kilometres north from

---

88   Laffoley and Baxter 2019.
89   Sully, Burkepile et al. 2019.
90   Bednaršek, Tarling et al. 2012.
91   Heuer and Grosell 2014.
92   Frommel, Maneja et al. 2012; Stiasny, Sswat et al. 2019.
93   Rosa, Rummer and Munday 2017.
94   Wright, Pinnegar and Fox 2020.
95   Tanaka, Van Houtan et al. 2021.

waters off the southern California coast to Monterey Bay by Santa Cruz. One impact of this has been a significant decline in sea otter numbers in this more northern location. In some cases, species are also genetically adapting to environmental change. Kovach et al. examined a pink salmon population in Alaska and found evidence that there had been genetic change toward earlier migration times and a decrease in a genetic marker for later migration timing between 1983 and 2011, concluding that they had found "compelling evidence that recent climate change has influenced the evolutionary dynamics of salmonid populations and their adaptation via migration timing to their respective habitats".[96] The physiological effect on body size mentioned above in relation to fishes has also been reported in some bird species. For example, Prokosch et al. found a relationship between increasing temperatures and declining body size in a population of mountain wagtails in South Africa,[97] and Van Gils et al. discovered a similar relationship between Arctic warming and the red knot.[98]

To recap, much marine life has been badly affected by industrial fishing and habitat loss and now faces a triple climate threat of deoxygenation, rising water temperatures and acidification. A further impact of climate change is freshwater inundation into oceans. This occurs in two ways – either from the increasing frequency and severity of storm events, or via ice melt, including from the large ice sheets of Greenland and Antarctica. This alters the salinity of the ocean and alongside temperature increases consolidates the process of ocean stratification whereby there is far less mixing between different layers of the ocean. As Cavicchioli et al. point out, this reduces nutrient flow from deep to surface waters, which has subsequent impacts on "primary production", the production of biomass by phototrophic organisms, such as phytoplankton or plants.[99] Inundation from storm events is already contributing to the prevalence of skin disease in species of coastal bottlenose dolphins in southern and western Australia which are impacted by the declining water salinity caused by

---

96   Kovach, Gharrett and Tallmon 2012, 3875.
97   Prokosch, Bernitz et al. 2019.
98   Van Gils, Lisovski et al. 2016.
99   Cavicchioli, Ripple et al. 2019, 571.

storm events.[100] Similar skin lesions were found on bottlenose dolphins in the Gulf of Mexico when in August 2017 Hurricane Harvey inundated the Galveston Bay estuary in Texas with record-breaking rainfall.[101]

Freshwater inundation from ice melt is raising sea levels and can also alter ocean currents. The Bramble Cay melomys, a rodent endemic to a small island in the Great Barrier Reef, was confirmed extinct in 2016, becoming the first mammal to be reported extinct due to sea-level rise associated with climate change.[102] The melting of major ice sheets is accelerating and during 2019 the Greenland ice sheet lost a record 532 billion tonnes, or the equivalent of 1 million tonnes per minute across the year.[103] Using satellite data, Slater et al. determined that between 1994 and 2017 the Earth lost 28 trillion tonnes of ice, a mixture of ice sheets, sea ice and glaciers.[104] A 2020 paper has documented freshwater ice melt impact on the Beaufort Gyre, an anticyclonic sea ice–ocean circulation system in the Arctic Ocean.[105] Whether more systemic impacts to ocean currents occur, which could have significant impacts on, for example, Western Europe, remains to be seen. However, research suggests that the system of ocean currents in the North Atlantic, known as the Atlantic meridional overturning circulation (AMOC), is slowing and is at its weakest in 1,600 years.[106]

Meanwhile, melting ice has significant impacts on the animals of the Arctic region where temperature increases have been more pronounced than in temperate latitudes. For example, "Over the past decade, the Arctic has warmed by 0.75°C, far outpacing the global

100 Duignan, Stephens and Robb 2020.
101 Fazioli and Mintzer 2020.
102 Fulton 2017.
103 Sasgen, Wouters et al. 2020.
104 Slater, Hogg and Mottram 2020. The breakdown of 28 trillion tonnes of ice melt included Arctic sea ice (7.6 trillion tonnes), Antarctic ice shelves (6.5 trillion tonnes), mountain glaciers (6.2 trillion tonnes), the Greenland ice sheet (3.8 trillion tonnes), the Antarctic ice sheet (2.5 trillion tonnes) and Southern Ocean sea ice (0.9 trillion tonnes). The rate of ice loss has risen by 57% since the 1990s.
105 Armitage, Manucharyan et al. 2020.
106 Caesar, Rahmstorf et al. 2018; Thornalley, Oppo et al. 2018.

average".[107] The aforementioned polar bear, much used in the cultural iconography of climate breakdown, relies upon sea ice for the hunting of seals. As Boonstra et al. phrase it, "Changing sea ice phenology limits spring hunting opportunities and extends the period of onshore fasting".[108] They measured stress levels in a population of Arctic polar bears and concluded that this had become worse because of a warming climate. Molnár et al. analysed Arctic polar bear sub-populations and modelled the effect that fasting pressures will have under various emissions scenarios. They suggest that survival thresholds may already have been exceeded in some sub-populations and that rapidly declining reproduction and survival will threaten the persistence of most Arctic sub-populations by 2100.[109] Pagano and Williams assessed the effect of sea-ice loss on polar bears and narwhals, finding that both have species-specific physiological constraints which leave them especially vulnerable.[110] Another species that relies on sea ice for feeding and has inspired high-profile media coverage is the walrus. Mass "haul-outs", when many thousands of mothers and their young pull themselves onto shorelines to rest, are thought to be the result of declining sea ice,[111] decreasing their access to their food supplies, and exposing them more to terrestrial predators and conflict with humans. The case of melting permafrost also has broad consequences for Arctic wildlife. For example, Beardsell et al. found that permafrost destabilisation through temperature rises and increased precipitation on Bylot Island in the Canadian Arctic led to the destruction of a significant number of nesting sites of the rough-legged hawk between 2007 and 2015.[112] Increased Arctic temperatures have also promoted wildfires, an impact typically associated with warmer biomes. As well as affecting wildlife these fires constitute a positive feedback loop whereby further

---

107 Post, Alley et al. 2019, 1.
108 Boonstra, Bodner et al. 2020, 4197.
109 Molnár, Bitz et al. 2020.
110 Pagano and Williams 2021.
111 Kochnev 2019.
112 Beardsell, Gauthier et al. 2017. For a broad review of impacts of changing permafrost and snow conditions on Tundra wildlife, see Berteaux, Gauthier et al. 2016.

emissions are released. By late August 2020, Arctic wildfires had already emitted 35% more carbon dioxide than for the whole of 2019.[113]

Impacts on Arctic species will, in the short term, depend upon the ability of specific species to adapt and exhibit behavioural plasticity. In their analysis of ringed seals and white whales off the Norwegian island of Svalbard, Hamilton et al. found that both species had adapted their feeding locations and in the case of white whales were feeding on species that were new to the region owing to shifting ranges made possible by sea temperature rises.[114] Humpback whales in the Canadian Gulf of St. Lawrence may be less adaptable, with one study showing a significant decline in the number of calves born over the past 15 years due to the decrease of their food source, herring, itself caused by a warming ocean.[115] The issue of phenological mismatch may also come into play for Arctic species with significant research conducted into species such as reindeer and caribou on this and other climate impacts.[116] Shifting ranges is an important climate impact for species generally with a substantial amount of research looking at the poleward, elevational and other directional shifts of different species.[117] As Pecl et al. point out, "Marine, freshwater, and terrestrial organisms are altering distributions to stay within their preferred environmental conditions"[118] and that this global and systemic trend has considerable import for species, ecosystem health, human well-being and the dynamics of climate change itself. Although these can be seen as examples of animal agency in the face of climate change, it is unclear how sustainable such shifts are for particular species.

While the example of Arctic wildfires cautions against the assumption that certain climate impacts map simply onto specific biomes, in warmer biomes, not surprisingly, other effects of climate change are more pronounced. As well as seeing rising temperatures, temperate areas may also be subject to greater precipitation, storms

---

113 Thomas 2020.
114 Hamilton, Vacquié-Garcia et al. 2019.
115 Kershaw, Ramp et al. 2021.
116 Mallory and Boyce 2018.
117 Büntgen, Greuter et al. 2017; Freeman, Lee-Yaw et al. 2018; VanDerWal, Murphy et al. 2013.
118 Pecl, Araújo et al. 2017, 1.

and flooding events. Blöschl et al. examined European river flood discharges between 1960 and 2010 and found increases in countries such as the United Kingdom, Germany, (northern) France and Belgium.[119] Mediterranean nations, on the other hand, such as Spain, Portugal and Italy, have seen a clear decrease. They conclude that their analysis is broadly consistent with climate models and provide further evidence that climate change is already happening. In England, the network of Wildlife Trusts has become increasingly aware of the effect of sustained flooding events upon wildlife. For example, in February 2020 the twin impacts of Storms Ciara and Dennis inflicted substantial flood damage on Herefordshire. This significantly affected river ecosystems and wetland habitats, including the destruction of kingfisher nests, salmon and trout spawning areas, and many other plant and animal species.[120] These recent impacts compound a longer trend of decline for UK wildlife, with 41% of species having decreased in abundance since 1970.[121]

Droughts and heatwaves are a greater threat to biodiversity in the hotter Mediterranean chaparral, savanna and tropical biomes. Having previously noted the colossal scale of damage and loss of life during wildfire events, there are also documented events of droughts and heatwaves that have already led to considerable loss of life. While it remains an issue for climate scientists to attribute these events to climate change, their increasing severity and frequency, along with the continued setting of global annual and regional temperature records, suggests it to be the case. The worst-case scenario involves a biomic shift whereby tropical areas could reach a tipping point and become savanna.[122] In 2019 Zimbabwe was hit by its worst drought in 40 years, severely affecting the economy and food security. More than 200 elephants died, and the country's wildlife agency enacted an emergency translocation moving 600 elephants, 2,000 impala, two lion prides, a pack of wild dogs, 50 buffalo and 40 giraffes to northern parts of the country.[123] Translocation is increasingly discussed in conservation

---

119 Blöschl, Hall et al. 2019.
120 Weston 2020.
121 Hayhow, Eaton et al. 2019.
122 Staal, Fetzer et al. 2020.

science and policy as a measure to save wildlife as climate breakdown worsens. However, there are many ecological, practical, economic and socio-political barriers to its success as an adaptation approach.

Desert species adapted to a hot, dry environment have also been affected by drought and heat stress. Lovich et al. conducted a survey of Agassiz's desert tortoises found in southern California, discovering that the adult population had declined greatly between 1996 and 2012. Evidence from tortoise carcasses was consistent with death from dehydration and starvation and the authors expressed concern that climate modelling predicts longer-duration droughts in the future, threatening the viability of the area as a habitat for this species.[124] In the nearby Mojave Desert, Iknayana and Beissinger surveyed the desert bird population finding "strong evidence of an avian community in collapse. Sites lost on average 43% of their species, and occupancy probability declined significantly for 39 of 135 breeding birds".[125] Declines in precipitation associated with climate change were given as the chief reasons for the avian collapse. Spiny lizards are a further example of a species adapted to hot weather but also under threat from climate change. Sinervo et al. looked at 48 Mexican lizard species at 200 sites, finding that since 1975, 12% of local populations had gone extinct. As they further outline, during heatwaves lizards may retreat to cool refuges rather than risk death by overheating but in doing so spend less time foraging and reproducing, potentially undermining population growth rates and raising extinction risk.[126]

A heatwave in southern Queensland, Australia, at the start of 2014 led to the death of up to 100,000 fruit bats.[127] This also resulted in more people encountering bats and having to receive anti-viral treatment. It is worth noting that in many regions of the world night-time temperatures have warmed more than daytime temperatures,[128] which could impact more upon nocturnal animals such as bats. The

123 Chingono and Ndebele 2019.
124 Lovich, Yackulic et al. 2014.
125 Iknayana and Beissinger 2018, 8597.
126 Sinervo, Méndez-de-la-Cruz et al. 2010, 894.
127 Bavas 2014.
128 Cox, Maclean et al. 2020.

north-east Pacific marine heatwave of 2014–16 which caused elevated sea temperatures is thought to have killed 1 million common murres, a fish-eating seabird.[129] Their cause of death was not directly from the heatwave but from starvation due to its effect upon the marine interactions between phytoplankton and the seabirds' usual fishes food sources. Thousands of dead birds were discovered on beaches from California to Alaska. A similar dynamic is thought to be responsible for the death of tufted puffins from October 2016 to January 2017 on St. Paul Island, in the Bering Sea, close to Alaska.[130] Both these examples show some of the cascading ecological effects of elevated sea temperatures.

Heatwaves and droughts also bring clashes and tensions between differently socially constructed animals. For example, during Californian droughts, species such as coyotes, bobcats and bears may be more attracted to urban spaces in search of artificial sources of water.[131] This drought-related water scarcity not only increases "human–wildlife" conflict but brings "wild" and "companion" animals into closer proximity. This has included fatal coyote attacks on companion animals, responded to with the trapping and killing of the coyotes. This highlights the importance of considering different social constructions of species, which are also always differing valuations, when trying to anticipate and account for the impact of climate change. In this case, climate change shapes the coyotes' environment but they are marginalised further through the hierarchical human constructions of other animals. A similar kind of refraction of climate impact occurs with the guanaco (a mammal related to the llama) in central Chile. Increasingly arid conditions have brought the guanacos into conflict with ranchers who perceive them as competing with their commodified "livestock" animals for pasture.[132] Both examples illustrate the spatial and range constriction that animals are subject to due to both climate change and dominant human understandings of other animals, land use and space.

---

129 Piatt, Parris et al. 2020.
130 Jones, Divine et al. 2019.
131 Phillips 2014.
132 Vargas, Castro-Carrasco et al. 2021.

This review has mentioned disease as a climate impact in relation to vector-borne illnesses becoming a greater threat to humans and other animals. However, climate-related disease is also a specific threat to amphibian and avian biodiversity. The now extinct golden toad has become a much-discussed species as controversy has arisen over whether climate change was responsible for their extinction. Certainly, research has identified a risk factor between rising temperatures and fungal infections that can affect amphibians. An analysis looking at both the golden toad and the similarly extinct harlequin frog in tropical Costa Rica concluded with very high confidence[133] that large-scale warming had contributed to the growth optimum of the fungal infection and thus their extinction.[134] Other toads thought extinct have been rediscovered, notably the Ecuadorian Mindo harlequin toad, raising the possibility that some species may have become resistant to the fungal infection.[135] Rising temperatures can also contribute to the fatal incidence of botulism experienced by waterbirds,[136] with the Indian Veterinary Research Institute concluding that changing climatic conditions had been responsible for 18,000 botulism bird deaths in the state of Rajasthan, Northern India, in 2019.[137]

Comprehending extinction and loss

This assessment of the environmental and physiological impacts of climate change highlights the fact that the systematic unravelling of viable conditions for biodiversity is underway. It further makes clear that the climate crisis is inescapably a question of animal ethics, of the politics of human–animal relations, and of the intensification of the Capitalocene's war on nonhuman animals.[138] This underpins the need for climate scientists, activists and policymakers to appreciate

---

133 Pounds, Bustamante et al. 2006.
134 See also Cohen, Civitello et al. 2019.
135 Bittel 2020.
136 Anza, Vidal et al. 2016.
137 Sushma 2019.
138 Wadiwel 2015.

not only that there is an overlap between the climate and biodiversity crises but that the climate crisis cannot be approached on the basis of human welfare, or anthropocentric notions of justice, alone. For animal advocates, this should unequivocally nullify any conceptual distinction between pursuing veganism "for the animals" and "for environmental reasons". The summary of this review in Table 3.1, which shows only a very small number of the species affected, is intended as a heuristic reference to track ongoing threats to all species. Increasing defaunation and ultimately extinction constitute the current grave conditions for life on this planet heading toward the finality of a seriously compromised and denuded biosphere.

Reviewing Table 3.1, it becomes clear that individual nonhuman animals are increasingly helpless in the face of systemic environmental change which in some cases is becoming written on their very bodies. Biomes are shifting and their conditions for life are being undermined. For example, there is little that marine species can do in the face of rising temperatures, acidification and deoxygenation when their capacities for adaptation are overwhelmed by the novel temporality of change. Similarly, stark vulnerability and tragic helplessness in the face of extreme weather events such as hurricanes, heatwaves and wildfires are already too evident. An impact like phenological mismatch cruelly breaks the ecological webs of interdependence between species with potentially serious cascading effects upon populations. Several of these changes – ice loss, wildfires and permafrost melt – are positive feedbacks which produce further warming, be that the reduction of the albedo effect via lost ice or further emissions through forest fires and methane from permafrost. It is precisely the relative inattention to such positive feedback mechanisms which led some climate scientists to criticise the IPCC *Special Report on the Impacts of Global Warming of 1.5°C* for understating the severity of the climate crisis.[139] While there is an issue of how to accurately model for such unknowable feedback events it seems unwise to exclude them from future scenarios.

Amid what is unquestionably a dire situation there are some conservation victories with efforts preventing some extinctions to bird and mammal species including the Iberian lynx and the Californian

---

139 IPCC 2018. See also Lenton, Rockström et al. 2019; Harvey 2018.

Table 3.1 A summary of the environmental and physiological impacts of climate change on animal species.

Climate change environmental or physiological impact	Example animal species affected
Phenological mismatch	Pied flycatcher, West Greenland caribou
Shifting range	Mosquito, tick, white-footed mouse, sea bass
Rising land temperatures	Desert tortoise, spiny lizard
Rising sea temperatures	Coral, phytoplankton, common murres, tufted puffin
Rising sea levels	Bramble Cay melomys, human
Ocean acidification	Antarctic sea snail
Ocean deoxygenation	Phytoplankton, zooplankton, fish species in dead zones
Disease spread	Golden toad, harlequin frog, cat, dog, human
Ice loss	Polar bear, walrus, human
Permafrost melt	Rough-legged hawk, methanogen, human
Drought/precipitation decline	African elephant, coyote, desert birds, human
Heatwaves and heat stress	Fruit bat, farmed animals, human
Wildfires	Koala, kangaroo, human, horse, sheep ungulates
Floods	Sheep ungulates, kingfisher, human
Hurricanes/storms	Dog, seabirds, human
Declining body size	Mountain wagtail, red knot
Genetic adaptation	Pink salmon

condor.[140] Nevertheless, recognising the gravity of the situation is also to reflect that it is philosophically, culturally, emotionally and

---

140 Bolam, Mair et al. 2021.

scientifically challenging to make sense of the decidedly anthropogenic (and more accurately "capitalogenic") sixth mass extinction. Lawyers, criminologists and activists increasingly deploy the term ecocide, defined as "mass damage and destruction of ecosystems – harm to nature, which is widespread, severe or systematic"[141] and strive to have it established as an international crime. Finality and irreversibility are important components of what is taking place, encouraging reflection upon what it means to live in a time of serious extinction threat. Rose et al. underline that:

> The significance of extinction, what separates it from the singular death of an organism, is precisely this: the ending of an ongoing lineage cultivated over hundreds and thousands, perhaps even millions of years of evolutionary time; the abrupt termination of a whole way of life, a mode of being that will never again be born or hatched into our world.[142]

To this we can add, as Yong has emphasised, the loss of animals' ancestral knowledge,[143] something likely to be omitted if we defer to broadly Cartesian understandings of other species. In an analysis of mammal extinction, Davis et al. attempt to quantify years of evolutionary time lost through extinctions and the time it could take to attain a comparable degree of (mammalian) diversity in the future. They conclude – based on likely mammal extinctions over the next 50 years – that it will likely take millions of years for mammals to recover from biodiversity losses.[144] Though extinction has been the fate for the vast majority of species to inhabit the Earth, it is erroneous to use such a fact to naturalise the sixth mass extinction, the first mass extinction event to be caused by a stratified segment of one species. In other words, the context for the sixth mass extinction is the Capitalocene.

This is an evocative way of thinking how the Capitalocene is making its mark: a breathtaking temporal intensification shaping the

---

141 https://www.stopecocide.earth/what-is-ecocide.
142 Rose, van Dooren and Chrulew 2017, 9.
143 Yong 2018.
144 Davis, Faurby and Svenning 2018.

planet for millions of years henceforth. Moreover, it may be tempting to project extinction risk onto other species thereby reinforcing a human/animal distinction and viewing "ourselves" as relatively safe. That is not to imply flirtation with theories of imminent human extinction but it is to say that the defaunation and extinction of nonhuman others has direct import to humans, because of our interdependency with other species. It is also to be clear that climate breakdown is already taking the lives of many humans and other animals and the near-term loss of life is further unpredictable due to i) the complexity of how climate change enfolds into social and economic policy and risks of conflict and war, and ii) the potential of abrupt climate change. What is clearer is that, following the logic of the Capitalocene, human morbidity and mortality will be shaped by class, gender and "race". Therefore, climate breakdown and the sixth mass extinction is not the moment to pause and decry "humanity's shame", or to ask, what have "we" done? – implicit in the end scenes of countless nature documentaries. This is deferential to the generalised Anthropocene thinking critiqued in Chapters 1 and 2. For similar reasons it is not the time to individualise either the causes of, or solutions to, climate breakdown. Rather, the systemic unravelling of life outlined in this chapter must be approached on the scale of the social and the critical systemic framing of capitalism which for centuries has cheapened the more-than-human alongside humans. The climate crisis becomes impossible to theorise apart from a crisis of the normative anthropocentrism which inhabits our practices and institutions.

There is a sociological and historical lesson here for ecologists, conservation biologists and climate scientists. Bodies such as the International Union for Conservation of Nature and Intergovernmental Panel on Climate Change consistently fail to overtly grasp the political, economic and specifically capitalogenic construction of climate breakdown and the stranglehold of geopolitical, class, gender and racialised privilege upon the status quo. Indeed, it may be some consolation to feelings of eco-anxiety, loss, grief, hopelessness and climate melancholia to not have to essentialise the intractability of climate breakdown as an "inescapable failure of the human".

Rather, Rose et al. are correct to probe "Which forms of human life are driving these catastrophic processes of loss, and in what other diverse ways are humans drawn into and implicated in extinction – and its resistance?"[145] The larger answers to the first part of this question are those forms of life and practice engaged in capitalistic growth and maldevelopment based around a fossil economy and a meat-based culture, and as highlighted earlier, the many non-climate, but just as exploitative, contributing practices to biodiversity loss. The slightly narrower answer which more closely aligns with the scope of this book is found in the forms of human life reflected in the societal and transnational infrastructure of the animal-industrial complex that perpetuate and normalise practices of human exceptionalism. This places interrogating dominantly exploitative human–animal relations at the heart of the response to the climate crisis and locates resistance in those practices which give meaning to forms of life beyond taken-for-granted anthropocentrism. The next chapter turns to the inescapably important question of the role of animal agriculture in contributing to the climate crisis, the politics around emissions figures and the cultural omission of this.

---

145 Rose, van Dooren and Chrulew 2017, 6.

4
Animal omissions, animal emissions

When the Food and Agriculture Organization of the United Nations (FAO) published its *Livestock's Long Shadow* report in 2006 it significantly raised the profile of the role of animal agriculture in the climate crisis. Although the report was subsequently critiqued for both under- and overestimating that role, it put the contribution of animal agriculture at 18% of all greenhouse gas emissions.[1] The later 2013 FAO report, titled *Tackling Climate Change through Livestock*, revised this percentage down to 14.5% and this has since become an often-quoted figure.[2] At the end of 2022 the FAO revised the figure down further to 11.2%, meaning that between 2006 and 2022 FAO estimates (using different and non-comparable methodologies) decreased from animal agriculture being responsible for 1 in 5.5 of all emissions to 1 in 9, despite the animal-industrial complex continuing to grow in that time. It is noteworthy that this decline in estimates coincided with an FAO partnership partly with, and partly funded by, commercial animal agriculture interests, known as LEAP (Livestock Environmental Assessment and Performance Partnership). Indeed, the FAO have played a curious role in stoking the cultural fleischgeist, both defending and initially contesting meat culture. Later in this chapter I will argue

1   Steinfeld, Gerber et al. 2006.
2   Gerber, Steinfeld et al. 2013.

that it is important to also look at non-FAO analyses of this percentage. Some place the figure at around 20%. Yet even adopting a conservative approach of accepting the lower FAO figures, which I shall open to critical scrutiny later, nevertheless underlines a specific set of human–animal relations bound up in the agricultural exploitation of nonhuman animals, as constituting significant contributory practices to the overall climate crisis.

The first FAO report was well covered in the media and the animal agriculture–climate change[3] link has received quite consistent coverage from the United Kingdom's left-of-centre press such as *The Guardian*, *The Independent* and *The Observer* newspapers since then.[4] While the broader media engagement with the animal agriculture–climate change link may have been significantly less consistent, there is little justification to state that media omission has shaped the paucity of *academic* engagement with the link during the last 15 years. To be exact, when I speak of academic omission, I am referring to very specific sub-disciplines in the social sciences and humanities with a focus on climate change which one would reasonably expect to have more coverage of the link between animal agriculture and climate change. Later I highlight research (Table 4.3), predominantly from a broad range of the natural sciences, that *does* foreground the link and calls for reductions in meat consumption.

The omissions that I have become interested in extend beyond excluding the animal agriculture–climate change link and into the exclusion of human–animal relations generally. The sub-disciplines of primary interest are the sociology of climate change and climate ethics. Analysing key works in both sub-disciplines reveals little evidence of

---

3   Although not a focus here, this also extends into the use of farmed animals as feed for companion animals.

4   For example, *The Observer* (United Kingdom) ran a front-page headline on 7 September 2008: "UN says eat less meat to curb global warming" (Jowit 2008). *The Guardian* (United Kingdom) includes an ongoing section called "Animals farmed", as part of its environment coverage, which consistently covers the animal agriculture–climate change link. In May 2018 it gave prominent coverage to Poore and Nemecek 2018, under the headline "Avoiding meat and dairy is 'single biggest way' to reduce your impact on earth" (Carrington 2018).

any serious engagement with the link between climate change and animal agriculture or work that considers the impact of the climate crisis on other animals, or work that entertains changing human–animal relations as having mitigative import for averting the worst outcomes of the climate crisis. This would seem to suggest that sociology and philosophy, as supposedly key disciplines of critical thinking, have done little to reflexively examine their own legacy of anthropocentrism, at least when attention is directed at the climate crisis. After saying more about these two areas of knowledge, I note similar issues with discourses of climate justice, just transition and sustainability. Then I turn to broader cultural animal omissions by examining research suggesting the media and non-governmental organisations (NGOs) have also been less inclined to cover the animal agriculture–climate change link.

In the second half of this chapter I turn, in more detail, to the topic of animal emissions, and what the debate says about, firstly, my contention in Chapter 2 that the climate crisis ought to be read as a crisis also of human–animal relations, and secondly, what the evidence base might imply for how policy should proceed. The relatively high level of emissions associated with animal agriculture examined in this chapter ultimately supports the contention of this book that the climate crisis can be seen as a crisis of the anthropocentrism in our dominant human–animal relations whereas the various animal omissions outlined indicate a tendency to deny this and protect human exceptionalism.

Looking for animals in the sociology of climate change, climate ethics and sustainability discourse

I have examined the sub-discipline of the sociology of climate change which has emerged and developed since the turn of the century.[5] This involved surveying both key texts and journal articles for evidence of engagement with human–animal relations and the animal agriculture–climate change link. The evidence for omission of such

---

5   Twine 2023.

engagement was strong. Of the 12 texts analysed,[6] only 2[7] had any meaningful focus on the topics. Furthermore, 62 journal articles were examined: 24 from a broad range of social science journals, and 38 from the journal *Environmental Sociology*. These 38 constituted all the journal's articles that were focused on climate change from its inception in 2015 to early 2019. In total, 57/62 of the articles were published after 2006.[8] As with the texts, the articles were examined for framings that included a consideration of nonhuman animals in the emergences, impacts and mitigations of climate change. The 62 articles examined involved 92 authors: 62 were male and 30 female. Taken together with the text sample, the gender split was 74 male authors (69%) and 34 female authors (31%). From the overall sample of 62 articles only one contained any significant framings or sections considering how human–animal relations figure in the emergences, impacts or mitigations of climate change. Tucker focused on the food practices of environmentally conscientious New Zealanders and discussed meat reduction and vegetarianism.[9] Several articles (n=14) contained very fleeting mentions of nonhuman animals, either noting emissions from animal agriculture (5 out of 62 including Tucker) or that impacts of climate change are also experienced by nonhuman animals (10 out of 62).

The sociology of climate change can consequently be seen as progressing in an unbalanced and non-inclusive manner according to its overall inattention to human–animal relations and the animal agriculture–climate change link. This dimension as part of overall food and agriculture questions should constitute a significant focus of sociological concern, alongside, and interwoven with, a focus on transport (mobilities) and energy. There are rare examples of environmental sociologists who have not excluded human–animal relations in this manner,[10] yet such authors are already working at the

---

6   Adams 2016; Doyle 2011; Dunlap and Brulle 2015; Dryzek, Norgaard et al. 2011; Giddens 2009; Koch 2012; Urry 2011; Norgaard 2011; Shove and Spurling 2013; Shaw 2015; Gough 2017; Latour 2017.
7   Doyle 2011; Adams 2016.
8   For full details of the content and methodology of the samples, see Twine 2023.
9   Tucker 2019.
10   Gunderson 2011; Gunderson and Stuart 2014; Whitley and Kalof 2014; Stuart and Gunderson 2020.

interface of environmental sociology and critical animal studies (CAS) because they have realised the importance of doing so.

In explaining the overall omission of human–animal relations from the sociology of climate change, one possible reason is based on the gendering of knowledge and care. As noted, men constituted more than two-thirds of the authors of the works sampled, and the sub-discipline of environmental sociology may have been shaped by the social construction of masculinity which has traditionally been less open to including or considering nonhuman animals. Furthermore, it is worth remembering that environmental sociology emerged out of an insistence upon the incongruity of excluding human–environmental intersections and impacts from sociology. Yet it is equally unjustifiable to exclude nonhuman animals from sociological understandings of the "environment", especially when one's focus is climate change and there is ample evidence of a link between climate change and animal agriculture. Human/animal dualism is not sociologically an improvement upon separating out "culture" from "nature"; it is merely a reiteration of it. Giving human–animal relations their sociological place has important consequences. It forces the animal-industrial complex to become more overtly an object of critical sociological scrutiny, akin to, and alongside, practice complexes of, for example, automobility. This can shift not only the sociology of climate change but sociological work on food, agriculture and political economy broadly construed. Seen in this way, the omission of human–animal relations and the animal agriculture–climate change link from the sociology of climate change, which I have characterised as a form of denialism,[11] is protective of a historically embedded core anthropocentrism in sociology. Once breached it not only engenders more accurate social science but also inevitably forces the discipline to critically reflect upon its underlying normative investments in contesting "oppression" and "injustice", previously limited as self-evidently anthropocentric concepts. An impoverished sociological framing of climate change inhibits sociological conceptual strengths in the deconstruction of dominant social norms (in this case, most clearly, those that embed and naturalise animal consumption). Therefore, an exclusionary framing sells the

11   Twine 2023, 115.

discipline short in terms of what it can contribute to the social innovation of alternative, more sustainable, (food) cultures, as part of broader urgent endeavours to avoid the most dangerous impacts of climate breakdown.

For the purposes of extending this analysis of animal omissions, other areas where one would reasonably expect the inclusion of human–animal relations and the animal agriculture–climate change link may be examined. Climate ethics, which could be situated as an offshoot of the broader and more longstanding sub-discipline of environmental ethics, is a sensible choice to examine in this manner. For this I examined all major texts published on climate ethics as shown in Table 4.1. Of these eight texts, published between 2008 and 2018, all were monographs except for two edited volumes.[12]

As Table 4.1 illustrates, it was only in the Arnold volume that there was meaningful inclusion of human–animal relations, and only in one chapter.[13] This chapter, on the place of nonhuman animals in the ethics of climate change, is by Clare Palmer, one of the few philosophers who have worked to bridge the divide between environmental ethics and animal ethics/animal studies.[14] Indeed, she begins her chapter with an assessment of climate ethics which Table 4.1 suggests has continued to the present day:

> Ethical discussion about climate change has focused on two highly significant sets of questions: questions about justice between existing peoples and nations, and questions concerning the moral responsibilities of existing people to future people. However, given the likely planetary effects of climate change, one might also expect to find a third area of ethical debate: questions about the impact of climate change on the nonhuman world directly. But on this subject, very little has so far been said.[15]

---

12 Arnold 2011; Gardiner, Caney et al. 2010.
13 Arnold 2011.
14 Palmer 2011. See also Palmer 2018.
15 Palmer 2011, 272.

Table 4.1 Climate ethics texts (2008–18) and their inclusion/exclusion of nonhuman animals.

Climate ethics texts	Inclusive of a focus on nonhuman animals?
J. Garvey. 2008. *The Ethics of Climate Change: Right and Wrong in a Warming World*. New York: Continuum.	No (brief mention of animal ethics).
S. Gardiner, S. Caney, D. Jamieson and H. Shue eds. 2010. *Climate Ethics: Essential Readings*. Oxford: Oxford University Press.	No.
D.G. Arnold, ed. 2011. *The Ethics of Global Climate Change*. Cambridge: Cambridge University Press.	Yes. Includes chapter by Palmer entitled "Does nature matter? The place of the nonhuman in the ethics of climate change".
D. Brown. 2013. *Climate Change Ethics: Navigating the Perfect Moral Storm*. London: Routledge.	No.
J.C. Tremmel and K. Robinson. 2014. *Climate Ethics: Environmental Justice and Climate Change*. London: I.B. Tauris.	No (brief mention of contribution of animal agriculture to greenhouse gas emissions).
J. Broome 2014. *Climate Matters: Ethics in a Warming World*. New York: W. W. Norton & Company.	No.
S. Gardiner and D. Weisbach. 2016. *Debating Climate Ethics*. Oxford: Oxford University Press.	No.
B. Williston. 2018. *The Ethics of Climate Change: An Introduction*. London: Routledge.	No (brief citation of Clare Palmer's work).

Climate ethics should be an important interdisciplinary contribution to debates around the climate crisis but, like the sociology of climate change, appears hampered by its own unacknowledged anthropocentrism. When climate ethicists consider ethical responsibilities to future generations[16]

---

16 Brown 2012, 83–84.

they almost always mean future generations of humans. In contrast to such omission, Palmer's work has considered other animals in various important ways. Not only has she considered the omission of animals in climate ethics, she has reflected on the issue of what ethical duties humans may have toward "wild" animals impacted by climate change.[17] A further exception to the omission of animals in climate ethics is found in the work of Nolt who has explicitly critiqued climate ethics for its anthropocentrism, reintroducing, for example, the moral considerability of nonhuman animals in ethical deliberations over future generations.[18] Palmer and Nolt point the way to a more inclusive climate ethics that does not assume human sovereignty.

Alongside the sociology of climate change and climate ethics there are further academic discourses that also marginalise other animals, omissions which have been critically addressed by others. In Chapter 2, I spoke of the importance of moving beyond the Anthropocene framing for arriving at a more inclusive understanding of "climate justice". As a discourse that has a clear presence in activist and NGO politics, climate justice has largely come to be focused upon a humanist intersectionality that stresses the interconnections of the climate crisis with class, gender and racialised inequalities. Yet, as an overlap with climate ethics, climate justice could certainly be broadened to include arguments for the just treatment of animals, and better understandings of how the climate crisis unravels in systemic injustice for the sustainability and viability of other animals. Such a critique of climate justice can be found in the work of Pepper, arguing further that animals must be brought into considerations of climate adaptation.[19] This approach signposts an important corrective to conceptualisations of climate justice, and the broader notion of environmental justice, which reproduce the very anthropocentrism that has facilitated the climate crisis and marginalise other ethical traditions based around care, empathy and relationality. Similar remarks apply to the discourse of "just transition"[20] which, again, though well intentioned, tends to raise

---

17 Palmer 2021.
18 Nolt 2011.
19 Pepper 2019.
20 Wang and Lo 2021.

questions of justice-in-transition only in relation to humans, even though domains of food, transport and energy have clear effects on more-than-human lives.

Less surprising is how discourses of sustainability and sustainable development have tended to omit considerations of nonhuman animals. Although the influential definition of sustainable development as "development that meets the needs of the present without compromising the ability of future generations to meet their own needs"[21] was useful for identifying limits and boundaries, the focus was anthropocentric. I have previously rehearsed the critique of these discourses for embedding a managerialist approach toward "natural resources" and for foregrounding efficiency savings to business-as-usual practices rather than transformative change.[22] This was exactly the approach of the aforementioned 2013 FAO report (that I return to shortly) entitled *Tackling Climate Change through Livestock* (my emphasis) where the emphasis shifted away from a critique that might imply necessary changes to demand and consumption practices to one that focused on how animal production could be an answer to climate change by becoming more efficient. The discourse of sustainability has also had a fraught relationship with capitalism, often easily domesticated to such a reformist agenda associated with the notion of ecological modernisation or to nationalistic agendas eager to co-opt arguments for local production. Relatedly, Lélé's critique of sustainable development as a concept that "does not contradict the deep-rooted normative notion of development as economic growth"[23] underlines its lack of utility for climate mitigation and its attractiveness to neoliberal framings of development.

This contradiction is most visible now within the UN's 2030 Sustainable Development Goals,[24] with one termed "Climate Action"

---

21 WCED 1987, chap. 2, para. 1.
22 Twine 2010a, 122.
23 Lélé 1991, 618.
24 Rawles 2006 referred to animal welfare as the missing dimension of sustainable development. The Sustainable Development Goals have been criticised for their silence on animal welfare (Keeling, Tunón et al. 2019), and for having only a limited transformative impact (Biermann, Hickmann et al. 2022).

and another "Decent Work and Economic Growth", the problem being that capitalism is centred around growth and the extreme challenge of decoupling growth from increases in carbon emissions.[25] The longstanding environmental critique of growth is easily applied to the maldevelopment of the Global North but is equally applicable to the imposition of this template upon poorer countries seen in neoliberal structural adjustment policies and, more specifically, the meatification of global diets[26] presumed by the notion of the "nutrition transition".[27] Whether global capitalism could satisfy the other Sustainable Development Goals, such as "No Poverty", merely brings into stronger relief the tensions between such laudable aims and an economic system that is inherently unequal and dependent upon the exploitation of the more-than-human. Pointedly it could be asked just when the United Nations, continuing a legacy of failed goals (see also the previous Millennium Goals), is going to start understanding *why* such goals are repeatedly being thwarted.[28]

Whether sustainability remains an intellectually meaningful concept with discursive leverage is questionable. It is not surprising that there have been attempts within CAS to contest it in various ways and it is important to do so given the polysemic and illusive nature of the concept, and its cross-disciplinary and extra-academic use. For example, Boscardin and Bossert critique both sustainability and sustainable development concepts for their omissions of nonhuman animals and for their instrumental subsumption within notions of natural capital.[29] They are particularly critical of the historical lack of attention to the animal-industrial complex within sustainable development discourse given its broad exploitative shaping of humans, nonhuman animals and ecologies; and they develop an ethical case to include nonhuman animals in the normative basis of sustainable development. In later work, Boscardin is critical of the use of

25   Jackson 2009; Kopnina 2020.
26   Weis 2007.
27   Popkin 1998.
28   For a critique of the Sustainable Development Goals on grounds of their loyalty to both neoliberalism and anthropocentrism, see Adelman 2018.
29   Boscardin and Bossert 2015.

sustainability discourse by the animal production sector (such as the idea of "sustainable intensification") as constituting a form of greenwashing,[30] a critique that echoes my own explication of the molecularisation of sustainability in farmed animal genomics.[31] Just as "animal health" has long been used as a discursive cover for increasing productivity (antibiotic use, for example[32]), the same can be said of the capturing of sustainability discourse by the animal production sector, which furthermore becomes a means of absorbing ecological critique into pre-existing framings and business-as-usual practices such as breeding optimisation.

An important earlier interjection into the anthropocentrism of sustainability discourse came from Earnshaw, arguing that "interspecies equity – the consideration of nonhuman animals based upon their inherent self-interests – [w]as the embodiment and ultimate test of a truly sustainable system".[33] This was influential upon the 2016 Sydney Declaration that I co-authored in relation to theorising the possibility of the sustainable campus.[34] This foregrounded a similar idea of *interspecies sustainability* which "recognizes that animals, too, have a right to the social, material and ecological bases for flourishing lives, sustained over time".[35] Other areas of research where sustainability discourse has become embedded have also received criticism for omitting nonhuman animals. For example, the field of education for sustainable development has been contested within CAS[36] and others have explored the implications for environmental education more generally of the "animal turn".[37] CAS not only contests the animal omissions in education for sustainable development but critically explores the degree to which it can problematise the

---

30   Boscardin 2017; 2018.
31   Twine 2010a.
32   Twine 2013b.
33   Earnshaw 1999, 113.
34   Probyn-Rapsey, Donaldson et al. 2016.
35   Probyn-Rapsey, Donaldson et al. 2016, 136. For a fuller theorisation of the concept of interspecies sustainability, see Bergmann 2019.
36   Kahn 2008; Pedersen 2019; Dinker and Pedersen 2016; 2019; Kopnina and Cherniak 2015.
37   Lloro-Bidart and Banschbach 2019; Oakley, Watson et al. 2010.

exploitation of nonhuman animals given the history of educational institutions as sites for the reproduction of anthropocentrism and the normalisation of the animal-industrial complex.[38] Furthermore, the CAS critique of animal omissions in education for sustainable development has developed productively to the formation of critical animal pedagogy,[39] a topic I return to in Chapter 5.

Considering the animal agriculture–climate change link in media and non-governmental organisation contexts

In reviewing this range of animal omissions so far, I have included sites of knowledge production which are more academic (the sociology of climate change and climate ethics) as well as those such as climate justice and sustainability which extend into broader cultural spaces. At the start of this chapter, I also briefly noted potential media inconsistencies in covering the link between animal agriculture and climate change. It is relevant to probe this further via CAS work that has exactly examined this broader coverage both within media and by environmental NGOs (again another space in which one might justifiably expect critical voices against the animal-industrial complex vis-à-vis the climate crisis).

An array of research on newspaper media in different parts of the world has tended to find low coverage of the animal agriculture–climate change link. Bristow and Fitzgerald[40] analysed major newspapers *The Globe and Mail* (Canada) and *USA Today* (United States) for coverage of agriculture and climate change in the period after the publication of the FAO's *Livestock's Long Shadow* report. Of relevant news items (124 from *The Globe and Mail* and 46 from *USA Today*) only 1 article in *USA Today* and 5 in *The Globe and Mail* "indicated that global climate change could be mitigated through dietary changes"[41] and only 2 articles (both in *The Globe and Mail*)

---

38 Pedersen 2019.
39 Nocella, Drew and George 2019; Pedersen 2019; 2021; Dinker and Pedersen 2016; 2019.
40 Bristow and Fitzgerald 2011.

cited the FAO report. Lee et al. (2014) examined "livestock"-related articles in the *Los Angeles Times* (United States) newspaper between 1999 and 2010, finding that only 5% (19/380 articles) included climate change as a theme, occurring irregularly over the period. Friedlander et al. examined news coverage on meat and climate change in leading Australian newspapers published between 1 July 2008 and 30 June 2013, finding that just under 1% (107/11,190) of the Australian climate change news stories mentioned meat.[42] Almiron and Zoppeddu examined a sample of the top 10 Spanish and Italian newspapers for an eight-year period (2006–13) for coverage of the animal agriculture–climate change link, finding that 1.5% (Spain) and 3.6% (Italy) of all articles on climate change during the studied period mentioned "the impact of our meat-based diet on the environment".[43] When the FAO report was cited in these newspapers its conclusions tended to be downplayed and a higher degree of scepticism on the link between animal agriculture and climate change was found in more right-leaning newspapers.

In a 2020 study Kristiansen et al. analysed how much attention the UK and US "elite media" paid to animal agriculture's role in climate change, and the roles and responsibilities of various parties in addressing the problem, from 2006 to 2018.[44] The media analysed were *The Guardian* and *The Telegraph* (United Kingdom) and *The New York Times* and *The Wall Street Journal* (United States) to also compare left- and right-leaning newspapers. During the period, the authors found a total sample of 188 articles focused on the animal agriculture–climate change link, of which 27 were from *The New York Times*, 7 from *The Wall Street Journal*, 113 from *The Guardian* and 41 from *The Telegraph*, showing a clear majority (140) in the two left-leaning newspapers. For context, the authors found a total of 114,000 articles on climate change and just over 5,000 of these (or about 4%) mentioned animal

---

41   Bristow and Fitzgerald 2011, 217.
42   Friedlander, Riedy and Bonfiglioli 2014.
43   Almiron and Zoppeddu 2015, 314. Almiron, Rodrigo-Alsina and Moreno (2022) also analysed 110 European think tanks finding a similar lack of attention to the animal agriculture–climate change link.
44   Kristiansen, Painter and Shea (2021).

agriculture.[45] The research also found that within the 188 articles, responsibility for change was framed as residing more with consumers rather than other actors such as business or government. While overall this research concurs with the above analyses suggesting very low overall coverage, it reinforces my earlier point that left-leaning media such as *The Guardian* have relatively stronger coverage of the link between animal agriculture and climate change. However, a newspaper such as this has a far lower readership than the major UK tabloids – which do not typically cover the climate crisis – and it still regularly advertises meat, a factor that otherwise could potentially dissuade media outlets from going "too strong" on the link and advocating for the downsizing of the animal-industrial complex. Further research that probed the extent of social media sharing of the relatively small number of mainstream media stories alluded to here could provide a valuable additional picture of coverage and engagement.

Taken together these are significant findings which suggest that coverage of the animal agriculture–climate change link in major newspapers in the countries analysed is not even closely proportional to the contribution of animal agriculture to the climate crisis. Reflecting upon this high degree of media omission, Almiron, in common with my analysis of the same omission in the sociology of climate change, considers this in relation to climate denialism, a response more intricate than whether one agrees that the climate crisis is anthropogenic:

> Many people and organizations do not deny either the facts, evidence, or consequences of the current climate crisis yet they do deny the psychological, political or moral implications that conventionally follow. [This] … directly impacts the solutions adopted (or lacking) by non-denialists and shows that rejection

---

45   The apparent disparity between the figures of 188 and just over 5,000 was accounted for by articles that only mentioned the animal agriculture–climate change link in one sentence (and so were discounted from the more focused sample) or were about other related topics such as the impact of climate change upon animal agriculture.

in the issue of climate change is much more complex than simply pointing at the right-wing denial countermovement alone.[46]

Almiron goes on to specifically name this as a denial of moral anthropocentrism and as contributory to the climate crisis. She argues that the main aspect of moral anthropocentrism is a failure to accord full moral considerability to other species[47] and that it is a core value held by both right-wing climate denialists and "progressive" climate advocates who, despite being environmentalists often proficient in climate science, have largely failed to contest what she calls the animal-based food taboo and argue for transformatory dietary change.[48] Almiron's larger point is that moral anthropocentrism is significantly historically embedded, forming part of the taken-for-granted narrative of meat cultures, and is found in the journalistic codes uncritically adopted by mainstream media. Consequently, although some parts of the media report upon scientific evidence in favour of meat reduction as an option for climate mitigation they are far less likely to contest the cultural status quo of moral anthropocentrism.

Almiron's critique of this denialism *within* positions which otherwise would regard themselves as passionately engaged in climate politics is significant, finding that environmental NGOs have also been slow to foreground the animal agriculture–climate change link. As shall be noted later (Table 4.4), organisations such as Greenpeace have come late to the issue, while others such as the World Wildlife Fund for Nature have a longer history of engagement in the need for dietary change. Pointedly, however, this interest only goes as far as calls for meat reduction and shuns more transformative opposition to meat cultures. This critique of environmental NGOs formed a large part of the well-known documentary *Cowspiracy* (2014), which critiqued organisations such as Greenpeace, Sierra Club and the Rainforest Action Network for their inattention to the link between animal agriculture and climate change. Noticeably, all three of these organisations have since included the issue on their websites, albeit

---

46   Almiron 2020b, 1.
47   Almiron 2020b, 3.
48   Almiron 2020b, 4.

with a reduction framing. CAS research has also examined this in more detail.[49] Freeman's analysis of the websites of 15 US environmentalist advocacy organisations found a dominance of reduction discourse or advice to consume "more sustainably sourced" farmed animals. She also found organisations tentative to recommend a plant-based diet.[50] Laestadius et al. analysed a broader set of NGOs in the United States, Canada and Sweden. They found that "aside from animal protection NGOs, few organisations have encouraged more than modest reductions in meat consumption"[51] and there was a lack of formal campaigning on the issue. Further analysis of their qualitative data, involving interviews with NGO staff, found that "an internal NGO culture offering explicit or tacit support for continued meat consumption worked as a barrier to the active promotion of meat-free messages"[52] or that the issue of meat reduction was not always seen as part of their core mission.[53] It is possible that some NGOs did not want to risk alienating sections of their support base by foregrounding this issue.[54] Moreover, NGOs were reluctant to engage in campaigns focused on "personal behaviour change".[55] While there may be valid sociological and political reasons for this, framing "meat reduction" or "dietary change" through the lens of individual behaviour change is not a given, a point explored in Part II of this book. Notably, by 2020 the World Wildlife Fund for Nature were producing reports that gave serious attention to dietary shift *and* vegan transition[56] and Greenpeace[57] was partially critical of both the European and the British meat industries. In a further important inquiry Almiron has analysed two major EU lobbyist organisations for the meat industry, the UECBV (European Livestock and Meat Trades Union) and the CLITRAVI (Liaison Centre for the Meat Processing Industry in the European

---

49   Freeman 2010; Laestadius, Neff et al. 2013; Laestadius, Neff et al. 2014; 2016.
50   Freeman 2010, 269.
51   Laestadius, Neff et al. 2013, 36.
52   Laestadius, Neff et al. 2016, 98.
53   Laestadius, Neff et al. 2016.
54   Almiron 2020a, 167.
55   Laestadius, Neff et al. 2014, 36.
56   WWF 2020; Loken, DeClerck et al. 2020 (WWF and EAT); see Table 4.4.
57   Greenpeace 2020a; 2020b; see Table 4.4.

Union), arguing that they work to downplay and obfuscate the animal agriculture–climate change link.[58]

The 2014 Chatham House survey[59] found lower levels of public awareness of the role of meat and dairy production in contributing to climate change in comparison to other sectors which, in part, may have reflected the above media and NGO omissions. Although environmental NGOs remain within a reduction discourse, these recent reports are indicative of a position more overtly critical of the animal-industrial complex and closer to a transformatory vision of the food system. As the urgency of the climate crisis becomes further apparent, the role of the food system and especially the connection of the animal-industrial complex with the climate crisis has become simply unavoidable, and the previous taboo against contesting animal consumption has weakened. It remains to be seen whether this shift endures and translates into broader transformatory change.

The matter of emissions

The second part of this chapter moves directly to consider the animal agriculture–climate change link. Although an issue full of complexities, it serves to undermine and question the range of academic, public and policy omissions just noted. Moreover, it both centres the importance of human–animal relations in the climate crisis and calls into question the priorities of national governments and international bodies such as the Intergovernmental Panel on Climate Change (IPCC) (and the UN more broadly) when it comes to their climate change mitigation recommendations, policies and strategies.

At the outset of this chapter, I indicated my own dissatisfaction over the science base when it comes to a confident quantification of the percentage contribution of animal agriculture to climate change. It

58   Almiron 2020a.
59   Bailey, Froggatt and Wellesley 2014. The survey was conducted online in
      Brazil, China, France, Germany, India, Italy, Japan, Poland, Russia, South
      Africa, the United Kingdom and the United States, with a minimum of 1,000
      participants in each country.

is undoubtedly a scientifically complex question to settle. For example, is the percentage calculated historically? Does it adequately consider all historical land-use changes that have been made in service to the animal-industrial complex? Or is it intended as a contemporary snapshot of current greenhouse gas contributions? The debate over animal agricultural emissions also necessitates that insights from the sociology of scientific knowledge are included. These have become more difficult to speak about in the context of a perceived policy need to valorise a clear message from climate science. The sociology of scientific knowledge has been important for acknowledging the political and cultural contexts of scientific knowledge production and for better understanding its process, including its contingent or uncertain nature. It is unfortunate that with the cultural ascendency (relatively speaking) of climate science there has been an unhealthy retreat from the importance of such insights in favour of a naïve realism. Yet acknowledging scientific uncertainty or understanding the social construction of scientific knowledge production is not the repudiation of scientific knowledge and does not open the floodgates to climate change denialism.

From these points it is important here to aim to offer an analytical context to the scientific construction of the animal agriculture–climate change link that acknowledges both the potential political contexts of particular claims and the scientific uncertainty over the issue. To do so is also inevitably to explore the trustworthiness of various knowledge claims in the ongoing, unfolding and cultural consolidation of the link. Once covered I will move on to underline the broad range of peer-reviewed scientific papers (Table 4.3) and reports from NGOs, the United Nations and UK government advisory bodies (Table 4.4) that have argued for animal consumption reduction based upon the link, and other co-benefits. Although these calls for reduction may not constitute the necessary *transformative* change required to avert the climate crisis, they nevertheless represent an impressive body of knowledge that has begun to normalise a consensus, at least in their own spheres of influence, around a reduction discourse. Less impressive is their low level of tangible policy influence to date, and the reticence of most governments worldwide to entertain any sort of notable change to the animal-industrial complex as a climate change mitigation strategy. The end of this chapter sets the scene for Part II of this book with a discussion of three scenarios that speak to contemporary debates around reducing meat/dairy consumption and the possibility of more systemic, transformatory change.

The two FAO reports mentioned at the start of this chapter have played a key role in the debate on the link between animal agriculture and climate change, suggesting that animal agriculture contributes overall 18% of greenhouse gas emissions,[60] then revised down to 14.5%.[61] The two lead authors of each report were the same and while it may be tempting to trust the authority of the United Nations it is important to highlight the paradigm from which these reports emerged. The authorships of both reports are dominated by agricultural (or "livestock") economists who are part of the epistemological apparatus of the animal-industrial complex. Pierre Gerber, for example, is Professor in the Animal Production System Group at Wageningen University and has worked as a Senior "Livestock" Specialist with the World Bank. It is common for animal science departments with a focus on animal production to employ economists to help translate research on farmed animal genetics and genomics into industry profitability. Both reports operate within a paradigm that assumes that technological efficiency fixes are the way to lessen the greenhouse gas impact of animal agriculture and could be reproducing a form of epistemological bias.[62] Neither reports advocated reductions in meat and dairy consumption, and both naturalised the view that demand for meat, eggs and dairy will increase in the future. As the reports were written by people immersed in a particular paradigm which takes for granted the normality of animal consumption it is not surprising that they adopted a technological efficiency-based frame protective of the animal-industrial complex. It is fair to say that had the reports been conducted in a balanced manner they would have also entertained demand-side policy recommendations which at minimum would have

---

60 Steinfeld, Gerber et al. 2006.
61 Gerber, Steinfeld et al. 2013.
62 Twine 2021b. The FAO has operated in a very political context. Some FAO researchers clearly did want to underline the environmental impacts of animal agriculture. See Neslen 2023.

explored possibilities for reductions in animal consumption. Almiron also makes clear that in 2012 the two leading EU meat lobby groups – the European Livestock and Meat Trades Union (UECBV) and Liaison Centre for the Meat Processing Industry in the European Union (CLITRAVI) – became stakeholders in the public-private LEAP partnership coordinated by the FAO animal production and health division,[63] the same section involved in their two reports and the effort of the Global Livestock Environmental Assessment Model (GLEAM; see below) which has informed subsequent emissions calculations. There are certainly questions to be asked about the closeness of this FAO division to the animal production sector. This was underlined in October 2023 by an investigation in *The Guardian* that highlighted how FAO leadership caved in to pressure from the meat industry and subsequently censored and downplayed their own researchers working on the environmental impacts of animal agriculture.[64]

The scientific rigour of the two reports can also be questioned further. Additional research would benefit from an interdisciplinary approach and a research team devoid of a values affinity to the very sector under examination. Furthermore, the topic of peer review is important to consider in this case. The standards of peer review of FAO reports are not as high as a scientific journal and a review would have to ensure that it was not just drawing upon scientists favourable to animal agriculture to assess such work. Despite these issues both reports came up with percentages which highlight a significant animal agriculture–climate change link, and it should be noted that the first report, *Livestock's Long Shadow*, was a broader assessment of animal agriculture's environmental impact with a focus also upon water, air pollution and biodiversity. These further impacts must be factored into assessing animal agriculture, with its parallel role in the biodiversity crisis making the exclusion of demand-side changes all the more difficult to sustain.

The assessment of the link between animal agriculture and climate change was complicated further by a paper produced for the environmental NGO the Worldwatch Institute, which made the claim

---

63   Almiron 2020a, 173.
64   Neslen 2023.

that animal agriculture was responsible for 51% of all greenhouse gas emissions.[65] This paper was based upon a critique of *Livestock's Long Shadow* for underestimating the contribution of animal agriculture to greenhouse gas emissions. It arrived at the figure of 51% by arguing that *Livestock's Long Shadow* underestimated the impact of methane emissions and excluded carbon dioxide emissions from animal respiration. However, there is no peer-reviewed position which has argued that animal respiration should be considered in such calculations. Their paper created something of a back-and-forth debate with *Livestock's Long Shadow* authors[66] which was unsatisfactory for several reasons. While it is difficult to disagree with Herrero et al. when they say, "Global estimates of livestock GHG [greenhouse gas] emissions are most reliable when they are generated by internationally recognized scientific panels with expertise across a range of disciplines, and with no preconceived bias to particular outcomes"[67] it is harder to countenance that many of the FAO authors did not have a preconceived bias toward animal consumption given their "livestock" science backgrounds. In several ways the Herrero et al. critique of Goodland and Anhang is merited, but they do so by expressing various tropes which essentially confirm their epistemological and moral affinity with animal production. For example, it seems merited to critique the inclusion of animal respiration in Goodland and Anhang's 51% claim. Yet Herrero et al. also express the much-deployed tropes of nutrient and livelihood benefits in defence of animal production and refuse to entertain reductions to animal consumption.[68] In contrast, Goodland

---

65 Goodland and Anhang 2009.
66 Herrero, Gerber et al. 2011; Goodland and Anhang 2012. Herrero, Gerber et al. 2011 has 14 authors: 2 were co-authors of Steinfeld, Gerber et al. 2006 and 3 were co-authors of Gerber, Steinfeld et al. 2013.
67 Herrero, Gerber et al. 2011, 779.
68 Curiously, a set of overlapping authors, Herrero, Thornton et al. (2009, 117), did briefly entertain "managing the demand for livestock products … as part of the solution". Given that one of the same FAO authors (Gerber) was a lead author on both Steinfeld et al. 2006 and Gerber et al. 2013, it is an oddity that neither report explored this solution. The key sentence in Gerber et al. 2013, 85 is: "While demand-side mitigation approaches that directly target consumers of livestock products are also important, they are considered not within the scope of this report". Presumably, these would have detracted from

and Anhang, not constrained by such epistemological and moral affinities to animal production, recommended a 25% reduction in "livestock" products by 2017. Calling these aspects of Herrero et al. "tropes" is not to deny that animal agriculture does provide employment and nutrition. However, the livelihoods argument is often expressed in terms of developing world livelihoods. This is an important point, but it is arguably reproduced as a trope because it is intended to block further debate and imply a static economics unable to change. It also does not deflect calls for animal consumption reduction in richer nations. There are sensible just transition reasons to argue that the dismantling of the animal-industrial complex should begin in richer nations. It is therefore a distraction to imply that reduction or vegan advocates are arguing to compound the economic exploitation and food insecurity of a global underclass. Moreover, it is actually the expansion of animal consumption in poorer countries that is significantly responsible for their rise in incidence of diet-related disease.[69] The nutrient trope valorises animal over plant protein and speaks to a long history of conflating state development and economic progress with animal consumption.

Herrero et al. also self-cite *Livestock's Long Shadow* and refer to it as rigorous and yet it had already been critiqued elsewhere for a flawed comparison of animal agriculture with the transport sector.[70] Presumably, had *Livestock's Long Shadow* been adequately peer reviewed this would have been quite a simple error to spot. Ironically (though not without justification), Herrero et al. critique Goodland and Anhang for *their* lack of peer review, a charge that they later denied.[71] This saga highlighted how politicised and heated the debate over the

---

the overarching emphasis upon technological efficiency approaches to farmed animal production which, of course, are protective of the global animal-industrial complex.

69   Popkin 2006.
70   Pitesky, Stackhouse and Mitloehner 2009. One of this paper's co-authors, Frank Mitloehner, has been criticised for bias toward animal agriculture interests. See "Do critics of UN meat report have a beef with transparency?", https://tinyurl.com/4j3a393m. See also Stănescu 2019. Significantly, Mitloehner was chair of the aforementioned FAO LEAP initiative in 2013.
71   Goodland and Anhang 2012.

percentage number had become. Both FAO reports and the Goodland and Anhang paper ought to have been subject to double-blind peer review from non-animal production scientists. An unfortunate outcome of the Goodland and Anhang paper arguing for a 51% figure is that unsurprisingly it was seized upon by some animal advocates, including the aforementioned documentary *Cowspiracy*, as a tempting expression of confirmation bias. This was unnecessary given that even on the lower estimate of 14.5%[72] the critique of omission of the animal agriculture–climate change link remains entirely valid.[73]

Percentage complexity: Why both 18% and 14.5% are unreliable estimates

However, the figure of 14.5% can also be shown to be dated. This can be illustrated by probing in more detail why the FAO revised down their 18% figure to 14.5%. It is useful firstly to examine what each report argued were the ways in which animal agriculture produces greenhouse gas emissions. Both reports stated that the animal agriculture–climate change link is constituted by the following main emissions sources: land-use changes for feed production and grazing, methane emissions mostly associated with the enteric fermentation of ruminant farmed animals, nitrous oxide emissions mostly associated with farmed animal manure, fossil fuel use during feed and farmed animal production, and fossil fuel use in the production and transport of processed and refrigerated dead animals. These reports follow the convention of using global warming potential (GWP) weightings for the two major non–carbon dioxide greenhouse gases – methane and nitrous oxide – which enables a carbon dioxide equivalent calculation which amalgamates all major greenhouse gases despite them having different GWP values. For example, it is well known that methane, an important greenhouse gas to consider for the link between animal agriculture and climate change, is a far more potent greenhouse gas than carbon dioxide but that it does not last as long in the atmosphere. As

---

72   Gerber, Steinfeld et al. 2013.
73   Chivers 2016.

highlighted later, the problem of GWP incommensurability has complicated the task of producing reliable estimates of the link. Conventionally, and in these two reports, the GWP of methane is calculated over a 100-year period. As the IPCC acknowledge, "If a shorter, 20-year time horizon were used ... short-lived gases would rise in relative importance".[74] Consequently, the choice of time horizon is significant for the social construction of the animal agriculture–climate change link. Surprisingly, as the United Nations Environment Programme (UNEP) points out, methane has "accounted for roughly 30 per cent of global warming since pre-industrial times and is proliferating faster than at any other time since record keeping began in the 1980s".[75] This means that deep cuts to methane could buy some time with regards to arresting the pace of climate change. Although the importance of methane has risen up the policy agenda recently, this is yet to result in a meaningful focus on a coordinated policy to reduce beef and dairy production, two of its principal sectoral causes.

The second report used a new methodological tool called GLEAM, a modelling framework developed within the Animal Production and Health Division of the FAO.[76] The Appendix to the report explains the differences between the two reports and gives some indication why 18% became 14.5%.[77] Firstly, it is notable that despite its 2013 publication date the second report used data that were relatively old: from 2004 (for total global emissions) and 2005 (for total emissions from animal agriculture). The first FAO report[78] used data from a 2000–2004 reference point (2000 for total global emissions and 2001–2004 for their animal agricultural emissions estimate). The second report produced an estimate that "total GHG [greenhouse gas] emissions from livestock supply chains are 7.1 gigatonnes CO_2-eq [carbon dioxide equivalent] per annum for the 2005 reference

---

74 Edenhofer, Pichs-Madruga et al. 2014, 45. Furthermore, animal agriculture interests have been accused of dishonestly using a new global warming potential metric known as "GWP" to falsify the impact of animal agriculture. See Elgin 2021
75 UNEP 2021a.
76 http://www.fao.org/gleam/en/.
77 Gerber, Steinfeld et al. 2013, 206.
78 Steinfeld, Gerber et al. 2006.

period".[79] They arrived at the 14.5% figure simply by expressing 7.1 as a proportion of the 2004 IPCC figure of total global greenhouse gas emissions of 49 gigatonnes of carbon dioxide equivalent found in the Fourth Assessment Report of the IPCC.[80]

So how did this analysis differ from *Livestock's Long Shadow*? The second report makes clear that their 7.1 figure is "in line with FAO's previous assessment, *Livestock's Long Shadow*, published in 2006".[81] Yet for the second report the FAO used a different dataset for *total* global emissions. Whereas *Livestock's Long Shadow* used the total global emissions data of the World Resources Institute, the second report switched their data source to the IPCC.[82] The World Resources Institute data are from its Climate Analysis Indicators Tool database which includes all gases and sectors. This switch away from the World Resources Institute to IPCC is significant because there is a consistent discrepancy between total global emissions data when the institute and IPCC are compared. It is also significant because the second report used the IPCC Fourth Assessment Report data for 2004, published in 2007, but when the IPCC published its Fifth Assessment Report in 2014, they had themselves revised down their total global emissions data for 2004. It is worthwhile examining these two points in further detail.

World Resources Institute total global emissions data are consistently lower than those of the IPCC. Whereas the second report uses an IPCC figure of 49 gigatonnes of carbon dioxide equivalent for 2004, the World Resources Institute data only reached that level in 2018. In contrast to the IPCC, the World Resources Institute global

---

79   Gerber, Steinfeld et al. 2013, 15. Carbon dioxide equivalent (sometimes shortened to CO_2-eq) is a way to incorporate non–carbon dioxide greenhouse gases into the calculation. The second report broke down its 7.1 gigatonne estimate as follows (with a comparison with the first *Livestock's Long Shadow* report in brackets): 2 gigatonnes of carbon dioxide equivalent of carbon dioxide per annum, or 5% (9%) of anthropogenic carbon dioxide emissions; 3.1 gigatonnes of carbon dioxide equivalent of methane per annum, or 44% (37%) of anthropogenic methane emissions; and 2 gigatonnes of carbon dioxide equivalent of nitrous oxide per annum, or 53% (65%) of anthropogenic nitrous oxide emissions.
80   IPCC 2007.
81   Gerber, Steinfeld et al. 2013, 15.
82   Gerber, Steinfeld et al. 2013, 15.

emissions data for 2004 are 39.66 gigatonnes of carbon dioxide equivalent.[83] Had the FAO retained the institute as their source of global emissions data[84] for their second report they would have concluded that animal agriculture was responsible for 18% of emissions (not 14.5%), the same figure as the first report. As mentioned, by the time they had published their Fifth Assessment Report the IPCC had themselves revised down their global emissions data for 2004, from 49 gigatonnes to 45 gigatonnes of carbon dioxide equivalent.[85] Although these IPCC data were presumably not yet available to the FAO authors, it illustrates that even using their preferred data source for global annual emissions the percentage contribution of animal agriculture to overall greenhouse gases would have been 15.8% not 14.5%. Therefore, provisionally, depending on whether IPCC or World Resources Institute data were used, there was a range of between 15.8 and 18% as a quantification of the link between climate change and animal agriculture.

However, there is further complexity to probe. Firstly, is it realistic that the figure of 7.1 gigatonnes of carbon dioxide equivalent remained the same between the two reports? Secondly, were there any other methodological changes between the two reports that may have kept the figure at 7.1?[86] It is disappointing that the second report 7.1 figure was from 2005 data, already eight years old at the time of publication. That the 14.5% claim of the second report has achieved the status of stable scientific fact ignores both the fluctuations in global annual emissions data and that the 7.1 figure had become very dated. To answer the first question, yes it could be credible for the 7.1 figure to remain broadly static between the two reports because they were drawing upon data from adjacent years. However, when consulting the Appendix of the second report there is further important information about the change in methodology and accounting performed in the second report. Specifically,

---

83 https://tinyurl.com/mwa49mxz.
84 It is worth noting that the World Resources Institute has tended to retrospectively revise down its data for global total emissions.
85 Edenhofer, Pichs-Madruga et al. 2014.
86 Twine 2021b.

The Livestock's long shadow assessment includes GHG [greenhouse gas] emissions related to the production of feed (including pasture) fed to all animal species (for a total of 2.7 gigatonnes CO_2-eq [carbon dioxide equivalent]), whereas this report only accounts for feed materials fed to the studied species, i.e., poultry, cattle, pig, small ruminants, and buffalo (for a total of 3.2 gigatonnes CO_2-eq including rice products). All manure emissions were accounted for in the Livestock's long shadow assessment (for a total of approx. 2.2 gigatonnes CO_2-eq), but only emissions related to manure management and manure application on feed crops or pasture are accounted for in this report (for a total of 0.7 gigatonnes CO_2-eq and 1.1 gigatonnes CO_2-eq, respectively).[87]

Although the second report covers the main farmed animal species, this implies that their figure related to the production of feed would have been slightly higher (and also the overall figure of 7.1) had they, like the first report, included all farmed animal species. Furthermore, it is unclear why the second report changed its accounting of manure emissions which resulted in 0.4 gigatonnes of carbon dioxide equivalent less being included in the overall total. Even though the authors of the second report claim that their new GLEAM modelling software is more accurate this illustrates that, for example, the difference in the percentage number between the two reports *could* be accounted for by a) accounting for manure emissions differently, and b) switching from World Resources Institute to IPCC data for total annual emissions. Some will contest the importance of a few percentage points. However, it was symbolically important that the second report produced a lower figure than *Livestock's Long Shadow*. It allowed those invested in the animal-industrial complex to claim the problem had been overestimated. The difference between 14.5% and 18% is also the difference between saying that animal agriculture is responsible for close to one in seven emissions, or to nearly one in five of all emissions.

---

87   Gerber, Steinfeld et al. 2013, 106.

The analysis presented is already sufficient to illustrate uncertainty over the construction of the 14.5% figure.[88] This figure of 14.5% was based upon an IPCC overestimate of total global emissions which was later revised down by the IPCC, and this has never been mentioned subsequently by the FAO authors. The oldness of the data used also means that the task of understanding what proportion of total emissions are produced by animal agriculture should be not only undertaken by interdisciplinary scientists without a values affinity to the animal-industrial complex but repeated at more regular intervals.

However, the examination of the FAO work does not end there because further analyses have been published. The results of GLEAM 2.0 were released in 2017 with new data based on animal agricultural emissions for 2010. Only this time there was no major FAO report and no media reporting of a new percentage figure. Looking at this new FAO data, the total emissions from animal agriculture had increased by 1 gigatonne to 8.1 gigatonnes in the period from 2005 to 2010.[89] Again, this can be expressed as a proportion of total global emissions for 2010 for both the IPCC and the World Resources Institute data sources (49 and 44.88 gigatonnes respectively), resulting in overall percentages of 16.5–18%. Given that the IPCC was the FAO's new favoured data source it could apparently be concluded that from their perspective, between 2005 and 2010, the proportion of greenhouse gases contributed by animal agriculture rose by 2% from 14.5% to 16.5%.[90] However, as will be noted shortly, this 16.5% figure was also unreliable. The important

---

88   Further uncertainty can be gleaned from the IPCC Technical summary of the Fifth Assessment report where 2010 total emissions are referred to as 49 gigatonnes plus or minus 4.5 gigatonnes (Edenhofer, Pichs-Madruga et al. 2014, 42), pointing to some degree of doubt over annual total emissions data.
89   FAO 2017. In comparison to 2005 data (used by Gerber et al. 2013), the 2010 data for GLEAM 2.0 (published in 2017) broke down as follows: Methane represented about 50% of animal agriculture's total (or 4.0 gigatonnes of carbon dioxide equivalent). The remaining part was almost equally shared between nitrous oxide, with 24% (1.9 gigatonnes of carbon dioxide equivalent), and carbon dioxide, with 26% (2.1 gigatonnes of carbon dioxide equivalent). For the comparison between Gerber et al. 2013 and Steinfeld et al. 2006, see note 78 above. The main difference with Gerber et al. 2013 is that methane increased from 3.1 to 4 gigatonnes.
90   Twine 2021b.

point at this juncture is that since GLEAM 2.0 was released in 2017, the FAO knew the 14.5% figure to be inaccurate but did nothing to publicly clarify that. As late as 2018, FAO authors were still using the 14.5% figure.[91] It is still widely used in the media and by other contemporary studies[92] but should be retired.

Intriguingly, the FAO GLEAM website briefly updated its estimate from 14.5% to 15.6%, stating, "Total GHG [greenhouse gas] emissions from livestock supply chains are estimated at 8.1 gigatonnes of carbon dioxide equivalent per annum for the 2010 reference period. That amount represents 15.6 percent of all human-induced emissions, estimated at 52 gigatonnes of carbon dioxide equivalent for the year 2004 (IPCC 2014)".[93] I believe this to have been an FAO error – the IPCC did produce a further estimate of 2010 total emissions, based on different global warming potential weightings, of 52 gigatonnes of carbon dioxide equivalent for *2010* not for 2004 from which the FAO likely yielded their 15.6%.[94] It is probable the FAO also detected a global warming potential incommensurability issue between their own data and the IPCC in relation to how they had weighted methane and nitrous oxide specifically. Together this could explain why this quoted statement was subsequently removed from the FAO website, despite no explanation being provided. It is clear that the FAO GLEAM 2.0 analysis negated the validity of 14.5%. However, 15.6% and 16.5% were also wrong as new estimates due to global warming potential incommensurability for methane and nitrous oxide, as the FAO changed their weightings values used in the GLEAM 2.0 analysis[95] (see Table 4.2). This all led to the unsatisfactory situation of no reliable percentage estimate of the animal agriculture–climate change link (at least using FAO data).

---

91   Mottet and Steinfeld 2018.
92   Lazarus, McDermid and Jacquet 2021.
93   FAO 2017. Thanks to Daniel Braune, Head of Research at ProVeg International, for this information. The cited quotation is viewable via the internet archive: https://tinyurl.com/yc3rtdy5.
94   Edenhofer, Pichs-Madruga et al. 2014, 45.
95   FAO ambiguity on this ironically shaped my own incorrect claim of 16.5%, see Twine 2021b.

In October 2022 the FAO released GLEAM 3.0 data based on the year 2015 and involving a revised methodology.[96] Using IPCC AR6 GWP values, this suggested a further revision down to 11.2%. This was based upon an estimate that by 2015, total CO_2-eq emissions from animal agriculture had shrunk to 6.19 gigatonnes, seemingly a decline of nearly 2 gigatonnes over five years. Although the FAO published a methodology document to accompany the data release there was no comparative explanation for the scale of decrease being asserted and the work was yet to be peer reviewed. No doubt aware of how this new data could be interpreted by diverse stakeholders, the FAO published the following statement in its release notes in 2023:

> While all these estimates were simulated with FAO's Global Livestock Environmental Assessment Model (GLEAM) and for different reference years (2005, 2010, and 2015 for GLEAM v1, v2, and v3), the methods and underlying data sets are different, and the different figures should not be interpreted as time series. It is, therefore, impossible to draw conclusions like "emissions went up" or "livestock emissions are becoming less important compared to total anthropogenic emissions" from those data sets.[97]

This is an important qualification, but it still leaves unanswered questions over how emissions from animal agriculture could have "declined" so quickly when the sector has been growing, or how different methodologies could arrive at such different conclusions. Given these uncertainties it is useful to consider global animal agriculture since 2010 (and 2015) which may impact future estimates *and* to consider non-FAO research.

---

96   FAO 2022.
97   FAO 2022, GLEAM 3.0 release notes, https://bit.ly/3TWoJIg.

4 Animal omissions, animal emissions

Table 4.2 Food and Agriculture Organization of the United Nations' estimates of the percentage contribution of animal agriculture to total global emissions.

Analysis	LLS	TCCTL	GLEAM 2.0	GLEAM 3.0
Total estimated CO_2-eq emissions from animal agriculture (gigatonnes)	7.1	7.1	8.1	6.19
Reference year(s) used	2001–04	2005	2010	2015
GWP weightings used	CH_4: 23 N_2O: 296	CH_4: 25 N_2O: 298	CH_4: 34 N_2O: 298	CH_4: 27 N_2O: 273
Total global CO_2-eq emissions (gigatonnes)	40.00	49.00 (45.00)[c]	52.00	55.50
Reference year used for total emissions	2000	2004	2010	2015
Source for total emissions	WRI CAIT	IPCC AR4	IPCC AR5	IPCC AR6
GWP weighting commensurability?	Yes	Yes	Unclear	Yes
Estimated contribution of animal agriculture to total global emissions (%)	18.00[a] (19.81)[b]	14.50[a] (15.78)[d]	15.60[e]	11.15

Note: CH_4 methane; CO_2-eq carbon dioxide equivalent; IPCC AR4 Intergovernmental Panel on Climate Change's Fourth Assessment Report (IPCC 2007); IPCC AR5 Intergovernmental Panel on Climate Change's Fifth Assessment Report (Edenhofer et al. 2014); IPCC AR6 Intergovernmental Panel on Climate Change's Sixth Assessment Report (IPCC 2022); LLS *Livestock's Long Shadow* (Steinfeld et al. 2006); N_2O nitrous oxide; TCCTL *Tackling Climate Change through Livestock* (Gerber et al. 2013); WRI CAIT World Resources Institute's Climate Analysis Indicators Tool (cited in Steinfeld et al. 2006).

[a] Widely cited figures.

[b] Recalculated based on 2022 WRI CAIT figure for the year 2000 of 35.84 gigatonnes (see note 82).

[c] Revised 2004 figure from AR5 Working Group III (Edenhofer et al. 2014).

[d] Recalculated based on updated IPCC figure for 2004 of 45 gigatonnes (Edenhofer et al. 2014).

[e] Incommensurate GWP weightings likely used. 15.6% was removed from the FAO website implying an error.

Firstly, if the rate of animal agricultural emissions has grown faster than total global emissions since 2010, then the percentage will now be higher. FAOSTAT data show that farmed cattle ungulate production (the farming of which produces the highest amount of greenhouse gases owing to land-use changes and enteric fermentation) had grown globally from 1.411 billion live animals in 2010 to 1.529 billion live animals in 2021, an 8.4% increase.[98] While some farmers will have attempted to instigate some of the efficiency recommendations of the two FAO reports, authors involved in these reports have indicated that adoption rates are low.[99] Thus, it is hard to imagine that an increase of 118 million cattle ungulates alone has not markedly grown the 8.1 gigatonnes of carbon dioxide equivalent reported for 2010.[100] During the same period, the number of chickens grew from 19.709 billion to 25.856 billion,[101] a 31% increase,[102] increasing land used for feed production. Also noting the FAO's own data on deforestation, South America lost 26,000 square kilometres of forest every year in the 2010–20 period, and globally, 100,000 square kilometres were lost annually from 2015 to 2020.[103] Significant areas of that deforested land will have been used to cultivate animal feed and would need to be part of any new calculation of the link between animal agriculture and climate change. Furthermore, research by *The Guardian* and Aid Environment found that between 2017 and 2022, 800 million trees were felled in the Amazon by the beef industry.[104]

Although both FAO reports factor in fishmeal as an animal feed in agriculture, neither report includes the broader fishing industry in its calculations of greenhouse gas emissions.[105] There are also question

---

98 FAOSTAT, https://www.fao.org/faostat/en/.
99 Herrero, Henderson et al. 2016, 459.
100 FAO 2017.
101 FAO data at FAOSTAT distinguish between global populations of farmed animals per year and the much higher figure of animals slaughtered per year. In the case of chickens, the growth in annual slaughter between 2005 and 2020 rose from approximately 47.5 billion to over 71 billion, see Orzechowski 2022.
102 FAOSTAT.
103 FAO 2020.
104 Wasley, Mendonça et al. 2023.
105 Parker, Blanchard et al. 2018.

marks over whether either report fully accounted for land-use changes related to animal agriculture. The second report makes clear that "[b]oth assessments include emissions related to land-use change from deforestation for pasture and feed crops and limit the scope of the analysis to the Latin American region".[106] Specifically, the second report limits its analysis of feed crop expansion to soybean cultivation in Brazil and Argentina only, whereas *Livestock's Long Shadow* included all feed crop expansion in Brazil and Bolivia. This delimiting does not account for the global scale of carbon sink loss via deforestation for animal agriculture, and especially not in a deeper historical sense. Deforestation in other parts of the world is also linked to the economic expansion of the animal-industrial complex. For example, beef production in Australia has been linked to 94% of all forest clearing in the Great Barrier Reef catchment areas between 2013 and 2018 and some of the remainder due to farming sheep ungulates. During the same period, 16,000 square kilometres of forest were cleared in Queensland.[107]

Secondly, other non-FAO studies have implications for the percentage calculation. Searchinger et al. are critical of the second FAO report for its limited accounting of land-use changes.[108] Poore and Nemecek pay particular focus to the issue of land-use change and the degree of mitigation that could be achieved via a global switch to plant-based diets. They calculate that "the land no longer required for food production could remove 8.1 billion metric tons of CO_2 [carbon dioxide] from the atmosphere each year over 100 years as natural vegetation re-establishes and soil carbon re-accumulates".[109] They later clarified in an erratum that their "'no animal products' scenario delivers a 28% reduction in global greenhouse gas emissions across all sectors of the economy relative to 2010 emissions".[110] This estimate of 28% makes a fuller attempt to calculate the mitigation potential of land carbon sinks than either FAO report, constituting the largest (peer-reviewed)

---

106 Gerber, Steinfeld et al. 2013, 106.
107 Cox 2019.
108 Searchinger, Wirsenius et al. 2018, 251.
109 Poore and Nemecek 2018, 991.
110 Poore and Nemecek 2019.

percentage figure for the contribution of animal agriculture to total greenhouse gas emissions.[111] Critics of Poore and Nemecek might argue that global transition to plant-based diets and such high levels of afforestation are politically challenging. This is true, but it does not detract from the importance of producing more accurate estimates and it is necessary to underline how the animal-industrial complex produces denuded ecologies and involves a profligate use of land with a poor return on nutrition. Furthermore, the implication of studies stressing the opportunity costs of the animal-industrial complex suggests that the FAO work may not have fully accounted for this.[112] A further study found that greenhouse gas emissions from food systems account for 35% of global total anthropogenic greenhouse gas emissions, of which 57% (i.e., 19.95% overall) corresponds to the production of animal-based food (including "livestock" feed).[113] Such peer-reviewed studies provide an important alternative to FAO data. The most cautious estimate at the time of writing, which will only be improved by further impartial and rigorous peer-reviewed research, is within the 11–20% range, although potentially higher if one factored in the afforestation/carbon sink potential of dismantling the animal-industrial complex. Finally, comparative statistics are also useful to convey significance, with the 2021 Meat Atlas underlining how emissions from the top 20 meat and dairy corporations in the world exceed those of Germany, and those of the top 5 exceed those of either Exxon or Shell.[114]

Global greenhouse gas emissions have yet to peak. Despite the GLEAM 3.0 data from the FAO, it is reasonable to assume that emissions from animal agriculture are also yet to peak. Whether the

---

111 Care was taken to check with Poore (personal communication, December 2020) that this was what the article was suggesting. See also the presentation, Poore 2019.

112 For example, Hayek et al. 2021, 21, found that "shifts in global food production to plant-based diets by 2050 could lead to sequestration of 332–547 GtCO$_2$ [gigatonnes of carbon dioxide], equivalent to 99–163% of the CO$_2$ emissions budget consistent with a 66% chance of limiting warming to 1.5°C".

113 Xu, Sharma et al. 2021.

114 Chemnitz and Becheva 2021, 35.

proportion of animal agriculture–related emissions within the total is also increasing awaits further data. The broad range of possible percentage figures included here nevertheless conveys a conundrum for advocates of the "sustainable intensification" of animal agriculture. For the figures suggest that without regulation economic growth will outpace any gains that can be made to improving efficiency, a point I return to below. This puts the work of the FAO in a difficult position because they have acted to embed the efficiency framing as the solution that should be favoured by the UN and national governments.[115]

From efficiency to reduction to transformative change

The efficiency emphasis of the FAO now appears out of step with a broad range of peer-reviewed research which argues that improving the efficiency of farmed animal production is not the key to reducing emissions from animal agriculture, but that reducing demand and consumption levels are (Table 4.3). Poore and Nemecek[116] is an example of such work, arguing that a scenario of a 50% reduction in animal "products" targeting the highest-impact producers delivers a 20% reduction in global greenhouse gas emissions. The efficiency emphasis of the FAO can be criticised for an impoverished understanding of the political global economy. They are prone to naïve understandings of demand, for example, referring to the "livestock" sector as "demand-driven"[117] and stating that "[m]odern livestock production is essentially driven by demand for livestock products".[118] This gives the impression that the growth of the animal-industrial complex is a natural developmental trajectory instead of being secured by the power of producers, governments and cultural normalisation. The efficiency perspective is also guilty of being overoptimistic that

---

115 However, both FAO report lead authors have exhibited some ambiguity on the matter, even co-authoring a short paper together arguing for a dual approach of "consuming less and producing better", Steinfeld and Gerber 2010.
116 Poore and Nemecek 2018.
117 Gerber, Steinfeld et al. 2013, 1.
118 Steinfeld, Gerber et al. 2006, 31.

new efficiencies to animal production will not be dwarfed by forecasted increases in production and thus fail to deliver meaningful reductions in greenhouse gases. *Livestock's Long Shadow* was critiqued as an approach that merely anticipates demand without adequately considering what levels of food consumption might be needed.[119] Applying a sustainable intensification ethos to animal farming will also conflict with animal welfare considerations[120] and raise the degree of genetic control over animals.[121] While the efficiency framing may be the best approach for protecting the normativity of animal consumption and the economic power of the animal-industrial complex, for the climate (and biodiversity) crisis it is the wrong strategy to pursue.

The sense that this framing has been superseded by academic research is strengthened by the 50 articles listed in Table 4.3, which all recommend reductions in animal consumption for climate mitigation, including for other co-benefits such as human health and biodiversity. They span 20 years (2003–22) and are published in highly respected peer-reviewed journals. Furthermore, Table 4.4 shows 50 key reports that also recommend reductions in animal consumption for climate mitigation and other co-benefits spanning 15 years (2008–22). These are a mixture of reports from either United Nations agencies (thus in conflict with the UN FAO), UK government or UK government advisory bodies, and NGOs, charities and think tanks.

The strength of recommendation within these articles and reports varies, but in contrast to the FAO they all consider dietary change as part of the solution to the deleterious impacts of the animal-industrial complex on both greenhouse gas emissions and other environmental and health issues. They have begun to normalise and mainstream a reduction discourse. The need to move beyond efficiency policies is now a well-established discourse in climate policy, extending to the transport and energy sectors. As O'Rourke and Lollo phrase it, "To meet the scale of the sustainability challenges we face, interventions and policies must move from relative decoupling via technological

---

119 Garnett 2009, 493.
120 Garnett, Appleby et al. 2013.
121 Twine 2010a.

improvements, to strategies to change the behavior of individual consumers, to broader initiatives to change systems of production and consumption".[122] Undoubtedly, efficiency framings have been attractive to governments and industry, allowing for business-as-usual and delaying more radical contestation of economic and cultural practices and infrastructures.

---

122 O'Rourke and Lollo 2015, 233.

Table 4.3 Fifty key journal articles recommending reductions in animal consumption for climate mitigation and other co-benefits (2003–22).

	Author(s)	Title	Journal	Year
1	D. Pimentel and M. Pimentel	Sustainability of meat-based and plant-based diets and the environment	American Journal of Clinical Nutrition	2003
2	A.J. McMichael, J.W. Powles, C.D. Butler and R. Uauy	Food, livestock production, energy, climate change, and health	The Lancet	2007
3	S. Friel, A.D. Dangour, T. Garnett, K. Lock, Z. Chalabi, I. Roberts et al.	Public health benefits of strategies to reduce greenhouse-gas emissions: food and agriculture	The Lancet	2009
4	A. Haines, A.J. McMichael, K.R. Smith, I. Roberts, J. Woodcock, A. Markandya et al.	Public health benefits of strategies to reduce greenhouse-gas emissions: overview and implications for policy makers	The Lancet	2009
5	J. Powles	Commentary: why diets need to change to avert harm from global warming	International Journal of Epidemiology	2009
6	B.M. Popkin	Reducing meat consumption has multiple benefits for the world's health	Archives of Internal Medicine	2009
7	T. Garnett	Livestock-related greenhouse gas emissions: impacts and options for policy makers	Environmental Science and Policy	2009
8	A. Carlsson-Kanyama and A. González	Potential contributions of food consumption patterns to climate change	American Journal of Clinical Nutrition	2009

	Author(s)	Title	Journal	Year
9	H.J. Marlow, W.K. Hayes, S. Soret, R.L. Carter, E.R. Schwab and J. Sabaté	Diet and the environment: does what you eat matter?	American Journal of Clinical Nutrition	2009
10	E. Stehfest, L. Bouwman, D.P. van Vuuren, M.G.J. den Elzen, B. Eickhout and P. Kabat	Climate benefits of changing diet	Climatic Change	2009
11	H. Steinfeld and P. Gerber	Livestock production and the global environment: consume less or produce better?	Proceedings of the National Academy of Sciences	2010
12	N. Pelletier and P. Tyedmers	Forecasting potential global environmental costs of livestock production 2000–2050	Proceedings of the National Academy of Sciences	2010
13	S. Wirsenius, C. Azar and G. Berndes	How much land is needed for global food production under scenarios of dietary changes and livestock productivity increases in 2030?	Agricultural Systems	2010
14	J.A. Foley, N. Ramankutty, K.A. Brauman, E.S. Cassidy, J.S. Gerber, M. Johnston et al.	Solutions for a cultivated planet	Nature	2011
15	S. Wirsenius, F. Hedenus and K. Mohlin	Greenhouse gas taxes on animal food products: rationale, tax scheme and climate mitigation effects	Climatic Change	2011
16	D. Nijdam, T. Rood and H. Westhoek	The price of protein: review of land use and carbon footprints from life cycle assessments of animal food products and their substitutes	Food Policy	2012

	Author(s)	Title	Journal	Year
17	E.A. Davidson	Representative concentration pathways and mitigation scenarios for nitrous oxide	Environmental Research Letters	2012
18	I. Tomlinson	Doubling food production to feed the 9 billion: a critical perspective on a key discourse of food security in the UK	Journal of Rural Studies	2013
19	D. Tilman and M. Clark	Global diets link environmental sustainability and human health	Nature	2014
20	W.J. Ripple, P. Smith, H. Haberl, S.A. Montzka, C. McAlpine and D.H. Boucher	Ruminants, climate change and climate policy	Nature Climate Change	2014
21	F. Hedenus, S. Wirsenius and D.J.A. Johansson	The importance of reduced meat and dairy consumption for meeting stringent climate change targets	Climatic Change	2014
22	B. Machovina, K.J. Feeley and W.J. Ripple	Biodiversity conservation: the key is reducing meat consumption	Science of the Total Environment	2015
23	S. Stoll-Kleemann and T. O'Riordan	The sustainability challenges of our meat and dairy diets	Environment: Science and Policy for Sustainable Development	2015
24	K.-H. Erb, C. Lauk, T. Kastner, A. Mayer, M.C. Theurl and H. Haberl	Exploring the biophysical option space for feeding the world without deforestation	Nature Communications	2016
25	L. Aleksandrowicz, R. Green, E.J.M. Joy, P. Smith and A. Haines	The impacts of dietary change on greenhouse gas emissions, land use, water use, and health: a systematic review	PLOS One	2016

	Author(s)	Title	Journal	Year
26	M. Springmann, C.J. Godfray, M. Rayner and P. Scarborough	Analysis and valuation of the health and climate change co-benefits of dietary change	*Proceedings of the National Academy of Sciences*	2016
27	D. Bryngelsson, S. Wirsenius, F. Hedenus and U. Sonesson	How can the EU climate targets be met? A combined analysis of technological and demand-side changes in food and agriculture	*Food Policy*	2016
28	P. Behrens, J.C. Kiefte-de Jong, T. Bosker, J.F.D. Rodrigues, A. de Koning and A. Tukker	Evaluating the environmental impacts of dietary recommendations	*Proceedings of the National Academy of Sciences*	2017
29	R.R. White and M.B. Hall	Nutritional and greenhouse gas impacts of removing animals from US agriculture	*Proceedings of the National Academy of Sciences*	2017
30	H. Harwatt, J. Sabaté, G. Eshel, S. Soret and W. Ripple	Substituting beans for beef as a contribution toward US climate change targets	*Climatic Change*	2017
31	M. Swain, L. Blomqvist, J. McNamara and W.J. Ripple	Reducing the environmental impact of global diets	*Science of the Total Environment*	2018
32	T. Searchinger, S. Wirsenius, T. Beringer and P. Dumas	Assessing the efficiency of changes in land use for mitigating climate change	*Nature*	2018
33	M. Springmann, M. Clark, D. Mason-D'Croz, K. Wiebe, B.L. Bodirsky, L. Lassaletta et al.	Options for keeping the food system within environmental limits	*Nature*	2018
34	A. Shepon, G. Eshel, E. Noor and R. Milo	The opportunity cost of animal-based diets exceeds all food losses	*Proceedings of the National Academy of Sciences*	2018

	Author(s)	Title	Journal	Year
35	J. Poore and T. Nemecek	Reducing food's environmental impacts through producers and consumers	*Science*	2018
36	H.C.J. Godfray, P. Aveyard, T. Garnett, J.W. Hall, T.J. Key, J. Lorimer et al.	Meat consumption, health, and the environment	*Science*	2018
37	M. Berners-Lee, C. Kennelly, R. Watson and C.N. Hewitt	Current global food production is sufficient to meet human nutritional needs in 2050 provided there is radical societal adaptation	*Elementa: Science of the Anthropocene*	2018
38	H. Harwatt	Including animal to plant protein shifts in climate change mitigation policy: a proposed three-step strategy	*Climate Policy*	2019
39	N. Bowles, S. Alexander and M. Hadjikakou	The livestock sector and planetary boundaries: a "limits to growth" perspective with dietary implications	*Ecological Economics*	2019
40	R.C. Henry, P. Alexander, S. Rabin, P. Anthoni, M.D.A. Rounsevell and A. Arneth	The role of global dietary transitions for safeguarding biodiversity	*Global Environmental Change*	2019
41	G.M. Poppy and J. Baverstock	Rethinking the food system for human health in the Anthropocene	*Current Biology*	2019
42	B.F. Kim, R.E. Santo, A.P. Scatterday, J.P. Fry, C.M. Synk, S.R. Cebron et al.	Country-specific dietary shifts to mitigate climate and water crises	*Global Environmental Change*	2020
43	M.C. Theurl, C. Lauk, G. Kalt, A. Mayer, K. Kaltenegger, T.G. Morais et al.	Food systems in a zero- deforestation world: dietary change is more important than intensification for climate targets in 2050	*Science of the Total Environment*	2020

	Author(s)	Title	Journal	Year
44	W.J. Ripple, C. Wolf, T.M. Newsome, P. Barnard, W.R. Moomaw and P. Grandcolas	World scientists' warning of a climate emergency (signed by 11,258 scientists from 153 countries)	*Bioscience*	2020
45	M.A. Clark, N.G.G. Domingo, K. Colgan, S.K. Thakrar, D. Tilman, J. Lynch et al.	Global food system emissions could preclude achieving the 1.5° and 2°C climate change targets	*Science*	2020
46	M.N. Hayek, H. Harwatt, W.J. Ripple and N.D. Mueller	The carbon opportunity cost of animal-sourced food production on land	*Nature Sustainability*	2021
47	I. Hamilton, H. Kennard, A. McGushin, L. Höglund-Isaksson, G. Kiesewetter, M. Lott et al.	The public health implications of the Paris Agreement: a modelling study	*Lancet Planetary Health*	2021
48	M. Crippa, E. Solazzo, D. Guizzardi, F. Monforti-Ferrario, F.N. Tubiello and A.J.N.F Leip	Food systems are responsible for a third of global anthropogenic greenhouse gas emissions	*Nature Food*	2021
49	M.B. Eisen and P.O. Brown	Rapid global phaseout of animal agriculture has the potential to stabilize greenhouse gas levels for 30 years and offset 68 percent of CO_2 emissions this century	*PLOS Climate*	2022
50	Z. Sun, L. Scherer, A. Tukker, S.A. Spawn-Lee, M. Bruckner, H.K. Gibbs et al.	Dietary change in high-income nations alone can lead to substantial double climate dividend	*Nature Food*	2022

Table 4.4 Fifty key reports recommending reductions in animal consumption for climate mitigation and other co-benefits (2008–22). Note: Row shading indicates the following: unshaded – United Nations (13); grey shading – charities, think tanks, NGOs and other (28); dark grey shading – UK government or advisory to government (9).

	Organisation	Title	Year
1	World Society for the Protection of Animals	*Eating Our Future: The Environmental Impact of Industrial Animal Agriculture*	2008
2	Food Climate Research Network	*Cooking Up a Storm: Food, Greenhouse Gas Emissions and Our Changing Climate*	2008
3	Food Ethics Council (UK)	*Livestock Consumption and Climate Change*	2009
4	Worldwatch	*Livestock and Climate Change: What if the Key Actors in Climate Change Are Cows, Pigs, and Chickens?*	2009
5	Sustainable Development Commission (advisory to UK government)	*Setting the Table: Advice to Government on Priority Elements of Sustainable Diets*	2009
6	United Nations Environment Programme	*The Environmental Food Crisis: The Environment's Role in Averting Future Food Crises, a UNEP Rapid Response Assessment*	2009
7	UN General Assembly	*Report Submitted by the Special Rapporteur on the Right to Food*	2010
8	United Nations Environment Programme	*Assessing the Environmental Impacts of Consumption and Production: Priority Products and Materials*	2010

	Organisation	Title	Year
9	Foresight and Office for Science (UK government)	*The Future of Food and Farming: Challenges and Choices for Global Sustainability*	2011
10	United Nations Environment Programme	*The Critical Role of Global Food Consumption Patterns in Achieving Sustainable Food Systems and Food for All*	2012
11	Live Well for Life and World Wildlife Fund for Nature	*Food Patterns and Dietary Recommendations in Spain, France, and Sweden*	2012
12	Live Well for Life and World Wildlife Fund for Nature	*Adopting Healthy, Sustainable Diets: Key Opportunities and Barriers*	2013
13	Food Ethics Council and World Wildlife Fund for Nature	*Prime Cuts: Valuing the Meat We Eat*	2013
14	Department for Environment, Food and Rural Affairs (UK government)	*Sustainable Consumption Report: Follow-Up to the Green Food Project*	2013
15	International Development Committee (UK government)	*Global Food Security Report*	2013
16	United Nations Environment Programme	*Our Nutrient World*	2013

	Organisation	Title	Year
17	Chatham House (UK)	*Livestock: Climate Change's Forgotten Sector*	2014
18	Intergovernmental Panel on Climate Change	*Working Group III: Assessment Report 5*	2014
19	Food and Agriculture Organization of the United Nations and Food Climate Research Network	*Plates, Pyramids, Planet: Developments in National Healthy and Sustainable Dietary Guidelines: A State of Play Assessment*	2016
20	World Resources Institute	*Shifting Diets for a Sustainable Food Future*	2016
21	ETC Group (Action Group on Erosion, Technology and Concentration)	*Who Will Feed Us? The Industrial Food Chain vs. the Peasant Food Web*	2017
22	World Wildlife Fund for Nature	*Appetite for Destruction*	2017
23	Rural Investment Support for Europe Foundation	*What is the Safe Operating Space for EU Livestock?*	2018
24	GRAIN and Institute for Agriculture and Trade Policy	*Emissions Impossible: How Big Meat and Dairy are Heating Up the Planet*	2018
25	Intergovernmental Panel on Climate Change	*Global Warming of 1.5°C: An IPCC Special Report on the Impacts of Global Warming of 1.5°C above Pre-industrial Levels and Related Global Greenhouse Gas Emission Pathways, in the Context of Strengthening the Global Response to*	2018

	Organisation	Title	Year
		the Threat of Climate Change, Sustainable Development, and Efforts to Eradicate Poverty	2019
26	Intergovernmental Panel on Climate Change	Climate Change and Land: An IPCC Special Report on Climate Change, Desertification, Land Degradation, Sustainable Land Management, Food Security, and Greenhouse Gas Fluxes in Terrestrial Ecosystems	2019
27	Climate Change Committee (advisory to UK government)	Behaviour Change, Public Engagement and Net Zero	2019
28	Greenpeace	Feeding the Problem: The Dangerous Intensification of Animal Farming in Europe	2019
29	World Resources Institute	Creating a Sustainable Food Future: A Menu of Solutions to Feed Nearly 10 Billion People by 2050	2019
30	EAT-Lancet	Food in the Anthropocene: The EAT-Lancet Commission on Healthy Diets from Sustainable Food Systems	2019
31	Institute for Agriculture and Trade Policy	Milking the Planet: How Big Dairy is Heating Up the Planet and Hollowing Rural Communities	2020
32	World Wildlife Fund for Nature	Bending the Curve: The Restorative Power of Planet-Based Diets	2020
33	Greenpeace	Farming for Failure – How European Animal Farming Fuels the Climate Emergency	2020
34	Greenpeace	Winging It: How the UK's Chicken Habit is Fuelling the Climate and Nature Emergency	2020
35	World Wildlife Fund for Nature and EAT	Diets for a Better Future: Rebooting and Reimagining Healthy and Sustainable Food Systems in the G20	2020

	Organisation	Title	Year
36	UK Health Alliance on Climate Change	All-Consuming: Building a Healthier Food System for People and Planet	2020
37	The Lancet	The 2020 Report of The Lancet Countdown on Health and Climate Change	2020
38	Feedback	Butchering the Planet: The Big-Name Financiers Bankrolling Livestock Corporations and Climate Change	2020
39	United Nations Environment Programme	Emissions Gap Report	2020
40	Convention on Biological Diversity and United Nations Environment Programme	Global Biodiversity Outlook 5	2020
41	Intergovernmental Science-Policy Platform on Biodiversity and Ecosystem Services	IPBES Pandemics Report: Workshop Report on Biodiversity and Pandemics of the Intergovernmental Platform on Biodiversity and Ecosystem Services	2020
42	European Commission	A Farm to Fork Strategy for a Fair, Healthy and Environmentally Friendly Food System	2020
43	Climate Change Committee (advisory to UK government)	Reducing UK Emissions: 2020 Progress Report to Parliament	2020
44	Climate Change Committee (advisory to UK government)	The Sixth Carbon Budget: The UK's Path to Net Zero	2020

	Organisation	Title	Year
45	Treasury (UK government)	*The Economics of Biodiversity: The Dasgupta Review*	2021
46	Henry Dimbleby (advisory to UK government)	*National Food Strategy: The Plan*	2021
47	Chatham House (UK)	*Food System Impacts on Biodiversity Loss*	2021
48	United Nations Environment Programme	*Making Peace with Nature: A Scientific Blueprint to Tackle the Climate, Biodiversity and Pollution Emergencies*	2021
49	World Wildlife Fund for Nature	*Land of Plenty: A Nature-Positive Pathway to Decarbonise UK Agriculture and Land Use*	2022
50	Intergovernmental Panel on Climate Change	*Working Group III Contribution to the IPCC Sixth Assessment Report*	2022

The juxtaposition of research in Tables 4.3 and 4.4 (and Table 4.5) with governmental inaction on reducing animal consumption highlights a severe impasse between science and policy. In the case of the UK government, many of the calls for reduction have come from its own advisory bodies. For example, in late 2020 the Committee on Climate Change called for a 35% reduction in all meat and dairy by 2050[123] (Table 4.4 #44), which was ignored by the time the government published its "Net-Zero" strategy.[124] In June 2023 the committee expressed concern over lack of action on this policy and others.[125] There are questions here over responsibility for effecting change, with governments on the right less sympathetic to public health and climate change policies. In the US 2017–18 election cycle, more than 80% of the energy sector's US$8.5 million in donations went to Republican candidates.[126] Moreover, neoliberal sympathies make such governments less likely to regulate the economy and more averse to being seen to tell people what to eat, even though governmental policies (such as subsidies for the animal-industrial complex) already do that.

A taste of what can happen in the exchange between science, policy and the media occurred in the United Kingdom in 2009 when *The Lancet* launched its work on health and climate change (Table 4.3, articles 3 and 4). Then Labour Health Secretary Andy Burnham gave a speech at the launch of the report. Departments of Health, Energy and Climate Change and an international development minister all endorsed the health-environmental co-benefits narrative of the report. The media represented the report as arguing for the culling of 30% of UK cattle and sheep ungulates, and that this policy was being entertained but without consultation with the Department for Environment, Food and Rural Affairs (which oversees agriculture) or the farming community. This was represented by the media as an amusing farce.[127] This media construction served to obfuscate the important co-benefits narrative and deflect attention away from a policy of reducing animal consumption. However,

---

123 UKCCC 2020b.
124 UK Government 2021.
125 UKCCC 2023.
126 Aronoff 2019.
127 Landale 2009.

it did serve to highlight the political, economic and practical complexities of making meaningful change to a food system with present co-disbenefits for environment and human health.

A continuum can be identified in societal options ranging from the efficiency framing all the way to a scenario of societal vegan transition. In the middle are varying degrees of reduction discourse which at very low levels resemble more the efficiency framing but at higher levels of reduction come closer to more transformatory change. It also matters whether reduction discourse is framed to target the consumption patterns of the wealthy, whether it is a uniform level of reduction in animal consumption across a population, or whether it is business-as-usual for some, reduction for some, and vegetarian/vegan transition for others. It is important to note that much of the work represented in Tables 4.3 and 4.4 which favours reduction options does so in a thoroughly anthropocentric manner.[128] Exceptions to this may be those works which bundle their co-benefit argument with biodiversity conservation. Conspicuously, none of these works consider the ethics of eating animals per se and therefore do not consider reduction in animal slaughter a further co-benefit.

The phrase "transformative change" has come to be used increasingly in work around sustainable transition[129] bearing relation to calls for systemic change. It is an open question what that might mean in terms of dietary and practice change regarding human–animal relations. Some stakeholders would see a 40–50% reduction in animal consumption as transformatory change. Others would prefer to reserve the phrase for a fuller plant-based transition. Others still would rather apply it to vegan transition and further fundamental shifts away from the main tenets of contemporary capitalism. The environmental argument for plant-based transition, differing degrees of which are necessitated by policies of animal consumption reduction, is bolstered by the further evidence base noted in Table 4.5, featuring 21 peer-reviewed studies which demonstrate that a plant-based diet can be the lowest emissions diet; a sample which includes five systematic reviews (studies 11, 13, 15, 16 and 20).

---

128 See Arcari 2017.
129 Grin, Rotmans and Schot 2010; Díaz, Settele et al. 2019.

Table 4.5 Twenty-one studies demonstrating that a plant-based diet can be the lowest emissions diet (2009–23).

[a] Also appears in Table 4.3.

	Authors	Title	Journal	Year
1	E. Stehfest, L. Bouwman, D.P. van Vuuren, M.G.J. den Elzen, B. Eickhout and P. Kabat	Climate benefits of changing diet	Climatic Change	2009
2	H. Risku-Norja, S. Kurppa and J. Helenius	Dietary choices and greenhouse gas emissions: assessment of impact of vegetarian and organic options at national scale	Progress in Industrial Ecology	2009
3	A. González, B. Frostell and A. Carlsson-Kanyama	Protein efficiency per unit energy and per unit greenhouse gas emissions: potential contribution of diet choices to climate change mitigation	Food Policy	2011
4	M. Berners-Lee, C. Hoolohan, H. Cammack and C.N. Hewitt	The relative greenhouse gas impacts of realistic dietary choices	Energy Policy	2012
5	P. Scarborough, S. Allender, D. Clarke, K. Wickramasinghe and M. Rayner	Modelling the health impact of environmentally sustainable dietary scenarios in the UK	European Journal of Clinical Nutrition	2012
6	T. Meier and O. Christen	Environmental impacts of dietary recommendations and dietary styles: Germany as an example	Environmental Science and Policy	2013
7	A. Joyce, J. Hallett, T. Hannelly and G. Carey	The impact of nutritional choices on global warming and policy implications: examining the link between dietary choices and greenhouse gas emissions	Energy and Emission	2014

	Authors	Title	Journal	Year
8	L. Baroni, M. Berati, M. Candilera and M. Tettamanti	Total environmental impact of three main dietary patterns in relation to the content of animal and plant food	Foods Control Technologies	2014
9	P. Scarborough, P.N. Appleby, A. Mizdrak, A.D.M. Briggs, R.C. Travis, K.E. Bradbury and T.J. Key	Dietary greenhouse gas emissions of meat-eaters, fish-eaters, vegetarians and vegans in the UK	Climatic Change	2014
10	C. Van Dooren, M. Marinussen, H. Blonk, H. Aiking and P. Vellinga	Exploring dietary guidelines based on ecological and nutritional values: a comparison of six dietary patterns	Food Policy	2014
11	E. Hallström, A. Carlsson-Kanyama and P. Börjesson	Environmental impact of dietary change: a systematic review	Journal of Cleaner Production	2015
12	K.-H. Erb, C. Lauk, T. Kastner, A. Mayer, M.C. Theurl and H. Haberl[a]	Exploring the biophysical option space for feeding the world without deforestation	Nature Communications	2016
13	L. Aleksandrowicz, R. Green, E.J.M. Joy, P. Smith and A. Haines[a]	The impacts of dietary change on greenhouse gas emissions, land use, water use, and health: a systematic review	PLOS One	2016
14	G.M. Turner-McGrievy, A.M. Leach, S. Wilcox and E.A. Frongillo	Differences in environmental impact and food expenditures of four different plant-based diets and an omnivorous diet: results of a randomized, controlled intervention	Journal of Hunger and Environmental Nutrition	2016

	Authors	Title	Journal	Year
15	M.E. Nelson, M.W. Hamm, F.B. Hu, S.A. Abrams and T.S. Griffin	Alignment of healthy dietary patterns and environmental sustainability: a systematic review	Advances in Nutrition	2016
16	B.C. Chai, J.R. van der Voort, K. Grofelnik, H.G. Eliasdottir, I. Klöss and F.J.A. Perez-Cueto	Which diet has the least environmental impact on our planet? A systematic review of vegan, vegetarian, and omnivorous diets	Sustainability	2019
17	M.C. Theurl, C. Lauk, G. Kalt, A. Mayer, K. Kaltenegger, T.G. Morais et al.[a]	Food systems in a zero-deforestation world: dietary change is more important than intensification for climate targets in 2050	Science of the Total Environment	2020
18	M. Pieper, A. Michalke and T. Gaugler	Calculation of external climate costs for food highlights inadequate pricing of animal products	Nature Communications	2020
19	A. Rabès, L. Seconda, B. Langevin, B. Allès, M. Touvier, S. Hercberg et al.	Greenhouse gas emissions, energy demand and land use associated with omnivorous, pesco-vegetarian, vegetarian, and vegan diets accounting for farming practices	Sustainable Production and Consumption	2020
20	A. Kustar and D. Patino-Echeverri	A review of environmental life cycle assessments of diets: plant-based solutions are truly sustainable, even in the form of fast foods	Sustainability	2021
21	P. Scarborough, M. Clark, L. Cobiac, K. Papier, A. Knuppel, J. Lynch et al.	Vegans, vegetarians, fish-eaters and meat-eaters in the UK show discrepant environmental impacts	Nature Food	2023

Table 4.5 studies are important reinforcements for the many reports and studies in Tables 4.3 and 4.4. When discussing veganism during the Introduction, I underlined the distinction between veganism as a transformative ethico-political philosophy, which far exceeds dietary change, and a plant-based dietary transition. Strictly speaking, Tables 4.3 and 4.4 studies are discussing a plant-based diet rather than the potential of vegan transition, which represents a stronger critique of anthropocentrism and the animal-industrial complex. Consequently, there is a difference between plant-based eating, veganism and intersectional veganism, as possible strategies to help combat the climate crisis. Furthermore, the findings of Table 4.5 studies could be considerably enhanced by further specifying the *type* of plant-based diet because inevitably there are potentially significant differences between various ways of eating plant-based and their greenhouse gas emissions. For example, if a plant-based food system further comprised lower levels of food processing, more wholefood and raw consumption, more local production, and less waste, the co-benefits to environment and health are typically enhanced. This, however, would still not be synonymous with a vegan food system, and certainly not an intersectionally vegan food system, which I turn to in Chapter 8.

Questions around plant-based transition need to be contextualised by an ongoing trajectory of an expanding animal-industrial complex. For example, the FAOSTAT statistics cited above show farmed animal production on the increase. While this is not inevitable, the FAO typically forecasts considerable future rises; indeed, it works toward achieving that increase. As their website articulates:

World meat production is projected to double by 2050, most of which is expected in developing countries. The growing meat market provides a significant opportunity for livestock farmers and meat processors in these countries ... The FAO programme in meat and meat products aims to assist the member countries in exploiting the opportunities for livestock development and poverty alleviation through the promotion of safe, efficient, and sustainable production, processing and marketing of meat and meat products.[130]

While the FAO role here may appear remarkable given its involvement in producing the two aforementioned reports, the guiding aim is stated as poverty alleviation. This does beg the question of whether the FAO might achieve better results for "developing" countries by promoting a more plant-based diet, which as the research above illustrates (Tables 4.3–4.5) has better co-benefits all round. It is not clear why meat and dairy consumption is the FAO's go-to solution for poverty alleviation. It has become common in pro-reduction studies to see recommendations that cuts to animal consumption should be higher in richer countries that have normalised a high level of consumption quite unusual even to their own recent histories, and that poorer countries should "be allowed" to see their rates of consumption increase. This mirrors climate justice discourse around geopolitical responsibilities for greenhouse gas emissions generally. While I concur that the transformation of food systems in rich nations should take precedence due to these reasons of historical emissions, this FAO view naturalises the Western nutrition transition as a pathway that will *inevitably* be mimicked by poorer countries and disallows such countries from choosing their own path that could be more plant- and evidence-based.

Plant-based transition is also contextualised by the *urgency* to address the climate crisis. In Part II of this book, I argue that the main constraints to such transition are cultural, economic and political. It is a transition already made by millions of people but not yet on a scale to make a tangible difference. The need to make urgent cuts to emissions justifies such a rapid food transition advantaged by not needing any new specific technological innovation. The cultural constraints to plant-based transition are undoubtedly seen in the part omission of nonhuman animals from climate discourse with which this chapter began. These illustrated that even when attempts are made to research the climate crisis there often remains a residual anthropocentric framing which largely excludes a contestation of animal consumption. Essentially, every single "radical" discourse of climate politics in recent decades – be that sustainability, just transition or climate justice – has avoided a critique of anthropocentrism. This aversion is retained by the reports in Table 4.4 that, although partly contesting animal

130 FAO 2016.

consumption, settle upon a reduction framing. It was also seen in the Anthropocene idea covered earlier.

This chapter has highlighted and reinforced the policy consensus of Tables 4.3 and 4.4 by showing that emissions from animal agriculture make a significant contribution to the climate crisis, such that specifically commodified human–animal relations should be seen as constitutive of this crisis. It is neither credible to bracket out this sector from climate policy and societal scrutiny, nor to prioritise an efficiency framing as the policy, as the FAO work has attempted. The biodiversity and health co-benefits (the urgency of which were underlined in Chapter 3), reinforced by research in Tables 4.3–4.5, and reductions to water consumption, water and air pollution appreciably strengthen the case for transformatory change.

Three scenarios

Moving forward, there are three possible scenarios. I refer to these as a goal of plant-based transition, vegan transition or intersectional vegan transition. Plant-based transition is being called for in much of the work in Tables 4.3 and 4.4 and describes a scenario where plants come to dominate our diets and lower levels of animals are consumed. In this scenario, as food cultures shift, more people adopt vegetarian and vegan diets than at present, but most people would be "reducitarians" or "flexitarians". The social presumption of meat and dairy consumption would erode but remain. Although the work in Tables 4.3–4.5 discusses veganism (especially Table 4.5), there are almost no recommendations for a full plant-based transition, for either rich or poorer countries. Moreover, this work tends to conflate vegan and plant-based, and any discussion of veganism as an ethico-political position is excluded because there is no stated interest in not killing animals as a further co-benefit. Plant-based transition is not antithetical to capitalism; indeed, it imagines transition taking place *within* capitalism.

In the second scenario, which I term vegan transition, there is an ultimate goal of abolishing the animal-industrial complex. This pursues a clearer vision of vegan culture and brings to the fore the ethical and political dimensions of veganism underlining how it pertains to more than just dietary practices. Decisively, this entails an explicit cultural critique of anthropocentrism. Transition occurs embedded in the critique of the instrumentalisation of other animals, including their environmental habitats because, as Chapter 3 underlined, a coherent veganism must include broad considerations of ecology and biodiversity. Transition also foregrounds scientific knowledge about the richness of nonhuman animal life and decentres human exceptionalism. This imaginary of vegan transition contains no specific anti-capitalist narrative and does not consider exploitative human–animal relations as emerging from a specific historical confluence of capitalist, patriarchal and colonialist relations. Rather, it tends to operate an understanding of animal exploitation in terms of a generalised atheoretical misanthropy or an individualistic understanding of flawed human behaviour. This imaginary is quite typical within mainstream animal advocacy organisations.

In the third scenario, which is occupied by forms of intersectional veganism, animal exploitation is seen as inseparable from, and enabled by, the development of capitalism, patriarchy and colonialism, as outlined in Chapters 1 and 2. This embeds veganism within a broader understanding of the emergences of the climate and biodiversity crises and sees their resolution in terms of addressing these interconnected inequitable relations and practices. This imaginary simultaneously complicates strategies for change but introduces potentially new tactics and alliances. The depth of societal transformation increases moving through scenarios 1 to 3. While all three scenarios could potentially deliver meaningful reductions to greenhouse gas emissions, intersectional veganism[131] goes further and, in line with the

---

131 I recognise that not all vegans endorse an intersectional veganism. Furthermore, I know of no serious position that has argued against subsistence animal consumption in some communities such as the often-mentioned example of Amazonian tribes. Intersectional veganism should not simplistically be wholly conflated with global veganism, instead recognising the intersectional complexity and situatedness of food ethics. For more on debates around vegan universalism, see Twine 2014a. Some readers will note a resemblance between my three scenarios and the ideal/non-ideal theory distinction in political theory (Rawls 1971). However, I do not see scenario 3 as "utopian" because it is already practised.

Capitalocene theory outlined earlier, contests the root causes of the climate crisis: capitalist economic growth and consumerism, patriarchy, racism and anthropocentrism. This final scenario imagines the abolition of the animal-industrial complex as unattainable unless also wedded to the elimination of these parallel root causes.

Veganism is not a panacea for the climate crisis but is part of an alternative transformatory imaginary which is concerned not only with rapidly reducing greenhouse gas emissions but also with engendering new just societies that are truly respectful of other species. Rather than being a trojan horse to introduce animal ethics into climate debates, intersectional veganism positions the critique of anthropocentrism and the ethico-political question of the animal as inescapably central to averting the climate crisis. A crisis that, this and the previous chapter have shown, is intimately about how the commodification of farmed animals in the animal-industrial complex is contributing to global heating, which, in turn, is now imperilling ecosystem viability for multitudes of other animal species worldwide. In both cause and effect, other species suffer. The ethical treatment of other animals cannot be conveniently ignored by climate policy.

Although scenario 1 fails to contest the anthropocentrism which, I argue, is partly constitutive of the climate crisis, Part II of this book keeps all three scenarios in play to further understand how transformative change can be achieved. The ambitiousness of all scenarios is matched only by their urgency. It is arguably unhelpful to view them as mutually exclusive; for example, scenarios 1 and 2 (and other collective movements such as intersectional feminism) could become pathways to intersectional veganism. Part II of this book is dedicated to understanding what needs to happen to transition our ostensibly meat cultures to vegan cultures. How might transition work? What presently negates it? Why is it already working for some? Fortunately, there is a wealth of social science research to draw upon and nascent cultural change that can prefigure ways out of the climate crisis. I begin with a topic central to any account of cultural reproduction: childhood.

Part II

Transforming meat cultures

5

A child's right to contest meat culture

In considering the potential for real transformative change in anthropocentric meat cultures, the second part of this book considers theories of change (Chapters 6 and 7) that may afford insight into how this can happen. There are already indications of transition which, although not systemic, provide prefigurative evidence of possible societal pathways. There are cultural differences worthy of further research, countries where animal consumption is especially ingrained, or others where plant-based transition may be more likely to gain a foothold. Yet all countries achieve their meat cultures and naturalise animal consumption via what could be called a generational universalism: the imposition of a set of meanings around animal consumption which construct it as normative for each new generation. This signals human childhood as an important topic for both critical animal studies (CAS) and any realistic attempts to avert the climate crisis and theorise transformative change. For the Capitalocene, childhood has been an important site for the reproduction of human–animal relations, securing majority acquiescence to animal agriculture and the profitability and cheap food it provides.

This chapter aims to open a rich vein which can further understanding of how anthropocentrism is woven into the fabric of societies. It begins by exploring child/animal intersections and then moves on to outline how these are involved in securing the reproduction of meat cultures. The last part of this chapter points to an alternative narrative which centres critical animal pedagogy, alongside the empowerment of children, to contest the generational universalism of compulsory anthropocentrism. The figure of vegan climate activist Greta Thunberg is employed to reflect on these themes. A focus on childhood also helps to address the overall omission of the social dimension of age from the discussions in Chapters 1 and 2.

Children, animals and the problem of innocence

That there are conceptual and political overlaps between CAS and childhood studies has only been recognised by a few,[1] despite Western cultures making varied symbolic associations between children and other animals. Childhood studies and C/AS (critical animal studies and animal studies) are quite recent fields despite now making clear contributions to understandings of society and social relations.[2] Both are interested in the ethical marginalisation of children or animals and investigate and advocate for children's or nonhuman animal agency. Furthermore, there are significant socio-historical intersections between children, childhood, animals and animality.

The field of childhood studies was influenced by the earlier famous work of Philippe Ariès who argued that taken-for-granted childhood is a relatively recent historical formation.[3] This has been important for the simple apprehension of the social construction of childhood and how diverse cultural meanings can change dominant understandings of children and childhood. How a society imagines childhood may tell us something about much broader social relations as well as shaping policies and social expectations toward children. Although not without critics,[4] Ariès argued that the concept of childhood emerged in Europe between the 15th and 18th centuries, which maps onto the periodisation of early capitalism noted by Moore earlier. Though debate

---

1   Cole and Stewart 2014.
2   Cole and Stewart 2014, 62.
3   Ariès 1962.
4   Prout and James 1997, 16.

takes place over the degree of absence of a child/adult dualism before this time, political, economic and cultural change during the emergence of capitalism led ultimately to a more pronounced child/adult distinction. This was partly shaped by the Western decline of child labour and the rise of mass schooling but also by shifting cultural constructions of childhood.

In certain respects, Western dualistic thought has coalesced to position children and animals similarly. That both children and animals may be seen as instinctual, unpredictable, unruly, needy, innocent, pure or cute, speaks, in part, to their mapping onto the devalued realms of nature, emotionality and the body. Shared meanings of childhood and animality may partly account for the ubiquitous presence of nonhuman animals in the lives of children, as toys and in children's media for example. This animal presence, which is often also infant animals,[5] may serve to reinforce notions of childhood innocence and an image of the child as not yet human. It is precisely the representation of children as human futures which has been a point of critique within childhood studies, as Lee outlines, because "identifying children with the future has tended unjustly to silence children as present-day members of society and to sanction their use as a resource instead of their inclusion as citizens with views and preferences of their own".[6] A process of objectification (alongside more complex anthropomorphic subjectification[7]) can also be seen in the infantilisation of animals in human childhoods, a process which may extend to actual companion animals in the lives of children and families.[8]

Myers discusses several ideological discourses of childhood animality which speak to the socio-historical intersections of children and animals in Western cultures.[9] Firstly, he discusses the idea of the "untamed child" in which children and animals share an original wildness, and children must overcome this "animal condition" through

---

5   Cole and Stewart 2014.
6   Lee 2013, 4.
7   This is not meant to imply all human/companion animal relations infantilise or objectify to the same degree. For a broad discussion of anthropomorphism, see Parkinson 2019.
8   Lewis 2020.
9   Myers 2007, 22–25.

socialisation. This idea was typically embedded within Christian discourse, valued "culture" over "nature", and assumed a negative image of the child. Secondly, Myers discusses the "child of nature", which in contrast views children and animals as morally good, juxtaposed against a fallen or corrupt notion of culture or civilisation. He points out that in the emergence of pet-keeping during the 19th century among the US middle classes, animals were seen as supporting the moral dimension of socialisation (especially for boys) acting as both "practice material for children learning to act kindly and ... as exemplars that could teach such virtues as gratitude, fidelity, and enduring love".[10] It is possible to make a discursive connection to more recent developments in animal-assisted therapy (AAT) used for children with trauma[11] and special educational needs.[12] Unsurprisingly, this discourse of the "child of nature" was shaped by romanticism and became an important influence on the view of children and animals as innocent, which I turn to shortly. It strengthens the discursive dualistic associations between children and nature, and between adults and culture. Significantly, Myers links this discourse of childhood animality to the contemporary positioning of children as "future redeemers", writing:

> Today, the corruption from which children may save society is environmental degradation. As in the earlier versions of the theme, society is problematic – only now in the thoroughgoing sense of destroying its own biophysical preconditions. Children themselves did not create this fallen condition and are not held responsible for it; in this they are seen to be like the animals of natural ecosystems.[13]

Myers mentions a third discourse of childhood animality that sees early childhood as a re-enactment of an early stage of human evolution. Although these three discourses of childhood animality all work to

---

10 Myers 2007, 26.
11 Signal, Taylor et al. 2017.
12 Fung 2017.
13 Myers 2007, 27.

associate children with animals and so strengthen a child/adult dualism, their different moral dimensions remain significant for how children and childhood are thought of today. Moreover, they also delineate what might be deemed normative for adult identities. For example, Cole and Stewart have noted how adult maturation is often predicated upon a disavowal of "sentimentalism" for other animals, with a cultural assumption that more instrumental relations emerge in a normative process of development.[14] This could suggest that particular constructions of childhood work to reinforce the dominant anthropocentric meanings of meat cultures. Were child development and adulthood predicated upon the *refining* of empathy (for the more-than-human), this could produce a quite different culture. Yet adult animal advocates are often implicitly infantilised for their views. This recalls the discussion on gender and care from Chapter 2, underlining that these contrasting assumptions around childhood and adulthood are also gendered.

The "child of nature" discourse has been bound up in one of the most pervasive mythologies of the Western child which has received considerable attention within childhood studies: the idea of childhood innocence. Myers is something of an exception in making this wider cultural link between innocent children *and* animals, with much of the critique of the idea more focused on how it institutionalises children's passivity. Initially, arguing against childhood innocence can seem counterintuitive, a move that might endanger child safeguarding. Indeed, a 1970s iteration of the left made the mistake of deconstructing childhood in a direction toward legitimising sexual access to children.[15] However, critics of the mythology of childhood innocence have argued convincingly that it makes children *less* safe by making a fetish of innocence, pathologising sexual children, and shielding children from knowledge.[16] Critical takes on childhood innocence then are not about

---

14 Cole and Stewart 2014, 62.
15 For example, famous names of 20th-century French intelligentsia: Sartre, De Beauvoir, Foucault, Barthes and Jacques Derrida whose work was later championed within C/AS, all signed a petition in the late 1970s which called for the decriminalisation of paedophilia using arguments from children's rights, see Henley 2001.
16 Kitzinger 1997, 164–65; Duhn 2012.

disavowing vulnerability[17] but are concerned with how this mythology has obscured children as social actors and the subjects of rights.[18] Here of course is another parallel between childhood studies and CAS, as both fields seek to raise the ethical considerability of children or nonhuman animals, sometimes (but not only) drawing upon rights frameworks.

The critique of the notion of childhood innocence becomes further interesting when it is highlighted that its 19th-century emergence was specifically tied to a romanticised view of white middle-class children. Garlen's critique of childhood innocence argues that it has operated to maintain white supremacy,[19] underlining that the construct of the white innocent child emerged at a time when white patriarchal culture was under threat from slavery abolition and first-wave feminism. By constructing childhood in this exclusionary way, it re-presented whiteness as virtue and purity, casting mothers as carers of children and protectors of innocence. Moreover, the historical association of women and children can be seen in the infantilisation of women and the construction of a normative white feminine innocence as in need of, and desiring of, male protection. Although increasingly antiquated, such constructions historically helped secure culture and knowledge as the domain of (white) men. These broader discourses suggest a limitation to an analysis of childhood innocence that focuses only on children. It may be a particularly important construct more broadly in the intersections of class, "race" and gender of the past 150 years, a myth that, as Garlen argues, "became the embodiment of human potential, and by association, a measure of humanness".[20] Accepting that the innocent child construct and its sharpening of child/adult dualism has intersected with understandings of class, "race" and gender, it is not a great conceptual leap to position it as also partially shaping ideas of the human and dominant ideas of anthropocentrism. It has already been noted that the innocent child (and woman) is permitted a sentimentalist attachment to other animals, but that "real" white male

---

17 Garlen 2019, 55.
18 Sarmento, Marchi and Trevisan 2018.
19 Garlen 2019.
20 Garlen 2019, 63.

human citizenship is achieved via a detached instrumental disregard for other animals. This may afford a relational role for myths of innocence in the construction of masculinity and normative ideas of human being which secure ideas of human exceptionalism because to care, to be empathic, is not understood as authentically male or human, whereas dominating animals often is.

Yet there is a further way in which childhood innocence acts to guarantee anthropocentric culture that may be probed by exploring how the mythology shields children from knowledge. Postman's well-known thesis on the disappearance of childhood is helpful to tease out these interconnections between childhood innocence and the child/adult binary.[21] He was interested in how transition from an oral to a written culture was accompanied by broader social changes and argued that the invention of the printing press in the 15th century and subsequent rises in literacy rates had significant impacts on the socio-historical construction of "childhood" and "adulthood". In an analysis in sympathy with Ariès' view that childhood is a relatively recent construct, Postman comments that:

> In a literate world children must become adults. But in a nonliterate world there is no need to distinguish sharply between the child and the adult, for there are few secrets, and the culture does not need to provide training in how to understand itself.[22]

This suggests that literacy and eventually formal education created a knowledge divide which inflects understandings of adulthood, childhood and human development, replacing an oral culture in which there is much less of a child/adult binary. Furthermore, this sets up a dynamic of secrecy between adults and children in which certain facets of life that may be deemed uncomfortable are withheld from children. This secrecy then becomes part of the cultural strategy of constructing childhood innocence. It is worth noting, for example, that adult literacy did not reach high levels in the United Kingdom until

---

21   Postman 1994.
22   Postman 1994, 13.

the 19th century, the time when the mythology of childhood innocence was consolidated. Postman continues:

> One might say that one of the main differences between an adult and a child is that the adult knows about certain facets of life – its mysteries, its contradictions, its violence, its tragedies – that are not considered suitable for children to know; that are, indeed, shameful to reveal to them indiscriminately. In the modern world, as children move toward adulthood, we reveal these secrets to them, in what we believe to be a psychologically assumable way.[23]

It is not just that secrecy has helped construct childhood innocence but that it also acted to defer for adults the potentially shameful disclosure of facets of life that might contest modernist narratives of progress and civility. Cultural secrets help to preserve a sanitised and de-animalised view of the human; they allow for a disingenuous dualistic association of the human with rationality, order and control. In this logic, proof of adulthood came to be seen in distancing oneself from both naiveté and the animalised elements of our cultural constructs of childhood. Unsurprising, then, that adult secrets centre around topics such as sexuality and violence – practices that may trouble not only the idea of childhood innocence but also the dominant cultural view of the human. Such normative shifts give intelligibility to cultural moments like "the talk", when parents are supposed to sit down with their children to discuss sex, if the "shame" has not been successfully outsourced to the education system. Sexuality and childhood remains a culturally contentious topic with ongoing controversies about sex education.[24] Other discoveries await that have also constituted controversy on school curricula – the Nazi holocaust, war, the inclusion of colonialism and slavery in history teaching – implying also that childhood innocence is further protective of nationalisms. Postman's analysis argued that childhood was beginning to disappear, that the

---

23   Postman 1994, 15.
24   For example, in the United Kingdom there has been faith-based opposition to LGBTQI+ sex education where homophobia exploits the childhood innocence discourse, see Giordano 2019.

child/adult dichotomy was weakening in part due to the diffusion of television and mass media giving children access to adult worlds. The emergence of the internet has also been accompanied by anxieties which seek to preserve a notion of childhood innocence, and while technologies may have contributed to a weakening of the child/adult binary it would be an exaggeration to say childhood has disappeared rather than shifted in meaning. More specifically, to talk of the binary eroding is too general since its contestation can either empower children or facilitate their exploitation.

The slaughterhouse as cultural secret, the climate crisis as adult shame

Devoting this space to a reflection on the mythology of childhood innocence is, of course, to suggest its relevance for thinking about the climate crisis. When the future is compromised childhood is more likely to become politicised as more people appreciate that business-as-usual modes of pedagogy are part of the problem. Since the climate crisis and the animal-industrial complex are associated with violence they are further examples of knowledge and practice which may be deemed both threatening to childhood innocence and shameful to adults. The animal-industrial complex constitutes a further "fact of life" that children eventually learn about. There is however no formalised cultural "talk", but many parents may ultimately discuss the topic informally.[25] The emissions discussion of the previous chapter underscored that the slaughterhouse itself is, like the coal mine, a site of the climate crisis. However, its contemporary form can already be seen as an example of sequestered shame, hidden out of sight. Mass animal slaughter is troubling not only for our constructs of childhood but also for societies generally. Contradicting modernist assumptions of civility[26] and rationality,[27] the animal-industrial complex is sharply antithetical to the typically romanticised and anthropomorphic

---

25 Bray, Zambrano et al. 2016.
26 Elias 1969.
27 Bauman 1989.

presentation of other animals in childhoods. This was poignantly captured in the 2013 performance art piece by Banksy entitled "Sirens of the Lambs".[28] A play on the 1991 movie *The Silence of the Lambs*, it involved a farmed animal transport truck stuffed with soft toys of chickens, pigs and cattle ungulates peering out while being driven around the traditional meatpacking district of New York, accompanied by looped pained crying audio. It is difficult to capture cultural hypocrisy and disingenuity more effectively with regard to how childhoods are suffused with other animals, which are simultaneously killed on a vast scale.

For parents to disclose the reality of societal animal killing is potentially fraught and shaming. Childhood innocence provides the rationale for not explicitly discussing the slaughterhouse with children. The possibility of traumatising children itself points to the disjuncture between sanitised human–animal relations and the harsh realities of commodification, slaughter and dismemberment. The slaughterhouse then acts, for a time, as a cultural secret between adults and young children. This secrecy protects not only "innocent children" from trauma but adults from shame and their anthropocentric norms from critical scrutiny. It also aids complicity with a major economic sector that contributes to the climate crisis. As highlighted in Chapter 2, the slaughterhouse is not just a symbolically disruptive site in which human exceptionalism is ritually reiterated, it is a point of intersection for classed, gendered and racialised dimensions of human–animal relations.

Despite these points, anthropocentrism and animal consumption are more or less successfully socially reproduced. Securing anthropocentrism may be a complex cultural accomplishment, but understanding it better and contesting its processes are key to addressing the climate crisis and enabling transformative change. There is resistance of course. Several CAS researchers have analysed incidents of animals escaping from farms or slaughterhouses and the ensuing media coverage,[29] attempting to understand why "when the slaughter

---

28   Banksy's "Sirens of the Lambs" can be seen here https://tinyurl.com/mvkehrm7.

29   Molloy 2011; Twine 2013a; Colling 2020; David and Stephens Griffin 2021.

of many millions of animals for the global food industries continues unabated, should some newsworthy animals be classified as morally considerable and allowed to live?"[30] Mercy, as in the Roman colosseum, is the ultimate expression of sovereign power, with these rare examples of animal resistance accommodated as curiosities and entertainment pieces within (the mediation of) the animal-industrial complex.

Children resist too. Imbuing childhoods with other animals to construct innocence can have unintended consequences. Shifting views on parenting mean that even omnivorous parents may better accommodate a child's agency if an aversion to animal consumption is expressed. Vegetarian and vegan families are likely to prefer to raise children similarly, which is important for the creation of alternative familial identities. However, challenges remain in the widespread use of dairy milk infant formula, contextualised by (in the United Kingdom especially) low breastfeeding rates, substantial marketing power from formula companies[31] and a lack of vegan alternatives. Resistance from children in their early years may also emerge once the consumption of other animals becomes apparent. Several videos of young children passionately opposing meat consumption have gone viral on YouTube which appear to convey a moral collision when the secret of the slaughterhouse is disclosed.[32] Furthermore, the National Farmers' Union of England and Wales reacted angrily in 2019 when supermarket TESCO aired a TV ad of a daughter declaring that she did not want to eat animals anymore and a compassionate father accommodating her preferences. It may be tempting to read such children, real or performed, in terms of the "child of nature" discourse mentioned previously wherein the morally valorised innocent child is elevated above the fallen adult culture, but not only does this reproduce the moral child/adult dichotomy; it falsely universalises an essential moral child. Research might be misinterpreted to support a valorised innate child morality, such as findings that children may not prioritise humans

---

30 Molloy 2011, 4.
31 Palmer 2009.
32 Perhaps the two best known examples are "Three-year old kid explains why he doesn't want to eat meat" https://tinyurl.com/adpy87xf and "'I won't eat animals,' girl tells her mother" https://tinyurl.com/2p8fentr.

over other animals to the same extent as adults. However, the conclusion for Wilks et al. is that the view of humans as "far more morally important than animals appears late in development and is likely socially acquired",[33] which is consistent with the perspective of Cole and Stewart above,[34] around adult maturation as commonly predicated upon a gradual moral devaluation of other animals. Children's violence toward other animals shows that they are just as morally complex as older people.[35] Furthermore, it is possible to respect the agency and moral decision making[36] of children who resist, to position them as potential educators of adults, and nurture their proto-veganism, without an advocacy buy-in to suspect constructs of the child, or a child/adult binary.

Vegan parenting constitutes a dazzling antagonism for meat culture, a sociologically potent marker of the "fleischgeist". To raise a child vegan is one of the clearest pre-emptive and prefigurative moves that can be made against the human, dominantly constructed in terms of human exceptionalism. Child development is infamously contested knowledge[37] partly because it speaks to dominant narratives of what it means to be human. Contesting and reconstituting child development in such a way that challenges nutritional fallacies presents a radically new way of doing human. Interestingly, opposition to vegan parenting on such grounds comes up against a broad range of scientific expertise which has pronounced upon the healthiness of plant-based diets.[38] Nevertheless, as animal consumption is increasingly problematised and veganism increasingly visible it is not surprising to see pushback in the media where vegan parenting has been represented as akin to child

---

33   Wilks, Caviola et al. 2021, 27. A finding reinforced further by McGuire, Palmer and Faber 2023.
34   Cole and Stewart 2014.
35   McDonald, Cody et al. 2018.
36   Hussar and Harris 2010.
37   Burman 2007.
38   British Dietetic Association (2017); Australia's National Health and Medical Research Council (2013); American Dietetic Association (2009). All have recognised a vegan diet as healthy. Specification within the category of a "plant-based diet" is important for optimal health.

abuse, although recent representations are more reflective, and supportive, of nascent vegan normalisation.[39]

Bray et al. in a rare study that looked at adult/child conversation around where meat comes from found that many conversations took place in early childhood and that caregivers who were most conflicted about such conversation tended to be female and urban.[40] Cairns and Johnston identify a paradox in expectations around contemporary maternal foodwork and meat consumption in which children should know where their food comes from but also be protected from the harsh realities of animal slaughter.[41] It is also a paradox increasingly present in educational settings and one that is difficult to resolve without either contesting animal consumption, childhood innocence, or both. Cairns and Johnston are correct to underline the limits of knowledge in achieving dietary change – the notion that the defetishisation[42] of industrial food production and animal commodification will simply achieve shifts in practice. This is a point I will return to in the next chapter, but here it is relevant to note that although the slaughterhouse can be identified as a cultural secret that offers up precarious and dissonant pedagogical moments that may provoke resistance, the overall societal picture is the successful reproduction of normative animal consumption.

This has led several CAS researchers to explore how this cultural reproduction is secured in childhood.[43] Cole and Stewart offer a four-quadrant conceptual map (Figure 5.1) to better understand how dominant cultural understandings of other animals place them into

---

39 This is not to distract from rare cases where parents (vegan or otherwise) feed their infants nutritionally inadequate diets (Geanous 2019) which means that vegan transition is likely to require new competences around nutritional knowledge (see Chapter 6). Supportive media features have also been published, Gander 2017; Marsh 2016. See also a related cluster of articles by moral philosophers on the topic of vegan parenting, Alvaro 2019; 2020; Hunt 2019; Milburn 2022.

40 Bray, Zambrano et al. 2016.

41 Cairns and Johnston 2018.

42 Cairns and Johnston 2018, 571.

43 Pedersen 2010; 2019; Cole and Stewart 2014; 2017a; 2017b; 2020; Schoonebeek 2015.

various categories of utility (e.g., food, companion animals, experimental resources, entertainment, wild, animal representations) according to an axis of both subjectification–objectification and sensibility–non-sensibility.[44] The latter axis refers to the degree to which other animals tend to be sensed (e.g., visible/audible) to humans. They argue, for example, that in the quadrant of objectification and non-sensibility are found animals socially constructed as "vermin", "laboratory animals" and "farmed animals", whereas in the quadrant of subjectification and sensibility are found "pets" and "working animals". Unsurprisingly, human animals occupy the far north-west zone. Cole and Stewart make clear that some animals – they discuss rabbits – are especially ambivalent in how they are constructed in Western cultures, in that they can be found across all four quadrants. The same can be said of rats and mice. Their point is more than just that children are socialised into anthropocentrism via the learning of different socially constructed categories of nonhuman animals. Cole and Stewart argue that one of the socially expected learned competences in the transition to adulthood consists precisely in coming to "inhabit a cultural milieu that mixes up affect, objectification and denial, often in respect of the *same* species of animal".[45] The conceptual map contributes to an understanding of how the majority of people learn to acquiesce to a morally arbitrary and incoherent positioning of other animals which normalises their exploitation within capitalist projects of accumulation.

---

44 Cole and Stewart 2014, 22. This map is most applicable to Western cultures.
45 Cole and Stewart 2014, 23 (original emphasis).

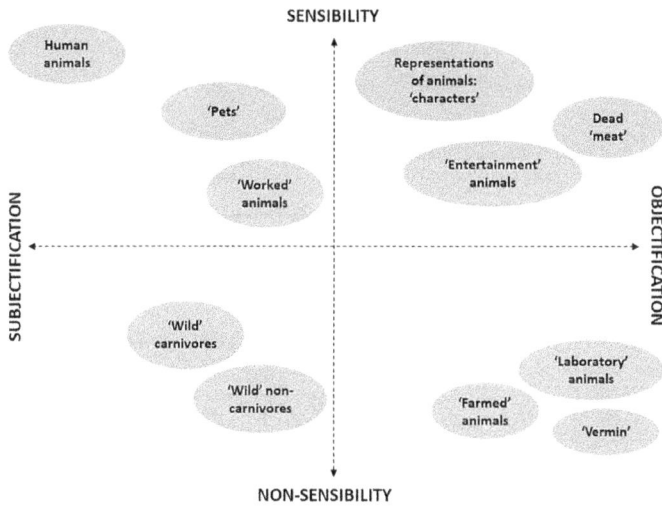

Figure 5.1 Cole and Stewart's (2014) conceptual map of the social construction of "other" animals.

They further identify four domains of particular importance where children learn meanings about other animals – the family, education, mass media and digital media. Such sites undoubtedly comprise an important part of the conceptual and affective apparatus of the animal-industrial complex whereby anthropocentrism weaves into practices, technologies, images, identities and markets.[46] These sites also roadmap points of resistance, being suggestive of areas in which to experiment with alternative narratives and practices.

In their subsequent work Cole and Stewart delve further into these domains, contesting the legitimation of exploitative human–animal relations in online "farming" simulation games,[47] in the comedy animation film *Sausage Party*[48] and in Hollywood CGI films.[49] As

---

46 Twine 2012.
47 Cole and Stewart 2017a.
48 Cole and Stewart 2017b.
49 Cole and Stewart 2020.

others such as Parkinson writing on the genetically modified CGI pig in the film Okja[50] (2017) or Plumwood writing on the film *Babe*[51] (1995), have shown, films using anthropomorphism can also be points of disruption and resistance against animal commodification.[52] Work on media which is especially targeted at and consumed by children complements necessary work on families and education in understanding the reproduction of cultural meanings about other animals. Pedersen has also examined how secondary and tertiary educational institutions further this social categorisation of animals and help confirm normalised anthropocentrism.[53] This can be noted in everything from school catering, animal science training and animal experimentation, nutritional education, the ties between the animal sciences and the animal-industrial complex, and the omission of critical animal pedagogy, which I turn to later. This is in affinity with Cole and Stewart who underline the normalisation of milk consumption in early childhood both at home and in schools, and how school food practices form part of the hidden curriculum that helps shape omnivorous eating as normative.[54] Such work within CAS constitutes the beginnings of clarifying the reproduction of anthropocentrism and the generational universalism of animal consumption. Animal consumption is a cultural imposition which survives under the guise of a naturalised part of being human. Though often simplified as a free choice of individuals, it is rather a social practice which people are typically steered into. Emphatically, this focus on children, childhood, child/animal relations and the child/adult binary is a necessary part of explaining how

---

50 Parkinson 2018.
51 Plumwood 1997.
52 There is often a knee jerk disavowal of anthropomorphism as a way to defend human/animal dualism. From a CAS perspective anthropomorphism can be objected to on the grounds of being anthropocentric and marginalising of nonhuman animal subjectivities. Parkinson (2019, 2) acknowledges these criticisms but also explores how it can be "a disruptive force, a capacity for imaginative appreciation of another's perspective; it opens the opportunity for cross-species intersubjectivity, and it can play a role in the development of empathetic relationships with other animals".
53 Pedersen 2010; 2019.
54 Cole and Stewart 2014.

anthropocentrism has also shaped the climate crisis. This is not just because human–animal relations contribute to greenhouse gas emissions but because the politics of childhood also intersect with and shape the climate crisis.

Understanding the slaughterhouse as a cultural secret of early childhood which initially protects childhood innocence and then is gradually disclosed through the learning of distinct categorisations of nonhuman animals illustrates an important aspect of meat culture which protects adults from a potentially difficult and shaming experience. The climate crisis, which meat culture is part of, presents a similar new dilemma for adults. How to explain to a child that their future is likely compromised? Unlike the slaughterhouse, which is a clear, albeit hidden, site of violence, the association of the climate crisis with violence is less overt. The slow violence of the climate crisis discussed earlier contrasts starkly with the intensely fast visceral violence of animal slaughter. In thinking this through via Postman's list of "facets of life" which structure adult secrets noted above, the climate crisis is perhaps not difficult to talk to children about because it is a violence (though of course it is), but more so because it is a tragedy, in at least two senses.

The crisis is tragic in the more abstract sense that Western modernity and the Capitalocene have been built upon exploitation, appropriation and are ultimately unsustainable. This punctures the Western sense of progress and superiority just as the slaughterhouse mocks pretensions of civility. The second sense of tragedy is more immediate, capturing how new generations in this century, who will now attempt to live into the next one, have a shadow cast over their lives. This also works with Postman's dynamic of secrecy and shame because not only is this an affectively difficult subject to talk to children about, but the everyday social practices of adults and parents – consumption related to energy, travel and food – are contributing to this tragedy, and most importantly, parents and teachers increasingly know this. This is not to generalise blame to "adults", or to individualise it away from larger scales of social organisation, but to underscore the potential precariousness of parenting in the Capitalocene which manifests itself in mundane activities including decisions over holidays, food practices and so on. The affective challenges of talking about the climate crisis not only disclose a failure of culture and a compromised future but also may intersect with the child/animal associations noted above as it is realised that animals cherished during childhood are increasingly under threat from the climate crisis. Contemporary extreme events linked to climate change such as wildfires and floods make it harder to conceal from children what is unfolding, making more likely a shameful disclosure of a diminished future.

Greta Thunberg as a vegan killjoy of childhood innocence

Nick Lee notes in an early book on children and climate change that in the context of modernity and its assumptions about human progress, future generations are assumed to have the opportunity for better wealth, health and longevity.[55] The climate crisis usurps this progress narrative creating a new expectation that future generations will have less abundant lives. The shadow cast by the crisis constructs an anxious future and constitutes a profound injustice against all future generations. It is noteworthy that sociological approaches to childhood which advocate for children's rights and agency have emerged at the same time that children's futures have been imbued with uncertainty and pathos. Chapter 2 noted how the climate crisis can be viewed as an injustice along lines of species, class, gender and "race", but it is also an injustice of age. Children and young people are often marginalised and disempowered from political decision making that affects the climate crisis. While young children are the humans most likely to experience the results of policy toward climate change, they are largely excluded from its formulation. Embedding his argument within childhood studies, Lee states that conceptions need to shift from seeing children as human embodiments of the future, to future makers. Without this shift, he argues, children are cast as "passively awaiting their role as 'janitors of the future'",[56] as inheritors of adult failure.

Championing the rights of children in relation to the climate crisis means a systemic critique of democracy as currently constituted and

55 Lee 2013.
56 Lee 2013, 130.

dovetails with longstanding ethico-political debates around the rights of future generations.[57] The politicisation of children and childhood that can now be noted in climate politics, thanks to a broad range of child activists, necessarily contests the mythology of childhood innocence. Markedly, the mythology may be present both in the exclusion of children from threatening climate knowledge and within environmental advocacy itself wherein a compromised future may be contested because it is seen to threaten assumed innocence. The movement for children's rights and progressive critical pedagogy in contesting innocence also challenges a sharp adult/child dichotomy on the grounds that children need citizenship and have a right to know that the climate crisis threatens their future. Casting children in the role of redeemers of the climate crisis is encouraged by the aforementioned "child of nature" discourse which itself, as already noted, valorises the child as good and innocent. There needs to be a distinction between the empowerment of children and the valorisation of the saviour child in this way because the latter runs up against the problematic of withholding knowledge from children in the mythology of innocence. In imagining children as agents of environmental change the child of nature discourse is both enabling and constraining. A further trap deserving of caution has been the critique of reproductive futurism.[58] Just as ideas of the animal have been used to construct key ordering concepts of modernity, Edelman argues that childhood serves "as the repository of variously sentimentalized cultural identifications, the Child has come to embody for us the telos of the social order"[59] and that ethico-political projects are often couched in terms of fighting for a better future for children. Edelman views this as normalising of a heteronormative notion of kinship and marginalising alternative queer forms. Climate crisis anti-natalism is perhaps the ultimate repudiation of reproductive futurism, a discourse which arguably also embeds an anthropocentric rationale, marginalising other kinship relations, such as those humans have with other species or those which take place between other species. Conceptualisations of the future tend to assume

---

57 De-Shalit 1994.
58 Edelman 2004.
59 Edelman 2004, 11.

both the heteronormative and the anthropocentric, and ultimately the humanist romanticism of valorising children as the future may exclude the idea that one could advocate to safeguard the flourishing of more-than-human kin.

Now a young adult, Swedish vegan climate activist Greta Thunberg is a clear cultural catalyst of the politicisation of children in climate politics and in the possibility of children as agents for change. Instigator of the *Skolstrejk för klimatet* (school strike for climate) movement, Thunberg can be seen as a classic killjoy figure. In her work on the feminist killjoy Ahmed argues for the political importance of the killjoy as a social agent for change who speaks out to critically contest the social order.[60] For Ahmed this also involves destabilising a "happiness order" preserved by social norms which act together to maintain patriarchal hierarchy. The feminist killjoy "makes sense if we place her in the context of feminist critiques of happiness, of how happiness is used to justify social norms as social goods";[61] for example, the assumed happiness of traditional gender performance, the assumed order of the patriarchal heteronormative family. Specifically, Ahmed evokes feminist critiques of the "happy housewife" as a docile caricature of complementary patriarchal society. This accentuates that oppositional politics advance through a contestation of happiness and that this has consequences for how others perceive activists and how activists experience themselves. Ahmed continues, "To be willing to go against a social order, which is protected as a moral order, a happiness order, is to be willing to cause unhappiness, even if unhappiness is not your cause".[62]

I have applied Ahmed's framework to thinking about the figures of the vegan, and vegan-feminist, killjoy.[63] The vegan killjoy questions the commensality of family eating based around animal consumption, disrupting the anthropocentric dimension of the social order. They contest the assumed happiness of animal consumption found especially in the marketing of the animal-industrial complex – notions of happy

---

60   Ahmed 2010a; 2010b; 2010c.
61   Ahmed 2010a, n.p.
62   Ahmed 2010a, n.p.
63   Twine 2014b.

meat, happy cows, of happy birthday parties in fast-food outlets. Killjoying this happiness construct means defetishisation and contesting the omnivorous denial of violent reality. Killjoying is not only critical but also creative practice, foregrounding the joy of emancipation from constraining ideology. That might be feminist or queer camaraderie in its original theorisation,[64] or the sharing of empathy for, and with, nonhuman animals in the vegan remaking of joy. I turn to these alternative narratives of happiness shortly. A climate killjoy potentially does even more systemic work, calling into question the Capitalocene and its misguided narratives of progress.

Greta Thunberg, as a vegan-feminist climate activist, is one of the leading cultural killjoys of our time.[65] However, she is also a killjoy to the mythology of childhood innocence itself. Her emergence as a political subject and agent of change illustrates the possibilities of resistance for children and young people. Defying notions of innocent ignorance, Thunberg became empowered partly through her grasp of climate science and politics. Thunberg's killjoying of childhood innocence also positions her against conservative instigators of moral panic over its loss – be that in areas of sex education or right-wing refusals to her calls to improve climate education in schools. Paradoxically Thunberg and the school strike movement also benefit from the innocence mythology because their resistance positions ineffective aged (typically white male) adults unavoidably as dispassionate destroyers of children's future. "How dare you!" Thunberg passionately decried in one of her most famous speeches,[66] delivered to the UN Climate Action Summit in New York on 23 September 2019:

> This is all wrong. I shouldn't be up here. I should be back in school on the other side of the ocean. Yet you all come to us young people for hope. How dare you! You have stolen my dreams and my

---

64   Ahmed 2010b.
65   In the media construction of Thunberg less attention is given to her feminism and veganism, but it is sometimes present in interviews. See Goldstein and Majsa 2019; and in Thunberg's speech here: https://tinyurl.com/mw3v5bfr.
66   Thunberg's speech can be viewed here: https://tinyurl.com/5n72nbxr. For an analysis of some of her earlier speeches from a childhood studies perspective, see Holmberg and Alvinius 2020.

childhood with your empty words. And yet I'm one of the lucky ones. People are suffering. People are dying. Entire ecosystems are collapsing. We are in the beginning of a mass extinction, and all you can talk about is money and fairy tales of eternal economic growth. How dare you!

This opening to her speech is of interest because it rhetorically positions ineffective world leaders as "stealing her childhood" and it cuts to the heart of one of the central capitalist contradictions – eternal economic growth. It is morally and culturally difficult for leaders and apologists to contest this child-led movement without appearing both callous *and* threatening to "childhood innocence". However, this has not prevented a male right-wing reaction against Thunberg that was entirely expected given her growing platform. Killjoying is risky behaviour. Ahmed says, "You cause unhappiness by revealing the causes of unhappiness. And you can become the cause of the unhappiness you reveal".[67] Hickman captures Ahmed's affective point well:

> The Australian Prime Minister Scott Morrison accused Greta Thunberg – the Swedish teenager who started the global strikes – of creating "needless anxiety" in children. So, it's Thunberg's activism that is responsible for the anxiety children feel about their future, not the climate crisis itself?[68]

The critical acerbity of the vegan-feminist climate activist killjoy will face backlash and Thunberg has triggered some other infamous men including Aaron Banks and Donald Trump.[69] This tricks such exemplars

---

67 Ahmed 2010b, 591.
68 Hickman 2019, n.p.
69 Banks, funder of Brexit, joked about her yacht capsizing during Thunberg's trip to New York, Busby 2019. Scott Morrison, the former Australian PM, reacted angrily to the school strike movement saying, "What we want is more learning in schools and less activism", Australian Associated Press (2018), and ex-US President Trump famously tweeted sarcastically after Thunberg's UN Climate Action Summit speech, "She seems like a very happy young girl looking forward to a bright and wonderful future. So nice to see!" Trump's sarcasm no doubt revealed how he thought children should present themselves, not as

of hegemonic masculinity into appearing vulnerable and fearful, again taking advantage of dominant cultural narratives of child/adult relations that shape expectations that grown men do not publicly attack children in the media. Such men struggle with the notion of a (then) 16-year-old girl having power. Right-wing male commentators have also used ableist language to try to pathologise[70] Thunberg in light of her having Asperger's, again another predictable response noted in work on feminist[71] and vegan killjoys.[72] All this has undoubtedly called for much resilience from Thunberg as she became educationally fast-tracked into an adult political world.

The potential mental health toll of comprehending the "adult knowledge" of the climate crisis disclosed to a child, and in the case of child activists the potential threat of backlash and misogyny, could all be mobilised as a rationale for re-protecting "childhood innocence". This brings us to an important point about trauma, mental health and a child's right to know, which has inspired debate in the literature on children and climate change. Sobel's mantra for (environmental) education was "no tragedies before fourth grade",[73] the age of 9–10, a position reiterated by Winograd.[74] I think this view falls victim to childhood protectionism. The mental health of children is important, but an age cut-off ignores the myriad ways in which topics can be taught. One of Thunberg's impacts has been to re-politicise the school curriculum, which I turn to shortly. It is also important to note that Thunberg is not the first child climate activist to gain prominence – Canadian Severn Cullis-Suzuki spoke at the 1992 Rio Summit – and watching Thunberg's rise could be suggestive again of children being afforded a limited profile and a cultural propensity to buy into the

---

critical or angry about the climate crisis. Trump would later tweet that Thunberg had an anger management problem, which in turn led to tweets from Thunberg mocking Trump, Luscombe 2020.

70 In the United States, Fox News had to apologise after one of their commentators referred to Thunberg as mentally ill in September 2019, Koerner 2019.
71 Ahmed 2010a.
72 Twine 2014b.
73 Sobel 1996, 27.
74 Winograd 2016, 10.

child of nature discourse. Through no fault of her own, Thunberg fulfils the accepted model of the white, middle-class saviour, even as she has resisted adult ideology of what constitutes acceptable behaviour for a child. Her profile may have been symbolically more acceptable to Western media and so marginalised child activists from the Global South;[75] however, she deserves credit for her internationalist outlook. Indeed, on 15 March 2019, a coordinated school strike day, 1.6 million people in 2,000 locations worldwide took to the streets to protest the climate emergency,[76] rising to 4 million people protesting in 163 countries on 20 September 2019.[77]

Thunberg's agency has both centred the possibilities of the child-as-educator (turning her father vegan but also agenda-setting globally) and illustrated how children's agency is also actualised in relation with peers and adults – she has a small team of advisory climate scientists and has sympathetic parents. A further example of this relational agency is seen in the *Juliana et al. v. United States* lawsuit in which climate scientist James Hansen has supported the case filed by 21 youth plaintiffs. Accusations that child activists are being manipulated by adults fail to countenance the possibility of children being invested in issues such as animal rights and the climate crisis. There is also a strong argument that participation in activism is a prerequisite for citizenship and democracy; as Winograd writes, "[Y]oung children's participation with their families, teachers, and peers is important in helping to develop the skills, dispositions, and confidence to become engaged in their communities, as 'agents of change'".[78] Building on this, Biswas and Mattheis contend that by striking from school, children take ownership of formal education, challenging curricula; and school strikes are themselves educative via learning from others, teaching others and learning about organised action.[79] Significantly, activism provides the setting for children to find their voice and self-confidence.

---

75 Unigwe 2019.
76 Holmberg and Alvinius 2020, 79.
77 Barclay and Resnick 2019.
78 Winograd 2016, 9.
79 Biswas and Mattheis 2022, 146.

Climate education, critical animal pedagogy and transformative change

Innovation in the theory and practice of early years education, informed by (early) childhood studies, has in recent years put the onus on ways to develop children's voice as a means of both empowering children and enhancing democratic structures in society.[80] Some have also argued for "childism", which goes further than including children, toward a societal restructuring of adult-centred norms.[81] Of further interest is the emergence of posthumanist perspectives in (early) childhood studies,[82] early childhood education for sustainability[83] and critical animal pedagogy.[84] Although these are overlapping areas there is less engagement with nonhuman animals in the first two. Chapter 4 already noted the various critiques of sustainability education for omitting human–animal relations. A more specific critique concerns how sustainability and related concepts are deployed in national early years curricula[85] and attempts, in line with CAS, to extend sustainability education into a critique of anthropocentrism and to move beyond an overfocus on children's agency toward a posthumanist appreciation of more-than-human agency.[86] Environmental education and education for sustainable development, which attempt to integrate social justice concerns, have longer histories but have faced obstacles gaining a foothold in curricula as education systems worldwide have become dominated by market ideologies and instrumental philosophies,[87] despite a supposed international agreement at the 1992 UN Rio Summit that "governments should strive to update or prepare

---

80   Moss 2014.
81   Wall 2022.
82   Murris 2016; Cutter-Mackenzie-Knowles, Malone and Hacking 2020; Malone, Tesar and Arndt 2020.
83   Davis 2005; Duhn 2012; Davis and Elliot 2014; Siraj-Blatchford, Mogharreban and Park 2016.
84   Dinker 2021; Dinker and Pedersen 2016; 2019; Nocella, Drew and George 2019; Oakley 2019.
85   Weldemariam, Boyd et al. 2017.
86   Weldemariam 2017.
87   Chapman 2011.

strategies aimed at integrating environment and development as a cross-cutting issue into education at all levels within the next three years".[88] Evidently this was not effective.

Climate education and critical animal pedagogy fall under the social reconstructionist educational philosophy that underlines the goal of education as personal and social transformation, tapping into a rich, varied history[89] which has informed critiques of traditional pedagogy based around adult-led hierarchies and assumptions of child docility. That this educational philosophy has been marginalised in the context of the rise of neoliberalism has contributed to the failure to establish effective climate education on national curricula. After all, climate education is inevitably critical of mobility, energy and food practices which still constitute major parts of national economies. An instrumental approach to education, fetishising credentials and employability, fits learners into the economy rather than entertaining critique of it. Therefore, although not all learning takes place in formal educational settings, the marginalisation of education-for-change philosophies can be seen as a barrier to effective climate education. If dominant educational philosophies constrain climate education, it is little surprise that they too must change. The movement inspired by Greta Thunberg has firmly put climate education on the agenda, shining the spotlight, for example, on the United Kingdom's poor record in this regard.[90] In late 2020 leading educators put pressure on new President Joe Biden to include changes to the US education system to both improve climate education and address the emissions of educational settings.[91] Climate education should be embedded across the curriculum and genuine fears over traumatising children can be addressed by incorporating mental health support within this theme. The UK national curriculum has also been recently critiqued for inadequately covering the history of British colonialism,[92] an issue that was amplified during the 2020 Black Lives Matter protests. Given the

---

88   UNCED 1993.
89   Dewey 1916; Freire 2017; hooks 1994; Giroux 2011.
90   Harvey 2020.
91   Milman 2020.
92   Goodfellow 2019.

Capitalocene framing in Part I of this book which made clear the role of colonialism in the climate crisis, it would be entirely appropriate to intersect these two curricular changes.[93]

Demands for better climate education are obvious; this analysis requires more. If anthropocentrism is a root cause of the climate crisis, then education systems must also be contested. Critical animal pedagogy, which seeks to move beyond the humanist focus of traditional critical pedagogy to also transform the lives of other animals, should be an integral part of climate education and the various strands of education for sustainable development and early childhood education for sustainability mentioned above. Critical animal pedagogy, for Dinker and Pedersen[94] and MacCormack,[95] must involve centrally, a leaving be of nonhuman animals, as a response to the historical exploitation of animals in educational settings. Furthermore, Dinker and Pedersen develop critical animal pedagogy as "an alternative education where students at all levels across the curriculum are invited to explore both a critical analytic and a radically transformative approach to animals and affect in education".[96] By critical analytic they mean that learners examine social norms, discourses and institutions (including education) that shape our affective responses toward animals in ways that increase our willingness to consume animal bodies rather than develop an ethical relationship with them. This is essentially bringing a critical sociological perspective on the animal-industrial complex into the classroom. It is pedagogically useful because it teaches skills about deciphering culture and the media, understanding society, economics and ethics. Some recent scholarship on posthuman childhoods favours contesting anthropocentrism via critical pedagogy,[97] though there is the usual tension over the degree to which a new field is inclusive of nonhuman animals. Critical animal

93 A London chapter of Black Lives Matter participated in an airport protest in 2016 to highlight interconnections between racism and the climate crisis, see Kelbert 2016.
94 Dinker and Pedersen 2016.
95 MacCormack 2013.
96 Dinker and Pedersen 2016, 418.
97 Kopnina, Sitka-Sage et al. 2020; Logan 2020; Saari 2020; Young and Bone 2020.

pedagogy is an approach in its own right, but can be seen as highly useful for embedding within climate education curricula. For without the critique of anthropocentrism, climate education is ineffectual in adequately contesting business-as-usual practices that have shaped the climate crisis, and in understanding the gravity of consequences for other species, as noted in Chapter 3. As per Calarco's definition of anthropocentrism[98] covered in the Introduction, pedagogy can contest human exceptionalism, a moralised binary account of human–animal differences, the tendency to dehumanise certain populations, and question institutions that privilege beings deemed fully human. In line with Wadiwel's work on epistemic violence[99] noted in Chapter 2, pedagogy can also contest the human sovereign right to dominate other animals by emphasising their agency, which is violated in their exploitation. Directly contesting anthropocentrism via critical animal pedagogy is one of the most powerful tools against the generational universalism of animal consumption discussed above, equipping children and young people with the agency to contest meat culture. Clearly this cannot be a didactic process, but it ought to be an enabling one which provokes real learning and engagement on human–animal relations even as it inevitably arouses resistance.[100] For at the moment, children are denied the knowledge to make informed choices about these relations, instead being compelled to submit to normative anthropocentrism.

Dinker and Pedersen also favour engaging students on the topic of intersectionality,[101] an appropriate framing for embedding human–animal relations within the broader framework of climate justice and education. They position vegan education, conceptualised as a critical intersectional framework, as the way to deliver the radically transformative aspect of their critical animal pedagogy, putting forward further practical suggestions. Significantly these include practices that are already common to school curricula such as growing vegetables but situated within reflection and the doing of (vegan) food practices,

---

98 Calarco 2021, 18.
99 Wadiwel 2015.
100 Darst and Dawson 2019.
101 Dinker and Pedersen 2016. See also Russell 2019.

nutrition, recipes, cooking skills and non-food aspects of vegan practice.[102] They further suggest class visits to vegan fairs, and one could add, animal sanctuaries. This would be one way to retain a degree of child/animal interaction in a refuge space external to the animal-industrial complex, despite their formulation of critical animal pedagogy as largely taking animals out of the educational experience. They also mention vegan picture books as useful educational materials for contesting normalised instrumental animal use in young children. Examples include *V is for Vegan: The ABCs of Being Kind*[103] and *Steven the Vegan*,[104] but the market has expanded in recent years, alongside children's vegan literature.[105] These could be used to provoke discussion about what different species mean to young children and why,[106] constituting early conversations about animal ethics. The radically transformative aspect of critical animal pedagogy is exactly an alternative narrative of happiness in a critical-creative practice of killjoy pedagogy.

There are already further indications of a developing vegan educational culture. This can be seen in challenges to school meal orthodoxy, such as the Scottish government adding non-dairy milk to its free nursery milk scheme in 2021 after a campaign by The Vegan Society,[107] Leeds City Council (United Kingdom) deciding to include vegan meals at 180 of its schools,[108] a Montessori Vegan Nursery in London,[109] and a United Kingdom–based campaign for vegan-inclusive education. This latter campaign has produced a guide for schools which is drawing upon the broader language of inclusion used in other social justice educational contexts and partly enabled by the characterisation of veganism as a protected belief under the *Equality Act 2010*.[110] Although the guide covers more than just food, it rightly makes a

---

102 Dinker and Pedersen 2016, 426.
103 Roth 2013.
104 Bodenstein 2012.
105 Koljonen 2019.
106 Schoonebeek 2015.
107 The Vegan Society 2021b.
108 Johnson 2020.
109 https://bodhitreenursery.co.uk/.
110 https://tinyurl.com/4xrjtvxw.

point of contextualising vegan-inclusive education in terms of climate education and mitigation.

These are examples of transformative change that are being promoted by a relational network of vegan children, parents, teachers and sympathetic allies. Also, in the United Kingdom there are early signs of institutional change on the reduction front, such as in 2020 school and hospital caterers vowed to cut meat served by 20%.[111] Systemic change perhaps not, but potentially seeds of change which could be accelerated by both critical animal pedagogy and vegan-inclusive education. The UK vegan-inclusive education campaign's use of equality legislation is illustrative of how employment and social rights extend the practice of veganism, further embedding societal change. For example, protections to ensure access to vegan food in institutions such as schools, universities, hospitals, care homes and prisons, leveraged justly via equality laws and codes of practice, can embed food-related vegan practices more widely in the social milieu. Overall, the contribution that critical animal pedagogy and allied approaches could make to both contesting anthropocentrism and creating vegan culture can be significant. Such promise, however, may be curtailed by incumbent conservative governance. An example was seen in the UK government's 2021 draft strategy for sustainability and climate change in education which included an aim to support schools to provide more plant-based meals by 2025. When the strategy was published in April 2022 this had been removed.

While it is unremarkable that transformative change to avert the climate crisis entails engaging with education systems (which have already been colonised by animal farming interests[112]), this chapter has asserted that our dominant understanding of childhood itself must also change. Conservative mythologies of childhood innocence act to maintain both climate and food knowledge as cultural secrets,

---

111 Carrington 2020.
112 A UK example of this is the Countryside Classroom project, https://www.countrysideclassroom.org.uk. Internationally, the dairy company Arla has an active classroom presence. In the United States 4-H is a youth program run by the Department of Agriculture which helps to normalise animal agriculture and consumption (Rosenberg 2016; Rosenberg and Dutkiewicz 2023).

preventing children from making informed choices and leaving the young as largely oblivious to how the climate crisis compromises their future. The food knowledge of animal slaughter is not overtly disclosed, protecting childhood innocence, adult shame, and the status quo of the animal-industrial complex.

This critique engenders an alternative transformatory narrative building upon the emancipatory work of both (early) childhood studies and CAS. In combining work that explores and values children's agency and sees early childhood pedagogy as the crucible of democratic renewal that must respond to the crises of our times,[113] alongside critical animal pedagogy's focus on a creative radical education beyond anthropocentrism, the marginalisation of both children and other animals can be addressed. That CAS would engage with childhood studies is inevitable given its concern with the politics of the human. These politics have significantly co-constituted the social construction of childhood because the ideological struggle over the meaning of the child is also the site in which narratives of human development have become stabilised. That a gendered human maturation has been seen in terms of a disavowal of care and empathy toward the more-than-human underlines the importance of opening up the contemporary politicisation of childhood to a questioning of normative anthropocentrism. This translates into a climate education inclusive of the nexus of issues around human–animal relations, animal consumption and veganism. Contesting the child/adult dichotomy through critical pedagogy can also serve the weakening of human/animal dualism, promoting leadership on the climate crisis among children and young people, empowered to understand the root causes of the climate crisis and the precarity it generates for all species.

---

113 Moss 2014.

.

6
Theorising transition

To better understand prospects for transformative change it is necessary to delve further into theories of transition. As should be clear from the previous chapters, the task of radically reducing emissions from animal agriculture via dietary transition while lacking any serious technological obstructions is nevertheless sociologically and politically challenging. Any intervention for transformative change has the weight of meaning, sedimented during childhood, to contend with. Although no social science understanding of transition offers a magic bullet to achieving rapid social change, some approaches offer insights in signposting pathways that may avert the worst of the climate crisis. A key aim of this chapter is to introduce a practice theory approach to transition as a leading candidate for a useful framework, one which I have employed in previous research on vegan and food-related transitions.[1] This and the following chapter will additionally consider critiques internal to practice theory concerning how it deals with issues such as power and the understanding of large-scale phenomena.[2] The approach is chosen in the belief that it can afford insight into both articulating and intervening in the animal-industrial complex. Furthermore, in what is an inclusive approach (albeit one that is

---

1   Twine 2014b; 2015; 2017; 2018.
2   Hui, Schatzki and Shove 2017.

sensitive to the incommensurability of contrasting theories), other theories will also be noted to provide a broader representation of social science research and the conceptual tools available.

A good place to begin when considering theories of transition is with the classic sociological and philosophical debate over agency and structure, especially because it is a discourse that tends to be implicitly reproduced in many discussions of strategy by activists invested in social change. This is seen in a dichotomous narrative whereby individual changes are sometimes devalued in favour of calls for systemic change. While the former would indeed be inadequate alone, and systemic (or structural) change is very much required, the narrative is arguably caught in this agency/structure dualism. An alternative approach is to see all scales as interdependent, co-constitutive and important.

For sociologists, the stakes of this debate are high because it speaks to the ontological demarcation of the discipline (much like the earlier discussions which contested the sociological exclusion of the more-than-human). For philosophers, it speaks to longstanding arguments over free will and determinism, and for political science, the debate is important because it informs reflection on what can be achieved politically in nominally democratic nations. Conceptualising "structures" for sociology has been an important part of disciplinary self-constitution that seeks to understand how, for example, institutions and infrastructures constrain and enable what humans can do. People are constrained, for example, by the class system, corporate power or by institutional racism. A contrary position finds this view potentially over-determining and aims to stress the agential capacities of social actors to change their own lives and society. The usual sociological "resolution" to this is to say that both positions are partly right, and this is reflected in much sociological research that tries to integrate understandings of social lives on the "micro scale" with broader "structural" considerations. Most sociologists acknowledge the agency of social actors, and their networks, but also appreciate how it may be limited. While the "structure" view may inadvertently foreclose opportunities to understand how societies change and the role of social actors within such processes, the "agency" view may be overly individualistic and fail to appreciate important sociological explanations of what might shape and constrain what people say and do.

Such reflections can be related to the notion of transforming meat cultures and the idea of vegan transition. So far I have noted that anthropocentrism is an entrenched belief system which gives rise to many dominant meanings helping to maintain a meat culture status quo. The agency view has not been particularly good at grasping how social meanings can be historically entrenched, if not immutable. As Walsh outlines:

> We are, as members, born into a social world which has been produced over centuries by the multiplicity of choices and decisions made by our ancestors, and we have to live with this inheritance and its effects in making our own choices and decisions.[3]

There are undoubtedly also culturally specific meanings that may further impede transition; for example, some countries more than others tie together masculinity and meat-eating.

In this vein I have noted the interconnected "structures" of capitalism, patriarchy and colonialism which have been argued to be inseparable from anthropocentrism and the formation of the animal-industrial complex in taking us on a pathway toward climate crisis. Norms of human sovereignty that naturalise an entitlement to the bodies of other animals have also been highlighted. In thinking through intersections between the social construction of childhood and anthropocentrism in the previous chapter it was argued that the discourse of childhood innocence contributes to the preservation of the anthropocentric status quo, and that neoliberal instrumental educational philosophies make climate education, let alone critical animal pedagogy, challenging. These are some hefty considerations for those tempted by an oversimplistic and optimistic narrative of the possibility of rapid social transformation premised around the notion of empowered individuals.

---

3   Walsh 1998, 13.

Despite such sobering considerations, animal consumption has become increasingly politicised, and veganism has become more visible. The increasing number of vegans, though still relatively small in the vast majority of countries (e.g., around 1% of the UK population[4]), does nevertheless underline resistance to prevalent anthropocentric norms. Furthermore, the previous chapter highlighted the influential power of individuals (and their networks) in the form of Greta Thunberg. Unsurprisingly, discourse among vegan communities tends to echo the structure/agency debate especially in terms of thinking through strategies for change. One might think of micro-level face-to-face interactional and social media vegan advocacy which often tries to popularise alternative ethical meanings in the hope that more people will decide to become vegan, and attempts to persuade social institutions, governments and corporations to promote a plant-based diet as more "macro-level" attempts to achieve change. This latter emphasis is noticeable in the climate justice movement's calls for "System Change! Not Climate Change", taken up by groups such as Extinction Rebellion and, relatedly, Animal Rebellion (later renamed Animal Rising), which have purposefully put forward a narrative and policy call for a plant-based food *system*, targeting institutions like universities.

As outlined later, one of the interesting aspects of practice theory is its refusal of the dichotomy between structure and agency and to argue against an ontological focus upon either the "individual" or "society" in favour of deploying a concept of practice as the unit of analysis, both for social science and for approaches to sustainability transition. However, before turning to practice theory, it is worthwhile paying attention to some other approaches within the social sciences, especially those which have already been applied to thinking about vegan transition or the reduction of animal consumption. Some of these approaches are influenced by social psychology or behavioural economics and have been specifically critiqued by practice theory. They tend to have a history of use in health-related behaviour change. Due to this

---

4   The UK Vegan Society commissioned research by Ipsos (2016; 2019) which found the level to be around 600,000 UK vegans or 1.16% of the aged 16+ population.

disciplinary heritage they often employ a terminology of "attitude", "behaviour" and "choice" which is avoided by practice theorists.[5] The more detailed theoretical reasons for this are explored later. The following approaches differ in the extent to which they might be conceptualised as coherent theories of transition (as opposed to being merely theories of behaviour) and whether, or how, they might approach the issue of scale. It is also useful to consider whether specific approaches are better oriented to understanding the hegemony of animal consumption and/or frameworks for understanding change, such as pathways toward vegan transition. Finally, it is also pertinent to ask whether the utility of approaches lies in their ability to act as explanatory frameworks, or whether they additionally offer possible ideas for policy intervention.

Social psychology and transition

A well-known psychological theory of behaviour is the theory of planned behaviour.[6] The theory is focused on the belief that behaviour is shaped by individual intention, and that intention arises as a result of the interplay between attitude, subjective norms (judgements of how others behave) and behavioural control (perceptions about one's ability to perform the behaviour). Applied in countless studies of consumer behaviour, the approach is rather mechanical, offers an overly rationalistic model of the social actor, fails to consider the gap between intention and behaviour (sometimes also called the values–action gap), and is certainly astructural and ahistorical. Nevertheless, judgements of how others behave introduces an important social component, and social psychological work is valuable for attempting to understand processes such as stigma as they relate to food practices and more generally how meanings around food are shared. Recent years have seen considerable expansion in social psychological research around meat consumption and veganism. This shall be referred to in the next chapter when meanings around animal consumption and veganism are

---

5   Shove 2010.
6   Ajzen 1985.

probed further. The usefulness of social psychology work can also be seen in further approaches that might inform thinking on transition, even though they might lack a capability to work at larger scales of analysis or be incommensurable with sociological approaches.

One such further approach is the idea of positive deviance. As Marsh et al. explain, "Identifying individuals with better outcomes than their peers (positive deviance) and enabling communities to adopt the behaviours that explain the improved outcome are powerful methods of producing change".[7] They further outline that "[p]ositive deviant behaviour is an uncommon practice that confers advantage to the people who practise it compared with the rest of the community".[8] Developed initially in the 1970s, the approach has subsequently been used, for example, in the area of children's nutritional health, rates of contraception use, and educational outcomes. It is also a community-based approach which stresses the importance of community engagement and the exploration of community-based norms, giving it some relevance for other scales of transition. The application of positive deviance to veganism is as yet unrealised but there are ways in which the practice could be classified as a positive form of deviance. A healthily and sustainably constructed veganism could be operationalised in the context of ecological public health programs. The practice encapsulates the norm-transgressing features of other forms of positive deviance, valuing a minority community practice which could have substantial benefits for all. A limitation in terms of climate action is that benefits may be difficult to apprehend in the short term since it could be difficult to equate lower carbon practice with the avoidance of climate crisis as benefits accrue for *future* generations. A vegan diet constructed to have advantageous health outcomes could have more potential to be constructed as a form of positive deviance.

The transtheoretical model (TM) is another widely used "behaviour change" framework. The approach emerged from work in psychotherapy during the 1980s.[9] It was specifically designed in relation to health-related behavioural change, being used in smoking

---

7   Marsh, Schroeder et al. 2004, 1177.
8   Marsh, Schroeder et al. 2004, 1177.
9   Prochaska and DiClemente 1982.

cessation. The main characteristic of the approach is to outline behavioural change as a series of changes between different stages. Mendes, who has briefly applied the framework to the process of becoming vegan, outlines the stages as follows: "The TM construes change as a five-stage process. The five stages of change are (a) precontemplation, (b) contemplation, (c) preparation, (d) action, and (e) maintenance".[10] Mendes considers how each of these stages may be applicable to the process of becoming vegan. It is a good example of an approach that is focused on the individual rather than the practice, or a broader notion of structure.

The behaviour change wheel framework is used by Grassian to analyse the effectiveness of different campaigns aimed at meat/dairy reducers, vegetarians and vegans.[11] Rooted in social psychology, it posits behaviour as the result of the interplay of capability, opportunity and motivation, a so-called COM-B model.[12] Capability is divided into physical and psychological, opportunity into physical and social, and motivation into reflective and automatic. This affords the framework the ability to account for social norms and the social environment to an extent. The wheel also has a layer named "policy categories", which includes considerations of more macro phenomena such as legislation, regulation and fiscal measures. It constitutes a novel framework in that its psychological roots are extended to consider larger scales of the social and economic. It is specifically aimed at having a multi-phenomenon relevance and is designed as an intervention tool for use with, but not restricted to, behaviour change in eating habits. For example, Grassian uses the framework to categorise potential barriers to meat/dairy reduction.[13]

The social norm approach has developed out of psychological research around conformity and consists of the idea that if people are told what lots of other people do, they are more likely to conform to that way of being (or buying).[14] This technique is used directly in social marketing

---

10   Mendes 2013, 142.
11   Grassian 2020.
12   Atkins and Michie 2013.
13   Grassian 2020.
14   Burchell, Rettie and Patel 2013.

campaigns, and more recently in attempts to change behaviour, notably in health-related campaigns. Burchell et al. assert that:

> In most sociological accounts, norms are explicit or implicit rules that guide, regulate, proscribe, and prescribe social behaviour in particular contexts. Thus, in sociology, norms refer to normative social influence with connotations of "ought" or "should". In social psychology, social norms also refer to patterns of group behaviour.[15]

Similarly, Bicchieri defines a descriptive norm as "a pattern of behaviour such that individuals prefer to conform to it on condition that they believe that most people in their reference network conform to it".[16] The expectation of social penalties (judgement, exclusion, stigmatisation) for non-conformity to social norms can constitute powerful social meanings which maintain status quo patterns of practice. Most research around animal consumption reduction and vegan transition relies upon ideas of social norms, usually to better understand their role in routinisation and why the majority of people persist with omnivorous practice.[17] When pro-vegan campaigns make appeals to shared social norms, those which many people would, on the face of it, seem to conform to, they are essentially using a similar approach without specifically naming it. Examples would include evoking social norms against direct violence to animals. Other techniques include using quantitative data to demonstrate that large numbers of people are reducing their meat and dairy consumption, in an effort to influence others and to encourage them along a pathway to veganism, or a more plant-based diet.

A further approach commonly used within marketing has focused on the diffusion of innovation by examining the role of early adopters in promoting new products.[18] This has explored not only why some

---

15 Burchell, Rettie and Patel 2013, 2.
16 Bicchieri 2017, 19.
17 Eker, Reese and Obersteiner (2019) specifically examine the social norm effect in relation to vegetarianism whereby a certain extent of normalisation can then accelerate further change.
18 Baptista 1999; Dedehayir, Ortt et al. 2017.

people are early adopters and others "laggards" but how early adopters impede or promote a wider diffusion. Riverola et al. have applied this approach to veganism outlining that:

> The "diffusion of innovations" theory categorizes the factors that drive the dissemination of innovations in the marketplace. The framework additionally distinguishes different adopter profiles that participate in the diffusion process, where "innovators" and "early adopters" are the first to adopt the innovation and are thus crucial for the innovation's further dissemination.[19]

Although veganism has increased markedly in recent years, there have obviously been vegans in the United Kingdom for many decades. It began to expand more in the 1990s and the first decade of this century. In contrast to the recent growth and social visibility of veganism, people who adopted veganism pre-2010, for example, could be considered early adopters. Such a time slice (The Vegan Society itself was founded in 1944) opens up potential comparative research between those who transitioned in specific decades. Research could focus on whether the salient meanings of veganism have shifted between decades. For example, Ploll et al. have also used a diffusion of innovation approach, finding differences between earlier and later adopters, with the latter more likely to report climate and environmental considerations.[20] If early adopters practise veganism in a certain way, it could, for example, slow social diffusion of the practice. As Riverola et al. put it:

> In some cases, innovators and early adopters act as opinion leaders in their respective communities and hence share the innovation and stimulate others to adopt it. Conversely, innovators and early adopters can differentiate themselves from the population, and in their deviance scare away potential groups of customers.[21]

---

19 Riverola, Ortt et al. 2017, 2.
20 Ploll, Petritz and Stern 2020.
21 Riverola, Ortt et al. 2017, 2.

Citing research by Centola,[22] they also argue for the importance of "homophily", or the tendency of social contacts to be similar to one another, as central to diffusion and adoption. Riverola et al. propose that "the perception of the characteristics of the innovation, as well as the image of innovators and early adopters, included in the communication message moderates the relationship between the receiver and the effect of adoption".[23] As will become clearer in the later discussion of practice theory, this approach speaks to interest in how early practitioners help create meanings around veganism, with implications for how non-vegans construct the practice. They continue:

> Omnivores perceive positively the vegan message, but they don't adopt it. Do omnivores believe that by adopting vegan behaviour they are belonging to a new group? Current research shows that omnivores perceived that adopting veganism implies a sense of belongingness to a different group. This negative sense of belongingness scares them away from adopting veganism.[24]

The notion that people may disidentify with the larger group associated with a particular practice is a relevant contribution to thinking about the diffusion of an innovative practice. An interest in understanding why a practice remains small-scale or becomes definitively embedded and normalised connects the diffusion of innovation approach to practice theory.

In considering transition I shift now to the more sociological theories that I advocate as especially useful because they are ostensibly better equipped for transgressing the agency/structure distinction. This is not to disparage social psychology; indeed, in the next chapter I return to the growth in work on the social psychology of animal consumption and veganism, as insightful for grasping food meanings relevant to transition.

---

22   Centola 2011.
23   Riverola, Ortt et al. 2017, 8.
24   Riverola, Ortt et al. 2017, 10.

Sociology and transition

Moving to more sociological theories such as social network analysis, the multi-level perspective and practice theory, a degree of synthesis may be possible. Though I shall place more emphasis on practice theory, their overlaps have been explored by others.[25] Practice theory is not intended as a theory of everything; as Schatzki, a key theorist, has argued,[26] it ought to seek out broader theoretical coalitions to improve its grasp of, for example, power[27] or larger social scales.[28] These issues are expanded upon in Chapter 7.

The first of these approaches – social network theory, or social network analysis – is a considerable area of research in the social sciences which studies social networks and relationships to better understand aspects of a practice, and multiple interlinked practices within social infrastructures. Although it has a long history, the social media connectivity of contemporary society affords new opportunities to better understand networks and their modes of influence. It is also an approach that overlaps with, and develops, the older sociological idea of social capital, the notion that one's social class position is partly determined by the characteristics of one's social network.[29] The areas of social analysis to which social network analysis has been applied are extensive. Research has suggested that health-related conditions such as obesity[30] or smoking may spread via social networks, as a form of "social contagion". One of the most important findings from this area is to suggest that "individual" health is socially interdependent, and that people are influenced in their health-related practices by those closely networked to them. Understanding broader national and transnational networks is also crucial to learning more about how large fossil fuel or animal agriculture companies maintain hegemonies within energy

---

25 Hargreaves, Longhurst and Seyfang 2012; Higginson, McKenna et al. 2015; El Bilali 2019.
26 Schatzki 2018.
27 Sayer 2013; Watson 2017.
28 Nicolini 2017.
29 Lin, Cook and Burt 2001.
30 Christakis and Fowler 2007.

and food systems, and how they link with lobbyists to deny their involvement in the climate crisis.

Although social network analysis has yet to be applied to vegan transition, prior interview data with vegans and vegetarians[31] make clear that people's constructions of veganism and their propensity to become vegan often emerges out of relations with vegan friends and family. A further way in which this approach could be applied to understanding the proto-normalisation of veganism would be to study the networks of vegan advocacy and to compare them with prior successfully normalised practices, or to chart the flows of vegan meanings on social media.[32] As veganism begins to become more normalised, there may be further benefit from other network-focused work associated with geographers and economists which has demonstrated the possibility of a network approach for comprehending larger scale systems by examining, for example, agri-food commodity chains[33] and alternative food networks.[34] The latter has tended to focus on the formation of fair trade, local production, organic food and farmer's markets as examples of resistance against elements of the capitalist globalised food system. That there is something of a gap between such visions of a just food system and that represented by a more radical shift to a plant-based food system is further explored in Chapter 8.

The second sociological approach is the multi-level perspective. Associated with the cross-disciplinary field of science and technology studies, like practice theory, it has been used to examine sustainable transitions.[35] The perspective argues that innovation and transition emerge from within the interdependency between three different levels of analysis. As Geels outlines:

---

31 Twine 2017; Parkinson, Twine and Griffin 2019.
32 Lawo, Esau et al. 2020.
33 Hubeau, Marchand et al. 2017.
34 Whatmore, Stassart and Renting 2003; Goodman, DuPuis and Goodman 2012.
35 Grin, Rotmans and Schot 2010.

The MLP [multi-level perspective] views transitions as non-linear processes that result from the interplay of developments at three analytical levels: *niches* (the locus for radical innovations), sociotechnical *regimes* (the locus of established practices and associated rules that stabilize existing systems) and an exogenous socio technical *landscape* ... Each "level" refers to heterogeneous configurations of elements; higher "levels" are more stable than lower "levels" in terms of number of actors and degrees of alignment between the elements.[36]

This is an attempt to connect different social scales for better understanding societal change. Although there are theoretical differences, another point of commonality between the multi-level perspective and practice theory is that both approaches often engage in historical analysis to understand how previous "socio-technical regimes" have transitioned. The three levels have been usually understood as a nested hierarchy wherein regimes are embedded in landscapes, and niches within regimes.[37] Regimes represent normalised and socially embedded clusters of practices something akin to constituent parts of the social order in classical sociology. However, regimes may be contested by niche-level social innovations and the broader socio-technical landscape, which can include a wide array of economic and political influences as well as events such as wars, immigration, environmental problems and pandemics, capable of putting pressure on existing regimes and opening up possibilities for change. An example is the climate crisis exerting pressure on pre-existing energy, transport and food regimes. However, regimes tend to attain a degree of stability because they consist of interdependent relations that have coevolved in a network of, for example, producers, users, government policy, infrastructure, science, technology and culture.

It is not difficult to see how the concept of niches could be applied to vegan innovation and to imagine a study which analyses the extent to which a vegan niche has been successful in influencing the

---

36   Geels 2011, 26.
37   Geels and Schot 2010, 18.

socio-technical regime, and to what extent the landscape is facilitating or curtailing this transition. For example, the climate crisis forms part of the landscape, but so do protectionist policies such as subsidies which benefit dominant, socially embedded agricultural and food practices. What is attractive about the multi-level perspective is that it allows the tensions and relations between the dominant socio-technical regime (what this book names as the "animal-industrial complex") and smaller niches such as proto-vegan or plant-based communities of practice to be better understood. Through developments such as food labelling and social visibility, vegan eating practice becomes increasingly stabilised within many societies. Amplified discourse on the climate crisis may create further opportunities for the vegan niche to embed. Although no research has yet applied the multi-level perspective to veganism, Vinnari and Vinnari applied a related, but broader, transitions management approach to plant-based transition.[38] Furthermore, research has applied the multi-level perspective to the emergence of plant-based milks,[39] and this is explored further in the next chapter as an example of how the perspective can improve understanding of the animal-industrial complex and its disruption.

The remainder of this chapter is devoted to a more in-depth consideration of a practice theory approach to understanding transition. Many different infrastructures require substantial change to address a meaningful societal transition to tackle the climate crisis. Levels of animal consumption across the world have increased markedly since World War II. The International Assessment of Agricultural Science and Technology for Development, commonly known as the World Agriculture Report, found that meat production almost quadrupled from 84 million tonnes in 1965 to a total of 330 million tonnes in 2017.[40] This underlines the importance of transition research to also attend to "negative trends",[41] because practices

---

38 Vinnari and Vinnari 2014.
39 Mylan, Morris et al. 2019.
40 https://tinyurl.com/2p9343py. The International Assessment of Agricultural Science and Technology for Development uses data from the Food and Agriculture Organization of the United Nations.
41 Antal, Mattioli and Rattle 2020.

effectively become "locked-in" and harder to dislodge. On average, every person in the world currently consumes 43.5 kilograms of meat per year. This average includes those who do not eat any meat. In 2013, US citizens consumed 115 kilograms of meat per year, UK citizens 81 kilograms, and citizens in India 3.7 kilograms.[42] Given that in 1960 the world human population reached 3 billion, 7 billion in 2011 and 8 billion in 2022, rates of meat production exceed rates of overall population growth. In reviewing similar historical meat consumption data, Sans and Combris conclude that richer countries have attained levels which exceed their needs.[43] While this ignores further questions around the politics of nutrition in which an argument could be made that there is no need for any animal consumption, it does draw attention to recent rapid rises in meat consumption and its unevenness between rich and poorer countries. Analysts have long pointed to a relationship between rising incomes and rising animal consumption under the rubric of the aforementioned "nutrition transition", and Cole and McCoskey have argued that an "environmental Kuznets curve" exists whereby meat consumption increases decelerate and potentially start to decrease.[44] They conclude that a deceleration of meat consumption takes place in high-income countries at a per capita income of US$49,848 and that such an effect will not have a *significant* impact any time soon. Consequently, they conclude that aggressive and controversial policy interventions will be needed to reduce levels of meat consumption. Although consumption levels have slowed or started to slightly decrease in the United States and some European countries, overall global increases are being driven by middle-income nations such as Brazil and China.[45] While this unsustainable transition to high levels of animal consumption has been partly shaped by rising population and income levels, it is important also to stress that it has been fundamentally promoted by the actors and practices of the animal-industrial complex itself, including corporations in the political economy and their interdependencies with governments and global

---

42   https://tinyurl.com/2p9343py.
43   Sans and Combris 2015, 106.
44   Cole and McCoskey 2013.
45   Ritchie, Rosado and Roser 2019.

bodies such as the World Bank and the Food and Agriculture Organization of the United Nations. Whether forecasts that Europe and North America will reach "peak meat" by 2025 are correct, remains to be seen.[46]

This negative transition, embedded culturally with new generations that may assume such levels of animal consumption are normal instead of the result of a specific and recent history, shows how new norms can become established in a relatively short time period and points to as yet unexplored questions around how these transitions were specifically achieved and to what extent they were contested in particular times and places. This post-war transition has constructed the dominant regime and the multiple interlinked practices which constitute the challenge for more climate-friendly ways of eating. In spite of this challenging context, sub-cultural niches of practitioners such as vegans already exist. The premise here is that vegans and those who have radically reduced their animal consumption hold important practice competences that should be explored and valorised culturally and sociologically as signposting transitions to more sustainable eating practices. In learning more about the "doing" and "saying" of these practices it ought to be possible to understand their reproduction, and broader social diffusion.

Treating practices, instead of individuals, as the primary unit of analysis, and examining how they consolidate and change, has been the focus of practice theory. Warde provides a useful genealogy of the concept of practice outlining its emergence as a focus of social theory in the 1970s and 1980s most notably in the work of Pierre Bourdieu.[47] However, Warde identifies the arrival of a second phase in practice theory associated with the work in particular of Schatzki[48] and Reckwitz.[49] This is of most interest here especially for how it has been taken up and refined by social scientists specifically interested to understand transitions toward more sustainable practices.[50] This phase

---

46 Carrington 2021a.
47 Warde 2016, 38.
48 Schatzki 1996; 2002.
49 Reckwitz 2002.
50 Shove, Pantzar and Watson 2012.

understood a practice as a patterned and routinised type of behaviour comprised of various elements,[51] and framed practices in two senses – as socially recognisable entities and as performances.[52] While broadly consistent performances of a practice maintain its social form, changes to the everyday performance of a practice (e.g., of eating), it is hypothesised, can engender change in the overall socially recognised practice-entity. In this way the approach aims to include a sense of mutual interdependence between different sites of practice.

Theories of practice are multiple, but this book, in addition to the theoretical writings of Reckwitz and Schatzki, is informed by the conceptualisation of Shove et al. of practice-entities comprised of the following three elements – competences, materials and meaning.[53] These are elaborated further in Figure 6.1. Their general argument is that "practices emerge, persist, shift and disappear when connections between elements of these three types are made, sustained or broken".[54] They also argue that connections are made between practices themselves, forming what they term "bundles"; an example might be driving and shopping. When such bundles become integrated parts of routine social infrastructure, they form what are referred to as broader embedded practice "complexes".[55] To underline, elements are "qualities of a practice in which the single individual participates, not qualities of the individual",[56] which has an implication for how the framework imagines both change and intervention and speaks to the theoretically posthumanist sympathies of the approach, foregrounding practice, backgrounding the individual. Exactly what connects elements of a practice together emerges as an important question, one which is explored below.

Shove et al. employ an understanding of change shaped by Giddens' structuration theory,[57] which was an important approach for moving beyond the aforementioned agency/structure distinction. This

---

51 Reckwitz 2002, 249–50.
52 Schatzki 1996; Reckwitz 2002.
53 Shove, Pantzar and Watson 2012.
54 Shove, Pantzar and Watson 2012, 14–15.
55 Shove, Pantzar and Watson 2012, 17.
56 Reckwitz 2002, 250.
57 Giddens 1984.

argues that the question of whether agency or structure is more important is moot because both are locked into an inseparable dynamic relationship mutually shaping each other. For Shove et al. this leads to two rejections, firstly, a) of "the notion that behaviours are driven by beliefs and values and that lifestyles and tastes are expressions of personal choice", and b) of the idea "that change is the outcome of external forces, technological innovation or social structure, somehow bearing down upon the detail of everyday life".[58] Practice theory can thus be seen as a rejection of both free will and determinism. In common with Giddens, these authors stress the routinised and habitual nature of social life which people accomplish by becoming enmeshed within multiple, overlapping, forms of practice. Therefore, although it is tempting to take the traditional anthropocentric (meant here literally as human-centred) approach to social change by focusing on either how to change the individual or how to understand wider social forces, the answer for practice theory instead is to attempt to understand exactly how practices emerge, evolve and disappear. As Reckwitz puts it:

> For practice theory, the nature of social structure consists in routinization. Social practices are routines: routines of moving the body, of understanding and wanting, of using things, interconnected in a practice. Structure is thus nothing that exists solely in the "head" or in patterns of behavior: One can find it in the routine nature of action. Social fields and institutionalized complexes – from economic organizations to the sphere of intimacy – are "structured" by the routines of social practices. … For practice theory, then, the "breaking" and "shifting" of structures must take place in everyday crises of routines …[59]

---

58 Shove, Pantzar and Watson 2012, 2–3.
59 Reckwitz 2002, 255.

COMPETENCY – Knowledge, skills and techniques

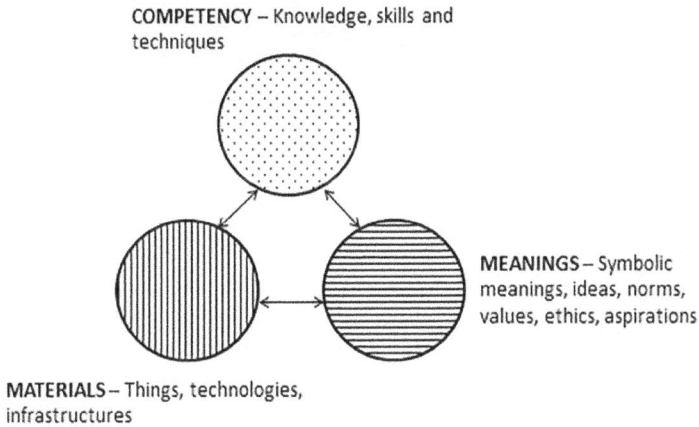

MEANINGS – Symbolic meanings, ideas, norms, values, ethics, aspirations

MATERIALS – Things, technologies, infrastructures

Figure 6.1 Elements of a practice. Adapted from Shove, Pantzar and Watson 2012.

In this way, practice theory approaches attend to the dynamism of elements in a practice, how socially shared and shifting meanings, materials and forms of competency change practices and render some redundant, or to use Shove's term, "fossilised". Such fossilised elements can become reincorporated into new practices, such as the use of many closed British railway lines for walking and cycling. These, like sub-cultural elements of practices, are potential resources for thinking about sustainable transitions. For example, one could note some benefits of a wartime rationed diet which was lower in meat consumption,[60] or of experiments with a New Nordic diet[61] or in re-engaging with the historical traditions of vegetarianism and veganism, often misunderstood as recent practices. In addition to changing elements, practices can change when others that they may be bundled together with change. Furthermore, although the primary conceptual focus on practices may suggest a posthumanist decentring of social actors, our agency, the networks social actors form remain

---

60   Cohen 2011.
61   Micheelsen, Havn et al. 2014.

important for understanding practice transitions. Practices presuppose relations. Thus, the way populations of carriers might change is important for the normalisation or the degeneration of a practice.[62]

In approaching food sociologically and through a practice-theory lens there are various conceptual issues. The everyday practice of eating intersects with many others, notably shopping, transport, storage and cooking practices, as well as its interdependency with cultural events and rites of passage. Moreover, invisible to most food consumers are the wide array of production practices, modes of distribution shaping and shaped by sets of international standards, trade relationships, governance, and the political economy of food. In the case of animal consumption, there is also the not inconsiderable question of practices bound up in the affective and spatial management of the killing of over 73.1 billion farmed land animals each year which are integral to the animal-industrial complex.[63] This underscores that there is also an ethical context to the prescription, meaning and organisation of eating practices.

This complexity and apparent weak degree of social organisation shapes Warde's view that eating is not in and of itself an "integrative practice" (this is Schatzki's term and is similar, if not identical, to the concept of a practice-entity mentioned earlier) but is better viewed as what he terms a compound practice (although all practices are arguably compound and overlapping): "Eating is formed from the articulation of different practices, including many in the long food supply chain, the domestic and commercial preparation of meals, and the organization of occasions for the consumption of food."[64] This is an understandable distinction given the diversity and complexity of what comprises eating practices. However, when examining a practice such as vegan eating, which during the last decade has, in many societies, taken on a degree of social intelligibility and organisation, an argument can be made that specific forms of eating are sociologically amenable, as identifiable, researchable forms of practice. For example, vegan eating practices may

---

62   Watson 2012.
63   See Introduction, note 17.
64   Warde 2016, 86. Indeed, veganism satisfies Warde's own test for constituting an integrative practice, based upon the existence of "how-to" books.

be more discernible than reducitarian or flexitarian eating due to their rule-based nature and relative degree of social organisation.

On the one hand, this introduces new complexities (over definitions of veganism and degree of adherence) but also presents a practice that is a more discernible and codified practice than omnivorous eating. Outside and inside the community of practice it is increasingly understood, at least in a dietary sense, as involving not eating foods, as far as is practical, from an animal source.[65] This is a shared norm. Vegan eating is largely consistent in this respect and delimited in a way that most eating is not. A wide variety of material artefacts are on hand to assist the everyday doings of veganism.[66] In the United Kingdom, The Vegan Society is an active presence acting as a reference for many in the successful performance of vegan eating (and non-eating) practices. This contrasting degree of social organisation to most culturally normative omnivorous eating practices makes veganism not only more likely to satisfy the conditions of a practice-entity but more amenable to study, and arguably, from a practice theory approach – though its degree of clustering or bundling with other practices means any social research on it is complex.

How then might the practice theory approach be usefully employed to understand the intricacies of vegan transition? Moreover, what affordances might the perspective offer in terms of interventions that could further embed the practice and expand the number of practitioners? While answers to the second question will seep into the next chapter, one answer to the first question is to deconstruct the practice of vegan eating in terms of the three elements of the practice and to offer explanations in terms of what makes the elements cohere together. Here I do so by drawing upon previous interview-based research that I conducted with 40 vegans living in the UK cities of Glasgow, Manchester and Lancaster.[67] This cohering of elements and the bundling of veganism with other practices is related to its nascent social normalisation, a process discernible in countries such as the United States, Canada, Australia, Germany and the United Kingdom.

---

65   See The Vegan Society definition of veganism discussed in the next chapter.
66   Twine 2018.
67   Twine 2017; 2018.

This construction of a useful framework is guided by the premise that vegan normalisation is necessary not just to recruit more people to the practice but to shift society overall so that more people eat diets lower in animal "products". A more vegan society catalyses opportunities for meat/dairy reduction. It would not be surprising if a discourse of "reduction" is more palatable to most people than one of "replacement" since the majority of people are fully invested practitioners of animal consumption, but the premise here is that vegan normalisation is co-productive of material infrastructure that also makes reduction more likely, even if that is not a satisfactory goal from an animal ethics perspective, and a less effective goal for meeting emissions and biodiversity targets.

The elements of a practice – competences, materials and meanings – are actively integrated and shape each other. An element can form part of several different practices. Competency includes forms of embodied skill, know-how and technique. It also includes the ability to evaluate one's own performance of a practice against shared understandings. Materials refer to things, technologies and infrastructures. Practice theory foregrounds an embodied ontology and so materiality here also includes the human body,[68] though I add here the bodies of other species. Nonhuman species are enmeshed in human practices and even more so when considering food. Importantly, Arcari has critiqued practice theory for an implicit anthropocentrism and ethical passivity in the way that nonhuman species are unquestioningly materialised in human practice.[69] The third element, "meanings", is used by Shove to "represent the social and symbolic significance of participation at any one moment"[70] but includes ideas, emotions, aspirations, values and norms. Practice theorists move beyond a causal understanding of many of these phenomena. Thus, when thinking about how people change it has often been assumed that attitudes, values, ethics, choices and norms should be individualistically targeted to initiate a linear outcome of "behavioural change" because individuals

---

68 Shove, Pantzar and Watson 2012, 23.
69 Arcari 2019. Arcari also examines whether nonhuman animals could be seen as practitioners.
70 Shove, Pantzar and Watson 2012, 23.

rather than practices have been seen as the agent of change. Instead, practice theorists view these phenomena as existing in a recursive relationship to practices and are continually "forged and reproduced through the process of doing" rather than acting as external drivers to particular behaviours.[71] Consequently, a target of intervention could be the socially shared meanings (or competences, or materialities) of a practice rather than the attitudes of an individual that have traditionally and erroneously been posited as existing in one's head. For practice theory, "attitudes" and "values" are social scripts of meaning that actors are recruited into. This is a significant difference with the aforementioned approaches from social psychology. One further point of conceptualisation – it is problematic to categorise values or ethics only within the element of meanings (contra Figure 6.1). This reinforces a cognitivist, abstract, disembodied understanding of what these are, jarring with scholarship that seeks to underline, for example, the lived, shared, embodied experience of animal ethics,[72] which points to the importance also of materiality and competences for ethical practice.

Competency and veganism

Methodologically using interviews to inform a practice theory approach may have some limitations. It risks collapsing practice theory back into something akin to the micro scale, and so potentially failing to transcend the agency/structure divide discussed earlier. Part of the next chapter will explore this issue of scale.

Certain themes related to acquiring competences were discernible from the interviews. The forms of know-how required to successfully perform vegan eating included a degree of new cooking skills, nutritional knowledge, ingredient-checking, new places of shopping as well as forms of social competency that made existing among normatively omnivorous culture easier. Related to these competences are various antecedent forms of know-how and knowledge around how to acquire these new competences. For example, with ingredient-checking one needs to know

---

71   Shove, Pantzar and Watson 2012, 143–44. See also Hards 2011.
72   For example, Acampora 2006.

which are of animal origin. Moreover, if one already has cooking skills the transition is likely to be easier. Prior internet access and competency also greatly aided the incorporation of vegan competences. Social skills in accessing both online and offline support groups and organisations proved very important to consolidating transition. Only one participant – Maggie, a 72-year-old who had become vegan in 1981 – did not use the internet in relation to her veganism. Access to material infrastructure also in many cases assumed an urban location. Therefore, for most of the sample there was a sense of ease in transition, though some of the antecedent forms of knowledge may assume economic privilege and technological proficiency.

The successful performance of veganism involves a set of competences which require initial work and reproduction. Misunderstood as a restrictive type of eating, most of the sample reported expanding their diet, eating foods that they had not chosen pre-transition. A practice theory lens makes clear that all eating habits tend to become "restricted" via food routines and habits. Vegan transition can potentially take one toward a greater diversity of fruit and vegetable consumption, even exploring the edibility of offshoot leaves from vegetables, common "weeds", and related food foraging expertise. Acquiring knowledge of new ingredients and foodstuffs (materiality) and potentially new ways of cooking was an important competency for this sample. A recurrent theme was adaptation and experimentation. Several participants embraced the challenge of cooking without animal "products", and many reported that veganism improved their cooking skills due to a greater need to "cook from scratch". Annie, 19, for example, took pleasure in such learning:

> I like experimenting with new recipes, I like getting recipes off friends that have tried stuff. I cook a lot of pasta using different sauces and stuff, I like cooking sauces from scratch. Yesterday I made vegan lasagne for the first time using a vegan white sauce and that was really cool. I've never done that before, and we used ground almonds and yeast extract on the top instead of parmesan cheese and that was really nice.

Knowledge of ingredients, labels and trademarks itself constituted a further vegan competency. The Vegan Society runs its own trademarking scheme, but this terrain is potentially complicated by the inconsistent use of vegan labelling among the major UK food retailers. Many participants reported that "ingredient-checking" was a new skill that had to be incorporated into their everyday shopping routines. As Steven, 24, outlined, "You quickly develop a thing of going straight for the ingredients and allergy advice that says milk, eggs, etc". In lieu of adequate labelling, vegans also need to know the names of obscure animal "products" often unexpectedly found in products. Since these interviews were conducted, UK supermarkets have acted to change store organisation to an extent, often including vegan products (at least those processed foods that tend to be symbolically derivative of animal "products") in dedicated aisles and sections. This, alongside increased vegan labelling, has partly addressed the need for ingredient-checking competences. As confirmed by the sample, vegans have created the category of "accidentally vegan" to describe products that are vegan but not labelled or sold as such. This is of interest for highlighting that everyone consumes "vegan food", a point that partly familiarises an assumedly "alien" practice.

The subject of labelling links to competency around food procurement and the sharing of knowledge (often online) between vegans on where to find particular products in a given locale, or which restaurant has just started offering vegan options. This was the norm among the sample, who turned to their new community of practice for knowledge and advice. I see this first-hand as an admin of the "Lancaster Vegans" Facebook group. Post-transition, people were more likely to augment their everyday supermarket shopping with visits to specialist shops which sell a higher proportion of vegan foods. While a small number of participants were dislocated from a broader vegan social scene, the three locations sampled all included events such as potlucks and vegan food fairs. These were experienced as being useful for sharing recipes, meal ideas, and promoting positive meanings of the practice and restoring commensality. Thus Tanya, 30, suggested that potlucks "have really helped me to try that bit more, try different ideas in cooking", and Rosemary, 64, felt that they had "increased my repertoire of dishes". Here social events can link together the three elements of vegan practice, a point I return to later.

A further finding asked questions about the boundary of vegan practice. In what I termed "non-practising practitioners",[73] several examples of friends and family of vegans taking on some of the elements of vegan practice were discerned. This included non-vegans cooking vegan meals for vegans and researching suitable restaurants that catered for vegans. Once a practice begins to socially embed it is not just direct practitioners that perform elements of the practice, but peripheral actors are partially enrolled into the practice, some of whom may eventually become full practitioners. Such people may be understood as making connections between practice competences and materialities but not yet integrating these with decisive food meanings integral to veganism.

A significant element of vegan practice is competency around nutritional knowledge. The majority of participants wanted to know about this during their transition to allay any fears they (or family) may have had. The supplementation of vitamins such as B12 into the materiality of foods has made this less of an issue. Post-transition, many vegans attain better competency around nutrition than the general population due to this learning process. Scientific knowledge claims around nutrition are pertinent to the terrain of sustainable food transitions. From a critical vegan perspective, assumedly scientific claims around the "balanced diet" as necessarily involving meat and dairy products have been particularly bound to cultural norms and economic interests; however, scientific nutritional knowledge advocating for veganism is now "catching up", with several national dietetic organisations endorsing plant-based eating.[74] Contested knowledge over nutritional quality inevitably colours constructed meanings of food practices with veganism located within that political field. This issue is returned to in the next chapter.

To illustrate the degree of social organisation at play, vegan communities of practice participate in and organise various activities that provide a conduit for competent performance. For example, the

---

73   Twine 2014b, 635–36.
74   See Chapter 5, note 38.

UK Vegan Society has had a buddy scheme for new recruits, which several of the participants had used, to aid performance of the practice. Websites such as Happy Cow[75] inform practitioners where they can locate vegan or vegan-friendly stores and restaurants. This international website was well known among participants, often featuring in their travel planning so they could identify places to eat. It served to partly address one of the trickier dimensions of transition.

The social context was experienced as the hardest part of transition. Friends, relatives and colleagues could aid or hinder transition in their response to veganism. This connects meanings to competences in that the contested meanings of the practice engendered a need for new vegans to devise forms of social competency to navigate a largely non-vegan social world. Carrie, 19, described having to "hold my tongue, so that I didn't offend people, even though I wanted to [laughs], learning restraint". Grace, 45, in reflecting on whether she had to learn any new skills during her transition answered, "social skills [laughs], assertiveness skills, yeah". This relates to asking about vegan options in such places as the workplace or when dining out. More generally participants reported having to deal with a significant degree of negative reaction from other people which had to be negotiated.[76] For most participants, it was the wearisomeness of this experience that made the transition occasionally difficult, rather than the more intrinsic elements of, for example, sourcing and preparing vegan food. Crucially, this highlights how food habits and routines are not constructed individualistically (highlighting that discourses of *individual* change can be misleading), but socially, in the form of friend, partner, familial and co-worker relations. Conflict is not surprising, and these relational ties can constitute significant blocks to vegan transition. On the other hand, contact with vegan friends or family helps to familiarise the practice, to learn alternative habits.

In thinking through the elements of practice it is important also to account for those of the dominant practice. The next chapter attends to this; however, here routine competences associated with animal consumption may be noted. Cooking and preparing meat can be a

---

75   https://www.happycow.net.
76   Twine 2014b.

highly skilled process. For example, it could be queried whether the UK tradition of the Sunday roast remains as strong, whether people still retain the preparatory know-how, or whether it is now "built in" to material kitchen appliances. It could also be explored which social groups remain most faithful to such practices, with such research potentially revealing early stages of practice fossilisation which could be further useful for theorising societal transition.

Materiality and veganism

As an illustration of the integration of elements in vegan practice, new forms of nutritional competency contest dominant meanings of health and ethics, in turn conveyed via new materials in the form of novel foods, vegan nutrition guides and charts, often placed in the kitchen. Such materials simplify the process of doing veganism, acting as aids to planning shopping and practising an understanding of "healthy eating".

Materiality is important to the practice theory ontology which stresses the role of objects in co-constituting social life. Again, this is a dimension that distinguishes practice theory from social psychological approaches to transition. Practices require materials, the theorisation of which has opened up the area of design as potentially important to sustainability transitions.[77] The animal-industrial complex draws upon a vast set of technologies and materials to bring animal "products" to the plates of consumers; for instance, the apparatus of molecular science pivotal to contemporary animal breeding,[78] specialised transportation for farmed animals, the material and temporal organisation of the slaughterhouse,[79] storage technologies and food preparation objects specific to the cooking of meat and other animal "products".

In thinking about how a dominant meat culture might be contested it is important here to explore the commonalities and differences between the materialities of meat culture and a counter-practice like

---

77   Scott, Bakker and Quist 2011; Shove, Watson et al. 2007.
78   Twine 2010a.
79   Pachirat 2011.

veganism. Moreover, how does the shadow of the dominant food culture shape alternatives? How materials are used and reinterpreted represents an area in which practitioners can innovate and contest the performance of (eating) practices. In the case of the kitchen environment there is not a marked difference between what an omnivore's kitchen might contain in comparison to a vegan's kitchen. A small number of kitchen artefacts are aimed at meat or egg preparation. Refrigerators used to be designed with fixed compartments for eggs; these now tend to be detachable or absent. A large roasting tray aimed at cooking a whole bird can easily be used instead to roast mixed vegetables. Vegetarian options are now mainstream in material food infrastructures and, in many locations, veganism has caught up. All participants were privileged in the sense of access to urban vegan or vegan-friendly restaurants and cafés, as well as supermarkets with a growing, if inconsistent, degree of vegan literacy. Veganism remains a less intelligible practice in more rural areas of the United Kingdom where norms of animal consumption are, on the whole, more immediate and tied to place.[80]

In approaching the data, vegan material adaptation and substitution of dominant practices were unsurprisingly significant themes. This highlights some of the ways in which the practice is not necessarily wholly distinct from the elements of the dominant animal-consuming culture. It also points to tensions within veganism over meaning and symbolism and the diverse ways of performing vegan eating. For example, some vegans take a position against the consumption of meat and dairy substitutes, preferring less processed alternative sources of nutrition. The so-called wholefood plant-based diet is recognised as a healthier mode of plant-based eating due to higher levels and diversity of fruit and vegetables, less fat, less salt, and no processed sugar. This is a materially different form of vegan eating which is not derivative of animal consumption. Thus Michael, 25, viewed substitutes "as something I enjoy as a treat, but I kind of still have this negative attitude towards processed foods". The

---

80 It is important not to reinstate stereotypes around the urban and rural. For example, in the United Kingdom several small rural tourist towns feature many eating options for vegans (e.g., Ambleside).

counterargument tends to view substitution as a convenient aid to transition away from meat and dairy. This latter view was dominant in the sample, with the majority of participants consuming substitutes during and after transition. Routines and habits perhaps receive the least disruption via the consumption of, for example, meat-free burgers and sausages, or plant milks. Natalie, 34, for example, explained:

> I always have loads of non-dairy milk and I quite like soya milk, coconut milk, all of those things, and yeah 'cos I'm a big hot chocolate fan so at least I didn't have to get rid of that. I'm a big cereal fan so at least I didn't have to change that, you know that was dead easy to substitute.

Although substitution might imply the possibility of a relatively simple food transition it could fail to do the work, as I discuss below, of the symbolic meanings of particular animal-based foods; in other words, a potential lack of integration between materiality and meaning. The proliferation of such substitutes in mainstream supermarkets, however, provides a visual presence for vegan foods: tangible evidence of the nascent normalisation of veganism. In recent years, the number of dairy and meat substitutes has soared in quantity and quality. Many different vegan cheeses are now available in the United Kingdom as well as soy, hemp, oat, almond, hazelnut and rice alternatives to dairy milk. UK supermarkets have now developed their own ranges of these foods which has lowered prices. Significantly, these products represent new supermarket spaces, the emergence of which also suggests new avenues for research. For example, food practices are intimately tied to the social organisation of time[81] but they also remake space[82] as in this supermarket example, or simply in how the word "vegan" becomes increasingly noticeable when negotiating urban space, or in virtual space. Thinking about the co-constitution of vegan practice and space may afford new opportunities for thinking about practice growth, especially in relation to the status quo spaces of animal consumption. The emergence of vegan butchers – already a number present in the

---

81 Southerton 2006; 2013.
82 Shove, Pantzar and Watson 2012, 133.

United Kingdom – could be exactly such an example. This nascent remaking of space is partly via this process of substitution, also being taken up by some caterers and restaurants, which is a significant step in normalisation, choice and availability. Corporate practices extend this process, illustrating the impact of resources on practice diffusion, but also how such commodification may modify specific practice meanings.[83]

In exploring food materiality, it is important to think of the foodstuffs themselves. The emergent part-normalisation of veganism in the United Kingdom is tied to the taste of vegan food, which in turn plays an important role in integrating positive meaning for veganism. There was a tangible sense of joy and enthusiasm among the participants when talking about the food they create and eat. Annie, 19, for example, used such experimentation to demonstrate veganism to family:

> I've experimented with a few different recipes that I found online and from friends, but mostly just cakes and different kinds of biscuits and I tried a cheesecake. I made that twice actually, I made it for my grandparents and my family and didn't tell them that it was vegan, and they all really liked it and then I dropped a bombshell [laughs].

> [What was their reaction?]

> They just didn't understand, they were like, how does it not have cheese in?

This suggests a playful material mockery of omnivores implying that if vegans can creatively fool them into believing that they are eating animal-derived food then it delegitimises their omnivorous diet. However, in demonstrating veganism like this, practitioners are able to build a bridge via the experience of vegan eating. Food materialities are used to communicate with non-vegan friends and family about veganism, to draw omnivores into the material, sensual experience

---

83   This is a focus of Chapter 8.

of vegan food. Here cooking for others and the agential taste of the food itself serves to positively change the meanings of veganism, to again reduce its sense of otherness. Elsewhere I have argued that this demonstrative veganism is a crucial form of vegan performance that can tempt non-practitioners into the practice within the context of friend or familial care relations.[84] Vegan competency in food materiality can shift the meanings of food for non-vegans.

Social events such as vegan potlucks and fairs are also designed to be demonstrative of taste to both practitioners and the curious, providing opportunities, literally, for tasting foods and the linked circulation of meanings, materials and competences. The network of vegan fairs in the United Kingdom has grown considerably since 2012, extending now to include smaller cities. Discussing the importance of events to practice theory, Birtchnell argues that they provide a celebratory context for a practice which is "given visibility, merit and institutional blessing".[85] Vegan fairs comprise numerous stalls that visitors explore which act to project meanings, introduce materials, and demonstrate competences in the practice, with cooking demonstrations an example of the latter. Vegan practice contests both a fixed view of taste and the dominant taste regime that celebrates the tastes of animal "products" divorced from the ethical treatment of animals.

Much of the talk between vegans centres on the tastes of new foods and recipes, stressing the corporeality of the practice, the co-production of taste[86] itself as a new competency and the agency of food materiality in recruiting new practitioners. Vegans have created their own verb, "to veganise", where traditional animal-based foods are re-created in vegan form. Certain foods take on almost iconic status among vegans, such as the vegan cupcake, or the vegan pizza, as a celebration of pleasure in the face of prevailing austere stereotypes of veganism. This is somewhat different to the aforementioned wholefood plant-based diet but again that should not be typecast as abstemious as practitioners use substances like date sugar to produce sweet tasting desserts. Food photography shared via social media such as Instagram

---

84   Twine 2014b, 636–37.
85   Birtchnell 2012, 500.
86   Hennion 2004.

also appears more common among vegans, perhaps again as a mode of communicating the retention of pleasure and commensality, as well as indicating the intensified social organisation of food practices in the everyday life of vegans. Since this interview research, veganism (or rather the capitalist marketing of products as vegan) has attained new visibility. From the Greggs (a UK chain) vegan sausage roll to the greater marketing of vegan products in supermarkets, the social intelligibility issues which may have beset some of the interviewees have been reduced.

Meanings and veganism

This intelligibility, in which veganism begins to become more culturally readable, involves a proliferation of meanings that, according to practice theory, includes the images, emotions, norms, ideas and aspirations that comprise eating practices, interwoven with their associated bundled practices. Veganism has to contend with a mythology around animal consumption, as well as the negative stereotypes of veganism[87] reproduced by a hegemonic meat-eating culture that operates normatively both to protect itself and to make defections to veganism less likely. However, it is within the "meanings-work" of veganism that anthropocentrism is most obviously contested.

Food is culturally abundant with meaning as it intersects with other practices related to gendered, classed, ethnic, familial, religious and generational identities. Food consumption is further bound up in cultural values pertaining to self-control, excess and pleasure. It is inescapably tied to meanings related to ethics, health and, more recently, environmental sustainability. As Sobal has argued, "Foods are objects inscribed with many meanings, representing ethnicity, nationality, region, class, age, sexuality, culture and (perhaps most importantly) gender".[88]

---

87   Cole and Morgan 2011; Twine 2014b.
88   Sobal 2005, 136.

Part of the economic success of the proliferation of Western meat consumption since the second half of the 20th century can be attributed to its symbolism, which has been a resource for the discursive practices of advertisers. Consuming meat has been associated with strength, health, power, status, virility, masculinity,[89] Westernisation and national progress. To consume meat is presented as the normal thing to do as a human; in some ways it is a practice that has become definitional for a certain view of what the human is, of an exalted position over the rest of nature. To consume animals is to perform the human sovereignty and entitlement discussed earlier. This may be one of the most resilient meanings surrounding meat consumption.

The association of meat with hegemonic masculinity and nation(alism) is important to consider when theorising the obduracy of animal consumption. If particular forms of masculinity[90] or nationalism[91] are socially dominant and are partly performed via eating practices, this points to what Shove et al. refer to as bundling, those "loose knit patterns like those based on co-location, sometimes turn[ing] into stickier forms of co-dependence".[92] Indeed, in another example pertinent to the climate crisis they discuss the bundling of masculinity with car driving and car repair.[93] With bundling, different practices reinforce each other – for example, the co-location of men, sports spectatorship and meat consumption. Yet forms of practice-bundling and practice elements such as certain meanings that transcend many different practices can also be instructive for the reinvention of practice, and new modes of bundling.

The practices of vegans can be read as showing much awareness of the social circulation of dominant meanings around animal consumption. The majority of interviewees, for example, were well versed in the cultural association between meat and masculinity. As a practice that could be read as intentionally challenging various hegemonies, including cultural anthropocentrism, it is not surprising

---

89 Adams 1990.
90 Adams 1990; Sobal 2005.
91 Nguyen and Platow 2021.
92 Shove, Pantzar and Watson 2012, 87.
93 Shove, Pantzar and Watson 2012, 36–37.

that veganism has been negatively represented as extreme, as a sacrifice or restrictive, as heroic in the sense of being assumed to be very difficult. Doubts may be raised about the vegan body as weak or potentially unhealthy. In response to such stereotypes practitioners called upon various elements of meaning to contest such claims and make veganism more socially intelligible and positively valued. Various oppositional strategies to reinvent the meanings of veganism focus on pleasure, health and naturalness, and attempt to erode the symbolism of meat as definitional to, and constitutive of, a meal. Laura, 43, stated:

> Being vegan, I think now how easy it is and actually how well it's made me feel, you know, healthy, I've got more energy, certainly definitely the things that have changed in my own sort of physiology over the last while, it's made me, well I wouldn't really want to go back.

And Bob, 20, similarly associated veganism with health and energy:

> I feel like my diet in itself is a lot healthier, I feel great, like my energy levels have never been higher. I am a runner and I thought it would have an impact on how I run but it's not. If anything, it has been a benefit really.

Contemporary Western urban veganism stands as a good example of what Soper terms "alternative hedonism" which occurs in tandem with a new negative aesthetic of animal "products"; "… commodities once perceived as enticingly glamorous come gradually instead to be seen as cumbersome and ugly in virtue of their association with unsustainable resource use".[94] To emphasise this hedonistic meaning many vegans routinely draw upon visual representation of their own meals and recipes, as mentioned above. Rachel, 42, reflected upon her photography:

> I keep thinking, why do I take photos? [laughs]. I think it is just a general obsession that vegans have with food and photographing

---

94   Soper 2008, 580.

food, evidencing food as being attractive [laughs] ... they feel that people see their diet as boring and lentil stew and so if they see a pretty pink cup cake or a lasagne it's like, wow our food can be sexy as well.

Such imagery often includes cakes and sweet foods which may serve to re-present veganism as pleasurable and as involving the innovative recombination of material elements rather than a sacrifice of omnivorous desire. The health benefits of low cholesterol food pleasures may also be underlined. Vegans draw upon their athletes (notably bodybuilders and long-distance runners) to emphasise an association of strength, health and endurance with veganism. The relative ease of veganism as noted among this sample is often stressed as a counter to its perception as difficult or awkward. Other central meanings circulated by vegans stress the health, environmental and animal ethics meanings of veganism, with these usually constituting the three pillars of vegan advocacy. For this sample, animal ethics framings for veganism were by far the most overriding initial factor but environmental and health discourses tended to come into play some time after transition and are expressed in the extracts below. Animal advocacy organisations have tended to foreground these three central meanings in an appeal to the rationality of the public. This has downplayed the everyday embedding of social actors in routine, habitual practices and led to an over-rationalised understanding of the process of practice recruitment. Practice theory can offer a more coherent view of food transition wherein meanings are dynamically co-produced in relation with materialities, competences and other practices; and a view of actors as socially and affectively situated and actively engaged in self-constitution via practices and their elements. This questions the common assumption of activists that non-practitioners are somehow deficient in information or in ethics. Social actors may indeed not have been exposed to certain knowledge and ethical meanings, but the explanations for why they practise what they currently do, according to practice theory, are more likely to be found in better understanding their current "practice portfolio", why it has "recruited" them, and how it has become embedded within their daily routine.

Participants were offered the opportunity to reflect at the end of their interview upon what veganism meant to them. This sample of answers below provides a rich narrative of self-reflection and ethical meaning, evidencing the importance of certain meanings in recruitment to vegan practice:

Treading lightly on the planet and respecting other sentient creatures. [Sonia, 67]

It means that I can walk on this planet and spend my life doing my utmost to not harm any living thing, that's probably it in its entirety really, it's not kudos for me or yeah, I'm better than you, it's just about my internal values and being able to stay true to them. [Lucy, 45]

I think it's just an all-rounded lifestyle towards aligning your beliefs and your ethics with what you're eating and what you're consuming in the world, moving away from violence towards not just the animals but your own body and the planet. So, it's a complete kind of harmony, it's congruent; it works together with your own ethics, with your own instincts. I think no-one really wants violence to be associated with their food. People don't go to Tesco and want to fight for their food or hunt for it, they want it to be peaceful, they want most of the interactions in their life to be peaceful and away from violence as much as possible. [James, 28]

I think it means living as good a life as I can really. About trying not to hurt anything I suppose, as much as that is possible. But yeah, just knowing that even though these horrible things still go on, it's kind of not done in my name and I hopefully haven't got a hand in those things. I can distance myself from them in that way, in a kind of useful way I suppose 'cos a lot of people distance themselves from them in a kind of let's kid ourselves, it's not really happening kind of way. But I suppose I feel it is a kind of a way of taking action to distance myself from being involved. [Grace, 45]

Table 6.1 Summary of the elements of vegan eating practice.

Competences	New cooking skills, new understandings of taste, ingredient-checking and label recognition, knowledge of animal-derived ingredients to avoid, nutritional knowledge, knowledge of new shopping and eating out infrastructure, forms of social and technological competency.
Materials	Shopping and eating out infrastructure (cafés, restaurants), vegan events (potlucks and fairs), websites and social media, cookbooks, vegan societies, kitchen space and technologies, vegan nutrition guides and wall-charts, vegan foods (including substitutes and plant-based foods formerly not part of everyday eating routines).
Meanings	The social construction of veganism associated with expanded choice, pleasure, progress, commensality, strength, health and fitness. Gendered meanings. Veganism as ethically meaningful; associated with peace, respect, non-violence, more ethical human–animal relations, and an environmental way of eating. Social distancing from, and transgression of, dominant meanings of animal consumption.

Such examples constitute earnest takes on the moral meanings of veganism. They are representations less often heard in the mainstream media, which tends to package veganism as a less threatening health issue. They are constructions of meaning in process that re-present and perform the practice positively and as personally rich in meaning. Within these perspectives are found forms of hope and expressions of self-liberation in the transgression of societal norms. I end this chapter with a summary of the elements of vegan practice (Table 6.1) which constitutes an initial mapping of those which this research found important for the performance and reproduction of vegan eating practice.

Setting out these elements of practice in this way and noting how their mutual coherence, their linking together, takes place – for example, via social relations, interaction and at events – goes some way to guiding how vegan transition occurs, as well as indicating why it does not. It is suggestive of various interventions which might better expose potential practitioners to these elements of practice. This provides a way to understand short-lived practitioners who may defect away from veganism, perhaps due to a lack of competency or not being enrolled into some of the central meanings of the practice that tend to keep vegans vegan. Since this summary is yielded from interviews, it excludes the broader elements pertaining to food systems. There are also "systemic" explanations for why more people are not enrolled into vegan meanings, from the dominance of meanings of animal consumption as a generational and cultural universal predicated upon human sovereignty discussed in Chapter 5, to the state subsidy of animal agriculture, the significant bundling of masculinity with animal consumption mentioned in Chapter 2 and here, and the more specific partial media omission of linking animal consumption to environmental damage noted in Chapter 4.

This sets up several aims for the next chapter. Firstly, it looks more closely at the meanings of veganism and animal consumption to better understand transition. Secondly, it considers how practice theory (and related approaches like the multi-level perspective) can be useful tools toward the better theorisation of the animal-industrial complex. Thirdly, this necessitates that such approaches are able to work at broader scales inclusive of infrastructure, so the chapter turns to some research on this question. Finally, the chapter offers suggestions toward the dismantling of the animal-industrial complex based on these approaches, premised upon the greenhouse gas mitigative effects that would have, but with co-benefits that speak to both the intersectional project of critical animal studies and the need to grasp the centrality of human–animal relations which this book has identified as bound up in the emergences, impacts and approaches to slowing and curtailing the climate crisis.

7
Toward the dismantling of the animal-industrial complex

To further engage with meanings around veganism it is appropriate to return to the different scenarios discussed at the end of Chapter 4. For eating practices generally, and for animal consumption and veganism specifically, the "meanings" element of practice is especially crucial in the theorisation of transition understood as practice normalisation and embedding. Vegan practice requires people to enrol into a new set of meanings around food which recast pre-existing normatively omnivorous practice in such ways which cannot help but question the animal consumer's practice affiliations, especially when the routine of eating is accomplished via so much social "bundling". Vegan transition is potentially challenging as it can entail undoing and redoing parts of the social fabric, questioning aspects of the routinised practices that make up our lives.

Contesting veganism

To recall from Chapter 4, plant-based transition was contrasted with vegan and intersectional vegan transition. I noted there is now much policy clamour for the former wherein plants would come to dominate eating practices with far lower levels of animals consumed. In spite of the evidence and scientific support for this scenario there is unsurprisingly strong opposition from animal-industrial complex interests and largely policy indifference from most governments to date. Arguments for this reducitarian transition are essentially anthropocentric. Conspicuously in this scenario there is no stated interest in not killing animals as a further co-benefit, and critical reflexivity around anthropocentrism is mostly lacking. There is certainly variation within this scenario, and it can encapsulate anything from quite meagre reductions in animal consumption to more substantial reductions in excess of 50% of current levels in rich nations. Proponents may also specify reduction calls only to ruminant consumption (especially beef and dairy). This may make superficial sense given the methane emissions associated with the farming of cattle ungulates, but it seems aloof to, for example, the deforestation and soya production for animal feed associated with chicken production (by far the most farmed animal numerically).

The EAT-*Lancet* report is an example of this approach arguing that,

> Transformation to healthy diets by 2050 will require substantial dietary shifts. Global consumption of fruits, vegetables, nuts, and legumes will have to double, and consumption of foods such as red meat and sugar will have to be reduced by more than 50%. A diet rich in plant-based foods and with fewer animal source foods confers both improved health and environmental benefits.[1]

It makes absolute sense not to call this plant-based transition scenario vegan and this would hold true even if the calls were for 100% plant-based transition due to the focus only upon food, which delimits all known and accepted definitions of veganism as both ethically oriented and extending to animal exploitation *beyond* food relations. Scenario 1 thus is not vegan because it provides no rationale not to consume non-food animal "products" even though doing so could also decrease consumption generally and benefit emissions reduction. In spite of the failure of the plant-based transition scenario to approximate veganism, it is arguably the most likely scenario and set of practices to actually take shape in the short term. Furthermore, it could be seen

---

1   Willett, Rockström et al. 2019, 3.

as counterproductive for vegan advocates not to support policies that could open a pathway to plant-based transition because they could break up the animal consumption status quo and make actual vegan transition more likely. Radicality is certainly relative here – for example, for vegans, EAT-*Lancet* recommendations are not radical, but for mainstream meat culture they are.[2] Given that they could downsize and destabilise the animal-industrial complex (though not abolish it) it *could* seem misguided not to support such initiatives, also for their health and environmental co-benefits.

In the second scenario, which I have termed vegan transition, there is indeed an ultimate goal of abolishing the animal-industrial complex. Definitively this *does* encompass a critique of anthropocentrism, bringing to the fore the ethical and political dimensions of veganism. This is not a narrative of efficiency savings but is premised upon meanings whereby consuming animals is wrong because the killing of animals for food is not justified either on the grounds of human need or health, or on the grounds of the value of nonhuman animal life. For vegans, such meanings contain valuable lessons for humanity, to transcend human exceptionalism and respect the right to life of other species. Many of the practice infrastructures that have created the climate crisis – overlapping systems of transport, energy and agriculture – have assumed the instrumental status of other animal species. Habitat destruction, roadkill, pollution and the systematic organisation of the mass killing of farmed animals have been deemed acceptable side effects of the progression of capitalist development. Proponents of this scenario likely concur with all the benefits of plant-based transition – lower emissions, healthier lifestyles, less pollution, better water conservation and biodiversity protection – but also see a societal benefit in no longer organising the global food system around the annual killing of billions of nonhuman animals. A vegan culture (as opposed to our hegemonic meat cultures) would be more protective against crises such as the climate crisis because energy, mobility and food systems (and technologies) would have to satisfy

---

2   As a stark illustration of this, the WHO pulled out of the launch event of the EAT-*Lancet* report in 2019, after an intervention from the Italian ambassador and permanent representative to the UN in Geneva (Torjesen 2019).

newly established norms that would strive to safeguard other species. However, this veganism could be deemed ineffective against the climate crisis because it has nothing to say about the interconnections between animal exploitation and capitalism, colonialism and patriarchy. It risks a veganism that reasserts whiteness and masculinity and uncritically celebrates plant-based capitalism.

Vegan animal advocates sometimes try to demarcate this veganism (scenario 2) from a broader critical animal studies (CAS) view (scenario 3) by self-identifying as "ethical vegans" or as being "vegan for the animals", as if such pronouncements can credibly keep separate "animal" and "environmental" meanings of veganism, a separation that was heavily contested in Chapter 3. Another connotation in self-identifying as being "vegan for the animals" has been to distance oneself from veganism as a broader social justice project. With its advocacy of *intersectional* veganism,[3] the perspective of CAS adds further complexity to the meanings of veganism and vegan transition. This conceptualisation, as noted in Chapter 4, articulates *systemic animal exploitation as intersectional*, as historically inseparable from, and enabled by, the development of capitalism, patriarchy and colonialism. Consequently, actual vegan transition is unlikely to be successful unless also wedded to a transformative social change beyond capitalism, patriarchy and white supremacism, because these power relations all provide blocks to transgressing anthropocentrism. It is difficult then to adequately theorise transition without intersectional competency because, for example, one will fail to comprehend how practice elements, practices and bundles overlap with each other to maintain the social order (intersections of class, gender and "race" with food practices to signpost the most obvious examples).

---

3   The naming of this type of veganism as "intersectional" could connote an appropriation of the idea from Black feminist thought (e.g., Crenshaw 1989; 1991). While it is true that care is needed here (Giraud 2021, 110), ecofeminism – a major influence upon CAS – represents a parallel strand of feminist theorising that addresses both specificities and conceptual overlaps between different categories of oppression. Certain ecofeminists (notably Adams 1994) also cited Crenshaw's work. It is possible to read the use of "intersectionality" in both ecofeminism and CAS more favourably as attempts to broaden and refine its focus to include the more-than-human.

This makes transition both harder and easier. For example, if one is already seeing the world through a feminist lens, then the attraction of eating in ways which may be read to contest the intersecting objectification of women and animals may for some be both intelligible and clear. A counterargument would say such bundling is in the wrong direction, by making a niche practice even more niche. Similarly, many scenario 1 or scenario 2 advocates would rather avoid associating either plant-based eating or veganism with the political left because it is reducing the pool of potential practitioners. Yet from a pro-intersectional vegan perspective the notion of plant-based transition under capitalism is highly problematic – to what extent would it really assist in the amelioration of the climate crisis if it is still within a capitalist economics which seems dependent on the ever-increasing exploitation of the more-than-human? If veganism is not intersectional, how could it be effectively linked to the broader climate justice movement which is exactly concerned with how the climate crisis is classed, gendered and racialised? Such noteworthy debates are left to the final chapter, but for now, the intention is to give a sense of the multiple meanings of veganism and how they may complicate transition.

So far I have identified three possible transition scenarios: plant-based transition within capitalism, veganism, and intersectional veganism. Within these scenarios are found the broad range of meanings that co-opt practitioners including those who may consume a plant-based diet primarily for health or environmental reasons. Nevertheless, research has suggested that initial salient meanings for transition are commonly reinforced by other meanings post-transition.[4] It is likely important to better understand the interplay and overlaps between these scenarios, especially if strategies emerge by which the more critical meanings and practices of intersectional veganism can be exported to the other two scenarios. Moreover, assumptions of linear progressive change are naïve in a modernist sense, and the sort of falsely optimistic narratives sometimes found within activist discourse may be liable to overestimate the rate or likelihood of change. The levels of ambitiousness increase moving from scenario 1 to 3, as does the

---

4   Twine 2017.

depth of social transformation which they imply. Mitigating the climate crisis does require systemic and urgent transformatory change, but the important points of critique that scenario 3 practitioners make of the other two need not be conflated with ignoring how the wider base of political (including dietary) change may contain useful practice dynamics and pathways for a broader oppositional movement.[5]

To build on the work on vegan meanings of the previous chapter, I firstly turn to research[6] which specifically studied how non-vegans construct veganism, and secondly, examine some of the social science debates around meanings and definitions of veganism (including the dominant practice of animal consumption). Both can aid understanding of what perpetuates animal consumption and what might attract and repel new vegan practitioners. This will then take us back to practice theory (and related approaches) to consider what insights the theoretical framework may afford in terms of intervention and transition. However, this also involves thinking critically about practice theory as an approach, and its ability to work at broader scales inclusive of infrastructure, and consequently to then consider how practice theory and related approaches can help improve the CAS theorisation of the animal-industrial complex. Finally, this chapter will consider how to best socially embed transition, and to work toward the dismantling of the animal-industrial complex.

The scenario 1 plant-based transition, as I have argued, is quite divorced from ethical meanings that question animal consumption on the grounds that it directly exploits animals. This distinction between "plant-based" and "vegan", however, is culturally vague with the terms being used somewhat interchangeably, for example, in the media or in retail. However, the suspicion is that "plant-based" is itself a linguistic reinvention which speaks to the capitalist accommodation of "veganism", cleansed of its threatening ethics and politics, and repackaged into a more consumable, domesticated space in which its health meanings constitute its most commodifiable value. It is within

---

5 This is not to gloss over CAS critiques of "vegan capitalism" which argue that it is not a pathway at all but merely an example of co-option (Best 2014). I rejoin this issue in Chapter 8.

6 Parkinson, Twine and Griffin 2019.

this context that assertions of "ethical veganism" become understandable statements against the dilution of veganism. The problem for vegan advocates is that ethical meanings can be strongly resisted by many non-vegans (more on this shortly) and so the question arises whether health or environmental meanings provide initial pathways to plant-based eating for those deterred or unsettled by the ethico-political meanings of veganism. The often-cited definition of veganism given by The Vegan Society is useful for further reflection:

> Veganism is a philosophy and way of living which seeks to exclude – as far as is possible and practicable – all forms of exploitation of, and cruelty to, animals for food, clothing or any other purpose; and by extension, promotes the development and use of animal-free alternatives for the benefit of animals, humans and the environment. In dietary terms it denotes the practice of dispensing with all products derived wholly or partly from animals.[7]

Several elements can be noted. Firstly, veganism is a philosophy and way of living, not a diet. Secondly, it recognises that the extensive commodification of animal bodies means *total* exclusion of animal use may be difficult. For example, if one is dependent upon medication that contains trace animal parts or was developed via animal research, then taking such medication ought not preclude one from defining as vegan (even though veganism, as this definition includes, must work to promote the removal of animals from medical research). As Giraud puts it, this acknowledges "structural difficulties that undermine scope for making these choices in particular contexts".[8] Thirdly, this is not really a definition of intersectional veganism. That would have to say something about the entanglements of animal exploitation with other relations of power and affirm the commitment of veganism also to feminism, anti- racism and anti-capitalism; however, the definition is general enough to include benefits to "humans and the environment". This does not preclude health benefits to humans, which I will shortly

---

7   https://www.vegansociety.com/about-us/further-information/key-facts.
8   Giraud 2021, 159.

argue can be consistent with an intersectional veganism. What it does preclude though is the notion of "health veganism", for while it is important to the practice that people may start consuming an entirely plant-based diet as a pathway to veganism, there is no initial rationale for such people to, for example, stop consuming leather and so on. Health veganism is an oxymoron because veganism is not a diet. However, that is not the same as saying health meanings are irrelevant to the practice of veganism.

In earlier work I offered a CAS inflected definition of veganism as "a systemic and intersectional mode of critical analysis and a useful lived philosophy counter to anthropocentrism, hierarchy and violence".[9] There is no doubt that ecofeminism and then CAS has worked to refine the meanings of veganism and to promote an intersectional understanding of the practice. In this definition I was also emphasising critique of large-scale systems such as the animal-industrial complex. It reflects some of the debates in CAS and Vegan Studies that have aimed to normalise intersectional understandings and to argue against campaigning strategies that, without a sense of intersectional competency, may be prone simply to discount the enmeshment of animal exploitation in a broader politics, or worse, uncritically reproduce veganism in sexist, racist, ableist or capitalist forms. Others have looked more closely at the historical formation of The Vegan Society to detect the ethical meanings of the practice.[10] For example, Cole's analysis makes clear not only that veganism was an ethical practice from the outset but that the opposition to exploitation of the early practitioners extended into a vision more generally of a compassionate and peaceful society.[11] This is not the same as saying that veganism was somehow always intersectional, but it does point out that the histories of veganism cannot be separated from their relations to other progressive causes and practices. Similarly, "intersections" between feminism and the anti-vivisection movement of the late 19th/early 20th century

---

9   Twine 2012, 19.
10   Stewart and Cole 2020.
11   Cole 2014.

manifested in figures such as Frances Power Cobbe (1822–1904) have been long noted.

With a degree of simplification, in the contemporary contestations of veganism just outlined, it is possible to summarise a central consistent meaning of veganism as an ethical social practice directed toward producing a society free of animal exploitation (scenario 2), but on the one hand there are accommodating and domesticating pressures from its commodification which seek to essentially dilute this history into a palatable "veganism" (scenario 1), and on the other, there are radicalising pressures from vegan activists and researchers who intend to intensify its broader political vision, under the rubric of "intersectional veganism" or similar terms (scenario 3).

Turning now to 2019 research which looked at how non-vegans construct veganism[12] is particularly valuable because the meanings and associations people have about veganism are as vital for thinking about how pathways to vegan transition emerge, alongside research focused upon vegans themselves. Parkinson et al. aimed to enhance understanding of barriers preventing transition to veganism, with a focus on vegan eating practice.[13] A key objective for the research was to gain insights into how non-vegans perceived vegans and messages about veganism. The research, located in the United Kingdom, comprised an online survey (n=1,435), semi-structured interviews (n=50: 10 single non-vegans, and 20 non-vegan couples) and 10 focus groups (n=72). Results were broad and detailed, but it is relevant to tease out findings related to possible health, environmental and ethical meanings. Of these the findings suggested that health meanings were the most likely to be received positively by non-vegans. Health messages were seen to have greater credibility than environmental or animal ethics messages. In total, 84.1% (n=1,207/1,435) of the survey sample thought veganism could be a healthy way of eating. This only dipped slightly to 79.5% (n=840/1,057) when including just omnivores. Furthermore, 78.4% (829/1,057) of omnivores thought that eating meat

---

12 Parkinson, Twine and Griffin 2019. The Pathways to Veganism project, funded by The Vegan Society, was carried out with colleagues from CfHAS, Edge Hill University. The report is available at https://tinyurl.com/56xvhrzu.
13 Parkinson, Twine and Griffin 2019.

was not essential to a healthy diet, and 90.1% (952/1,057) for consuming dairy milk. On the other hand, 52.2% of the survey sample said they would have health concerns about becoming vegan. Examining this in more detail revealed broad perceptions of nutritional inadequacy and uncertainty over how veganism would work with pre-existing medical issues. Unsurprisingly, vegetarians and pescatarians had a better perception of the healthiness of veganism than omnivores. This sample was not representative, but the high percentages remain noteworthy.

Although this research suggests that the practice of veganism has work to do in convincing others of its nutritional adequacy (or its nutritional *advantages*), it provided evidence that overall healthiness has become a generally positive meaning associated with veganism and has at least played a role in contesting previously negative media coverage.[14] This was reinforced in the focus group element which explored the traction of various vegan meanings. Groups ranked health meanings as most credible, and these were always ranked higher than both animal ethics and environmental meanings. In particular, quite strong resistance to animal ethics messaging and less awareness of the environmental and climate dimensions of food consumption were revealed. While it could be tempting to read these findings in terms of intractable human self-interest, they were bound up in discomfort around being asked to reflect upon human–animal relations. Rather than undermining transition strategies inclusive of a critique of anthropocentrism, they highlight the importance of framing in which such a critique could be shaped around multispecies benefits of, for example, land-use changes and alternative, richer relations with other animals. Potentially these findings further reflected how "veganism" has been commodified and presented in mass culture where the emphasis has been on health messaging. Since this research, TV ads in the United Kingdom have started to include food retailers communicating the environmental benefits of their plant-based ranges.

This is only one piece of research but could be taken as underlining the importance of including the health associations of plant-based eating as a gateway for either plant-based or vegan transition. It adds

14   Cole and Morgan 2011.

to the argument for co-benefit framing when arguing for changes to the food system to help combat the climate crisis. The Vegan Society definition has long had the element of "benefitting people", while intersectional understandings of veganism have a more complex relationship with health meanings. Since social class and racialised inequalities include inequalities of health, intersectional veganism cannot remain quiet on how poor diet is socially shaped along these lines and how access to healthy food may in some contexts be very constrained. While some vegans have championed a "junk food veganism" to either foreground pleasure or to make clear that their veganism is disinterested in health and purely "for the animals", this can reinforce our present food system's issues with cheap food, inequality and illness, mistakenly assuming that the only ethical questions pertaining to food relate to whether or not they include animal "products". An intersectional veganism can certainly take issue with a perfectionist notion of the healthy body to the extent that it can end up reinforcing ableism and the social judgement of individuals, but it is barely credible that it could ignore health meanings and the ability of a healthier diet to partly address intra-human inequality. Indeed, there is potential for intersectional veganism to be allied with a wholefood plant-based diet which varies according to cultural differences.

A larger transition strategy point here is that although vegan advocates are fundamentally arguing for a radically new ethical relationship with other animals – and increasingly understand environmental/climate crisis arguments as inescapably linked to this – the co-benefit meanings of human health have to be taken seriously if more transition is to occur. Moreover, if animal consumption is for most people "simply a part of everyday routine" rather than an ethical issue, the go-to vegan strategy of *leading* with ethical meanings may counterintuitively be counterproductive. This does not mean pandering to a meat culture that might be more amenable to a sanitised veganism. However, it does underline the need to think carefully about how ethical meanings are conveyed. A recognition of multiple and diverse pathways to veganism is likely necessary to further grow the practice.

At this point I return to and enhance the discussion in the previous chapter about the meanings of animal consumption so that those which sustain such eating practices can be brought into relief. I have noted the socio-historical associations between meat consumption and strength, health, masculinity, power, patriotism and virility, and how some vegan meanings try to contest these directly by moving into the "meaning territory" of animal consumption, or also by refuting stereotypes of veganism in meat culture. I also argued that practices of animal consumption perform meanings of the human, as dominantly understood, to involve a sovereign right and entitlement to the bodies of other animals. Veganism performs the human radically differently and its threat to such assumed rights and entitlements goes part of the way to understand backlash from meat culture and its practitioners.

In recent years work from social psychologists related to the meanings of both veganism and animal consumption has proliferated.[15] As noted in Chapter 5, meat cultures are contradictory, with social expectations around animal care coexisting with routinised violence against many species. This disjuncture has been conceptualised as "the meat paradox"[16] but does not just apply to animals that are eaten. Ambivalence and moral disengagement strategies have been identified as coping mechanisms used by omnivores to address these contradictions[17] and cognitive dissonance theory is well established as an explanation for the meat paradox.[18] Piazza et al. contend that individuals respond to the meat paradox in two ways – "one can reject meat consumption, bringing one's behaviours into alignment with one's moral ideals, or one can bring one's beliefs and attitudes in line with one's behaviour through various psychological maneuvers".[19] One such manoeuvre is what Adams describes as the construction of the "absent referent", a concept used to explain the reinvention of animals as "meat",

---

15 Dhont, Hodson et al. 2019. Social psychologists, like other social scientists, are similarly beholden to problematic historical disciplinary norms of "detached neutrality". The psychological engagement with the climate crisis and human–animal relations inevitably asks questions of the contemporary role of psychological knowledge and psychologists (Orange 2016; Adams 2016).
16 Bratanova, Loughnan and Bastian 2011; Loughnan, Haslam and Bastian 2010.
17 Buttlar and Walther 2018; Povey, Wellens and Conner 2001.
18 Bratanova, Loughnan and Bastian 2011; Loughnan, Haslam and Bastian 2010; Piazza, Ruby et al. 2015.
19 Piazza, Ruby et al. 2015, 114.

and the dissociation of "meat" from violence.[20] The "absent referent" at the meal table is, Adams argues, the someone, the animal, whose life has been taken, and suffering endured, to enable the omnivore to consume "their" meat. As Adams and others[21] have long noted, this dissociation is aided by the socio-historical embedding of meaning in language whereby the living animal is absented through words such as "ham" and "beef", or protective euphemistic killing verbs are used to describe human practices of killing other animals.[22]

The meat paradox may also be negotiated by claiming that animals eaten do not suffer[23] or that they are "unworthy" of moral consideration.[24] Loughnan et al. demonstrate how omnivores draw boundaries of moral concern in a motivated rather than absolute way, making it possible to legitimise animal suffering and meat consumption through the removal of moral status from consumed animals.[25] This means that individuals can, while purporting to care about the welfare of some animals, be motivated to deny or ignore the subjective lives of other animals, to remove, in their conscience, the capacity to suffer: "if animals lack moral status, then killing them is not a moral issue, and eating meat is not morally problematic".[26] While a willingness to eat meat can be reduced by heightening moral concern,[27] the Loughnan et al. study proposes that the *practice* of eating meat can reduce moral concern for animals.[28] Some social psychologists appropriately refer to this denial of nonhuman animal subjectivity as "dichotomisation",[29] which echoes the role that CAS and ecofeminism have given to the social construct of human/animal dualism in shaping anthropocentrism. As they underline, experiments that attempt to disrupt this process by making clear the subjective capacities of other

---

20 Adams 1990.
21 Dunayer 2001.
22 Franklin 2020.
23 Piazza, Ruby et al. 2015.
24 Bratanova, Loughnan and Bastian 2011; Loughnan, Haslam and Bastian 2010.
25 Loughnan, Haslam and Bastian 2010.
26 Loughnan, Haslam and Bastian 2010, 157.
27 Bratanova, Loughnan et al. 2011.
28 Loughnan, Haslam and Bastian 2010.
29 Rothgerber and Rosenfeld 2021.

animals can provoke greater levels of disgust at the thought of eating animals.[30] Adapting earlier work by Joy,[31] Piazza et al. found that omnivores justified meat consumption through "the 4 Ns": that meat is "natural", "normal", "necessary" and "nice".[32] Various studies have also suggested that right-wing people are more anthropocentric, eat more meat and are more resistant to progressive change in human–animal relations,[33] although such research is yet to explain why those who identify as left-wing mostly also conform to animal consumption.[34] What this work underlines is that animal consumption inescapably involves the direct funding of animal killing and a complicity in violence that is threatening to sense of self. Although this gives rise to strategies which aim to protect self-identity (literally the meanings one ascribes to oneself) it also potentially signposts alternative practices that would better situate eating *within* prevailing social norms of non-violence. That many people would rather not be complicit is a resource for vegan transition.

This social psychological work is useful, though one can note tensions between it and practice theory, not only because the unit of analysis switches back from practices to individuals but also because it is prone to over assume the rational (as opposed to routinised) character of behaviour.[35] Meanings that cohere around (animal) consumption are not necessarily reflected upon by practitioners, which might raise doubts over whether contesting meanings alone is enough to shift a practice. The sort of material substitutions noted in the previous chapter attempt to change a practice potentially with minimal disruption to meanings.[36] Practice theory is suggestive of further

---

30   Ruby and Heine 2012.
31   Joy 2010.
32   Piazza, Ruby et al. 2015.
33   For a summary, see Dhont, Hodson et al. 2019.
34   One theory could be the role of anthropocentric humanism in left politics wherein human/animal dualism has partly shaped aspirational political projects for recognition and rights along lines of class, gender and "race".
35   To be clear, the intention is not to generalise this to all social psychological approaches.
36   The example of cellular meat is considered in Chapter 8.

transition strategies, which are covered in the second half of this chapter.

Broader social meanings and norms around food, beyond those specific to animal consumption or veganism, also play a role in maintaining the status quo. These include interconnected norms of autonomy, privacy and choice which not only perpetuate the depoliticisation of human–animal relations but deflect critical thinking around food practices generally.[37] These sets of norms are environmentally damaging as they speak to the notion of the entitled sovereign consumer as well as the sovereign human. One can buy an SUV, one can fly as much as one likes and one can consume as many "steaks", only as money dictates, unconstrained by any consideration of labour or environmental consequence, or feelings of complicity in violence toward animals. Freedom to choose under neoliberal capitalism, to have one's choices protected by privacy norms, and to assume that one's practices are autonomous from "nature", rather than part of the capitalist reproduction of "nature", or, rather than being performative of particular human/more-than-human relations, together form a heady set of interrelated assumptions protective of the status quo. On an interpersonal scale, part of the omnivore's objection to vegan scrutiny is not just that they are having their morality, or happiness-order,[38] questioned but that these norms of autonomy, privacy and choice are also being transgressed. The intersectional veganism of CAS exactly aims to bring into relief the classed, gendered, racialised and anthropocentric dimensions of everyday consumption, which allies it partially to a climate justice agenda. That the food system is profit-driven instead of being defined by health, ecology or species flourishing is a further key taken-for-granted normative meaning that is fundamental for linking together the multiple practices of the animal-industrial complex and is examined more closely in the next chapter.

Mapping this terrain of meanings within a practice theory framework can potentially contribute to a more systematic approach to intervention in order to theorise the fossilisation of certain meanings

---

37   Jenkins and Twine 2014, 225.
38   Ahmed 2010b.

and the promotion of others. Moreover, the affordances of the practice theory approach now come into view. Growing veganism or plant-based eating and shrinking animal consumption practices likely works not just by theorising a general replacement, but by thinking through how the links between the elements of practices can be broken and made, as well as how practice-bundling and practice-complexes emerge.[39] In the previous chapter vegan competences and material substitution and creativity were noted as integral to vegan transition. The expanded mapping of meanings in this chapter is summarised in Table 7.1.

Table 7.1 Summary of major relevant food meanings.

Practice meanings	Estimated strength of link (for example, in UK society)	Potential interventions
Meat (consumption) as necessary	Strong but receding	Demonstrate healthy veganism.
Dairy as necessary	Medium but receding	Demonstrate healthy veganism.
Meat as normal	Strong	Contest essentialism of human as omnivorous. Erode conflation of meat and protein.
Dairy as normal	Strong	Contest essentialism of human as omnivorous. Demonstrate and embed non-dairy alternatives.
Meat as strength promoting	Strong	Erode conflation of meat and protein. Emphasise overconsumption of protein, especially in rich nations.
Meat as a nationalistic practice	Medium	Erode nationalism and/or bolster environmental/ animal ethics national identity. Question what it

---

39   Shove, Pantzar and Watson 2012.

Practice meanings	Estimated strength of link (for example, in UK society)	Potential interventions
		means to be an "animal loving" nation.
Meat as masculine	Strong but contested	Erode gender essentialism. Contest and redefine hegemonic masculinity. Embed pro-animal ecomasculinities.
Meat as healthy	Strong/ medium but contested	Amplify meanings from nutritional knowledge, most clearly around red/processed meat. Expand health meanings to include planetary health.
Dairy as healthy	Strong but contested	Amplify meanings from nutritional knowledge. Compare with plant-based milks.
Human sovereignty/entitlement to consume animals	Strong	Contest anthropocentrism. Highlight the unnecessary violence involved.
Meat as centre of a meal	Strong	Wholefood plant-based eating practice can be particularly effective at decentring "protein objects" from centre of meal. More diverse and healthy plate.
Meat as pleasurable and salient to memories and identity	Strong	Contest via alternative sensory food experiences and vegan creativity.
Farmed animals as having limited subjectivity	Strong but contested	Contest via knowledge from ethology/animal welfare science.
Animal-industrial complex as major employer	Strong	Transition should accommodate pre-existing animal-industrial complex labour. Just transition.
Veganism as against animal exploitation	Medium	Could be presented in line with pre-existing values of the majority such as pro-peace, anti-violence, and affective connection to companion animals. Further

Practice meanings	Estimated strength of link (for example, in UK society)	Potential interventions
		emphasis on non-food aspects of veganism.
Veganism as healthy	Strong but contested	Foreground wholefood plant-based eating, background "junk food veganism". Demonstrate healthy veganism, e.g., through bundling with other health practices and nutritional knowledge.
Veganism as restrictive	Strong	Contest via expansion of fruit and vegetable consumption. Emphasise narrowness of many omnivorous routines.
Veganism as environmental	Medium	Stress environmental (including greenhouse gas mitigation) benefits of a vegan food system. Bundle veganism with other practices such as localisation and anti-food waste measures.
Veganism as intersectional	Weak	Coalition building between pro-vegan groups and those working on climate justice, as well as on broader issues.
Food practices and systems as autonomous and private	Strong but contested	Autonomy contested by challenging anthropocentrism. Privacy contested by stressing the far-reaching impacts of consumption.
Food system as profit-driven	Strong but contested	Bundle veganism with food initiatives that seek to de-commodify food, e.g., community growing and sharing.

This summary of meanings potentially provides an overview of how to intervene, to weaken links within practices of animal consumption, and to strengthen various meanings of both plant-based eating and veganism. Vegan advocates will recognise many of these as already forming important parts of activist strategies for change. What practice theory provides is a sense of coherency to these but vitally sets them in a broader context of being linked to examples of competency and materials, and then practices themselves being bundled with others, eventually in larger forms of complexes, or infrastructure. Thus, one strategy for change is certainly to contest meanings, but it risks becoming limited if this broader context is ignored. Context and scale then become unavoidable issues. Even in considering these meanings, it is clear that some are more overarching than others. For example, if one could successfully challenge dominant ideas of masculinity that are still tied to society/nature dualism and find expression in practices such as eating meat or driving, or more fundamentally, challenge human sovereignty and anthropocentrism (as vegans do), then transition could likely be accelerated. As Shove et al. contend, this points to a more

> elaborate picture in which diverse elements circulate within and between many different practices, constituting a form of connective tissue that holds complex social arrangements in place, and potentially pulls them apart. To the extent that this is so, the attaching and detaching of meaning and signification sends ripples across the cultural landscape as a whole.[40]

Similarly, if international bodies like the Food and Agriculture Organization of the United Nations changed direction and started to advocate for a majority wholefood plant-based diet for people in poor countries (instead of their current "livestock" advocacy) it would fundamentally contest the association of animal consumption with ideas of Western progress and deliver the numerous co-benefits outlined earlier. Such high-level leadership can engender ripple effects.

Scaling up: Apprehending the animal-industrial complex

This brings us now to the need to apprehend larger scales of practice, and to explore whether practice theory and allied approaches can be useful for transgressing the dichotomy of agency and structure, and

---

40 Shove, Pantzar and Watson 2012, 36–37.

suggestive of ways in which to intervene at larger organisational scales. An interrelated question to that of scale is how practice theory theorises power. The two questions are related because the ability of practice theory to make sense of larger scales of social organisation is tied to its ability to "move beyond the localism of a focus on performances of practices"[41] to explain the operation of practice entities which are able to shape the doings and sayings of many social actors across time and space. The emergence of social norms in the space between practice performances and practice entities is one way in which practice theory envisions both scale and power. In one sense, as Watson notes, power has been implied in practice theory rather than explicitly articulated because it is present at all moments of its theorisation.[42] For example, mastering competences, deploying materialities, extending meanings and linking these elements in practice can all be seen as effects of power as particular dominant practices come to construct the social. The previous extended discussion on meanings also pertains to questions of how and why particular food meanings remain socially dominant over others. An approach to scale and power in practice theory briefly noted in the previous chapter has been to try to theorise the interplay between practices, how they link together in "bundles" and then in "complexes",[43] or as infrastructure.[44] For example, if the co-location of shopping and driving constitute a practice bundle, then the whole supermarket-led food system constitutes a highly intricate and powerful complex of practices. For Shove et al.:

> Complexes represent stickier and more integrated combinations, some so dense that they constitute new entities in their own right. Inter-practice relations of these and other kinds have emergent, cumulative, and often irreversible effects for individual practices, for the elements of which they are composed, and for the spatial and temporal texture of daily life.[45]

---

41 Watson 2017, 170.
42 Watson 2017, 181.
43 Shove, Pantzar and Watson 2012.
44 Shove and Trentmann 2018.
45 Shove, Pantzar and Watson 2012, 81.

I take it as coincidental that these authors use the word "complex", even though much research has used the term in the context of the animal-industrial complex, and there are longer standing uses in ideas of the military-industrial complex or entertainment-industrial complex.[46] Though practice theory writers have not tapped into this lineage, a practice theory approach to theorising the animal-industrial complex could be both possible and useful, and I make some rudimentary attempts toward this now.

Until this point much of how I have brought practice theory into thinking about vegan transitions has stemmed from my own interview research with vegans and non-vegans and so, in part, panders to the way "practice-based sensitivities are often pigeonholed as part of microsociology".[47] As a reaction against such theoretical branding, more recent practice theory has specifically tried to theorise larger scale social phenomena.[48] Yet before this, practice theorists were already considering both how practices interact[49] and how large-scale complexes of practices shape everyday practices with environmental relevance such as showering and laundry habits.[50] What practice theory is asking us to do is to suspend our usual way of thinking with regard to scale in the form of making distinctions such as micro/macro or agency/structure. For "large-scale phenomena" are inescapably present in practices that might be thought of as tangible, routine and mundane. Rather than being abstract and enigmatic, the multiple overlapping "large-scale" practices and relations that accomplish the consumption of, for example, animals in one's home, like those that accomplish domestic water consumption, *are* part of the practice-complex.

Most recent practice theorists subscribe to what is known as a "flat ontology" in which "so-called 'large-scale phenomena' are constituted by and emerge through the aggregation of interrelated practices and their regimes of reproduction".[51] Practice-complexes in turn interlink, such as

---

46   Twine 2012.
47   Nicolini 2017, 98.
48   Hui, Schatzki and Shove 2017; Shove and Trentmann 2018.
49   Shove, Pantzar and Watson 2012.
50   Shove 2003.
51   Nicolini 2017, 99.

the animal-industrial complex and complexes pertaining to water and energy, or to transport. Better understandings of how infrastructures "co-exist and mutually adapt"[52] constitutes another significant practice theory–inspired research direction for understanding how interventions can help steer practices in more sustainable directions. Within intricate contemporary societies certain practice-complexes take on the ability to transcend space and time as they become socially embedded, influential and have a very extended reach. The animal-industrial complex is particularly entrenched given its global character and its dominant generational universalism, noted earlier, which constitutes part of its "regime of reproduction". Relevant practices such as preparing, cooking, storage, eating, disposal and washing up may be quite commonly co-located[53] in the home; others are very spatially diverse yet still interconnected, such as domestic chicken and "beef" consumption in the United Kingdom existing in a practice-complex with logging practices in Latin America, or veterinary treatment on a farm that breeds animals for human consumption. In approaching the profound complexity of practice-complexes it is helpful to try to consider the host of sites in which relevant practices cluster, not only to better understand their interrelation but also to imagine or experiment with interventions.

Shove et al. suggest that bundles and complexes "arise and disappear as a consequence of competition and/or collaboration between practices".[54] Here two examples may be considered, one of collaboration and one of competition, of relevance to veganism and the animal-industrial complex. Firstly, as a case of collaboration and practice-bundling, the UK running club "Vegan Runners" formed in 2004 combines the practice of veganism and running. Socially visible through distinctive club kit and tapping into nationwide networks of races and the weekly Parkrun phenomenon of 5-kilometre runs, this case of bundling can be seen as one example of the normalisation and embedding of veganism. The club has grown to become (in 2023) the largest registered by Parkrun with over 4,500 members. Recent increases in membership have reflected the rising visibility of veganism

---

52 Cass, Schwanen and Shove 2018, 165.
53 Shove, Pantzar and Watson 2012, 84.
54 Shove, Pantzar and Watson 2012, 88.

since 2015. This bundling has the effect of contesting meanings which might associate veganism with weakness or poor health, presenting it instead as an attainable practice, associated with strength and fitness. Though for some this might take veganism "too far in a health direction", the club ensures involvement in other vegan ethico-political issues such as taking a stand against Parkrun's former sponsorship by the Happy Egg Company. It would be hard to think of a regular UK runner or volunteer who is not now aware of the club and anecdotally it has created a pathway for non-vegan runners to become vegan and non-running vegans to take up running. The embodied visibility of practitioners wearing the club vest and further via social media then becomes a significant part of the practice of running as a vegan, rendering the extensive local networks of races in the United Kingdom also, in part, vegan spaces. This bundling then affords one way in which veganism becomes socially embedded and is instructive for how that process could be intensified. Bundling is at the same time a diversification of a practice as it becomes more woven into the social fabric, something that is partly achieved via increasingly vegan visibility when dining out, on menus, and also within institutional (especially educational and healthcare) settings. A further example of cooperative bundling has been the combining of veganism with (food) fairs and potlucks, mentioned in the focus on events in the previous chapter. The institutional blessing (albeit concurrent with the dominant meat culture) afforded to vegan eating, if not to the broader meanings of veganism, again is relevant for embedding.

Secondly, and more competitively, are examples of meat, milk and egg replacements. As argued in the previous chapter, substitutes are attractive to transition advocates because they potentially offer a faster, less disruptive change to food habits, but they can also fail to do the work for pre-existing animal consumers of integrating meanings and materiality. Supermarket choice in the United Kingdom for meat alternatives has increased significantly since 2015, not only with burgers and sausages but in foods like veganised chicken nuggets, chicken Kiev, meatballs or bacon. The technological research and development devoted to such substitutes has intensified and the market expanded. Such novel foods are not excused from scrutiny over their environmental and health profile, or from questions over what kind of "nature" they reproduce, but in mimicking an established omnivorous regime of processed meats they compete against and potentially undermine the animal-industrial complex.[55] The potential of plant-based milks is a further case in point. Arguably, contemporary meanings of animal milks, while complex,[56] do not secure as much practitioner loyalty as meats. Their use as ingredients and in breakfasts is relatively easy to substitute, and the competition from plant-based milks in countries such as the United Kingdom is now well underway.

This can be explored further by returning to the multi-level perspective, noting its synergies with practice theory. Hargreaves et al. exactly attempt to bring the two approaches closer together, pointing out areas of overlap and tension.[57] The multi-level perspective and practice theory both focus on stability and change and have been heavily applied to the topic of sustainability. Some of their respective concepts can speak to each other; for example, competition between practices could be seen as taking place at the interface between niche and regime.[58] Although my approach here is to advocate for and to explore a practice theory approach to understanding and intervening in the animal-industrial complex, it is inclusive toward overlaps with other sociological perspectives like the multi-level perspective and social network analysis, as well as to insights from social psychology. The application of the multi-level perspective to thinking about plant-based milks by Mylan et al. which focuses on niche-regime interactions is of interest to this overall discussion.[59] One

---

55 Though Poore and Nemecek (2018, 990) found that "the impacts of the lowest-impact animal products exceed average impacts of substitute vegetable proteins across GHG [greenhouse gas] emissions, eutrophication, acidification (excluding nuts), and frequently land use".
56 DuPuis 2002; Stănescu 2018.
57 Hargreaves, Longhurst and Seyfang 2012. See also El Bilali 2019, for a critical look at applying the multi-level perspective to the food system.
58 However, there is a clear theoretical tension between the two approaches in the contrast between practice theory's "flat ontology" and the multi-level perspective's "nested hierarchy" of three levels. In noting this, multi-level perspective theorist Frank Geels (2011, 37–38) subsequently suggested dropping the notion of hierarchy from the perspective. In positing an open approach to different perspectives, it is important to be mindful of potential theoretical incommensurability.
59 Mylan, Morris et al. 2019.

of their contributions is to suggest a bidirectional niche-regime analysis including not only niche-to-regime processes but also regime-to-niche dynamics. This allows an analysis which accounts for both the sort of practices engaged in by niche advocates and also how incumbent actors and institutions respond to new innovations. They draw upon Smith and Raven[60] who proposed "two ideal-type empowerment patterns in niche-regime interactions: 'fit-and-conform', in which niche-innovations diffuse because they fit in with the existing selection environment, and 'stretch-and-transform', in which niche-innovations diffuse because advocates succeed in transforming existing regimes".[61]

Mylan et al. suggest hybrid pathways which combine elements of both patterns. They further argue that "cultural dimensions" likely play an important or leading part in sustainable food transitions, a position which echoes the importance I have given to meanings in this chapter. I also agree with their understanding of diffusion as a process of societal embedding either in relation to a novel "regime", or in practice theory language, new practice-bundles or complexes, because innovative performances need to become routinised practice entities to be a normalised part of the social order. They then orientate their analysis toward niche struggles in four environments: users/ markets, business, culture and policy. They suggest that in the United Kingdom dairy milk consumption was already a declining regime (since the 1970s) due to health reasons and shifting breakfast patterns but that plant-based milks may now be contributing to its problems. It is relevant to note that although certain practices may be deemed to be in competition it is often broader shifts and meanings that come into play. For example, from a vegan perspective, the ongoing decline in UK milk consumption as yet has little to do with a proliferation of ethical meaning over how the dairy industry treats cows. This is reinforced by their finding that most consumers of plant milks do so alongside continued use of dairy milks. In considering regime/niche dynamics across their four environments Mylan et al. are able to produce a nuanced analysis which shows how plant-based milks have begun to be socially embedded.[62] For example, they note the accommodation of

---

60   Smith and Raven 2012.
61   Mylan, Morris et al. 2019, 235.

plant-based milks by incumbent regime actors such as supermarkets in the business environment, alongside the resistance and defensiveness from dairy representatives, expressed in legal challenges against plant-based companies using the word "milk", lobbying of UK politicians, and policy and cultural environment struggles over the health claims of both dairy and plant-based milks.

This detour to the multi-level perspective can be read back into practice theory. It has provided an example of a "competing" practice but might require a reconsideration of what that means when many dairy consumers are simultaneously also drinking plant-based milks. Reminiscent of the non-practising practitioners noted in the previous chapter, it is important to bear in mind that elements of practice are shared across practices and practitioners. Are such people in transition? Or are they content to keep consuming both? This could be an issue for all plant-based meat or milk substitutes wherein they become normalised as a novel and additional food to consume, as an occasional buy-in to eating more sustainably, but do not actually end up substituting consistently for animal "products" (a point I return to in the next chapter). This example also asks of competing substitutes whether they are overly derivative of animal "products" and so may reinforce the normativity of animal consumption.[63] There is no need to consume any type of milk at all, one can obtain its nutritional profile elsewhere, and to assume so is to witness dominant omnivorous practices shaping their "competing" practices. For example, a vegan innovation in the performance of breakfast could radically redefine what that meal is, and one could highlight pre-existing cultural differences in breakfast practice that may not be so centring of dairy produce. Nevertheless, this implies an extra level of food routine creativity that might struggle to compete with the convenience of substitutes.

Returning now to the practice theory encounter with questions of power and scale, I have noted examples of practice collaboration and competition which for Shove et al. are integral to the formation of practice-bundles and complexes, that is, more complicated practice

---

62   Mylan, Morris et al. 2019.
63   Lonkila and Kaljonen 2021.

interdependencies.[64] Both processes are complex, but they are instructive guides for how something like veganism (or merely plant-based eating) may come to be socially embedded. Nicolini suggests several ways in which practice theory can apprehend larger phenomena. Firstly, connections between practices can be traced but also the understanding of how practices can influence each other should be expanded, including those that can potentially exert great influence across both time and space.[65] For Nicolini, "representing large-scale phenomena requires a reiteration of two basic movements: zooming in on the situated accomplishments of practice and zooming out to their relationships in space and time".[66] A second approach is to examine manifestations of the global within local everyday practices. In other words, it is a mistake to typecast practices which may be ordinary and tangible as not also being constitutive of large-scale phenomena. Finally, Nicolini advocates attending to particular practitioners and practices that are specifically integral to how large-scale phenomena are constructed.[67] These could include specific actors within, and practices of, corporate boardrooms, decision-making practices that solidify agricultural policies, and governmental procedures that might favour particular food sectors. This dovetails well with an important contribution from Welch and Yates who seek to extend practice theory so that it might better include forms of collective and strategic forms of activity.[68]

It is here again that practice theory deliberations on scale inevitably extend into how it understands power. The effects of certain key practices are discernible in the *coordination* of practice-complexes that strive and struggle to stabilise everyday conduct for multitudes of practitioners, be that employers dictating occupational practices or major actors shaping the food system. Maintaining a degree of control over the habits and routines of large practitioner populations is one

---

64 Shove, Pantzar and Watson 2012.
65 Nicolini 2017.
66 Nicolini 2017, 107.
67 See also Watson 2017.
68 Welch and Yates 2018. This paper is also significant for linking practice theory with social movement theory.

way of understanding power. As Watson outlines,[69] this brings practice theory into contact with Foucault, not only in the sense of his ideas of governmentality[70] and biopolitics,[71] concepts intimately related to the management of populations, but in his understanding of power "as effect rather than object".[72] This influential shift has understood power not as something that can be possessed, instead imagined as emerging as an effect of relations. Watson argues for a complementarity between this way of thinking and practice theory. He is correct that the theorisation of power within practice theory has been too implicit to date, but that "to be consistent with the ontological commitments of practice theory, power must be understood as an effect of performances of practices, not as something external to them".[73] It could also be added that power is an effect of the *elements* of practice (meanings, for example, can foreclose what can be said or done), and becomes sedimented within the connections between practices, within practice-complexes. It is within these complexes that the articulation of practice across time and space is generated, entailing notions of path dependency and embedded habits and routines. At the same time, practice theory is consistent with Foucault's assertion that power is not totalising (not a possession that can ultimately be controlled) since actors can change and intervene in the performance of practice that can ultimately shift the practice-entity.

Like Shove et al. and Nicolini above, Watson places the practice theory theorisation of scale and power within understanding relations between practices, asserting that "appreciating the ability of some practices to orchestrate and align others makes it possible to account for the appearance of institutional hierarchy and scale and for differential capacities to act, while retaining a flat ontology".[74]

People are situated differently within a nexus of practices, due to often longstanding practice-complexes that are imbued with, for

---

69   Watson 2017.
70   Foucault 1991.
71   Foucault 1990.
72   Watson 2017, 174.
73   Watson 2017, 171.
74   Watson 2017, 177.

example, classed, gendered, anthropocentric and racialised practice elements. Many practices themselves are exclusionary along these and other lines, especially those with "a disproportionate capacity for shaping action elsewhere".[75] Similarly, institutions and organisations are differently situated, for example, as part of influential pre-existing practice-complexes or as emergent forms of practice. Importantly, a "flat ontology", which for practice theory is to maintain the idea that practices are the key unit of analysis, should not be confused with an inability to theorise power or hierarchy. The final section of this chapter considers how these deliberations might be applied to thinking through the animal-industrial complex and ways to disrupt it.

Dismantling the animal-industrial complex

The animal-industrial complex remains successful because it recruits an enormous range of practitioners involved in the production and consumption of animals. It is a complex favoured by large-scale projects of governmentality interested in delivering cheap food to populations and achieving large profits. It consequently receives considerable patronage and investment, despite the fact that a plant-based food system would have its own biopolitical advantages. Highly diverse practices are linked together by the sort of shared meanings discussed earlier. This continuing practice theory conceptual work which tries to address prior limitations in thinking about power and scale also begins to afford new ideas for intervention and transition. If relations between practices are key, and certain practices more important than others, then research, and potentially interventions, need to be directed toward these relations and key practices. The analysis implies that key sites, events, practices, practice connections, institutions and intersections between practice-complexes are points at which to contest their ability to routinise everyday eating practice.

In my 2012 contribution to theorising the animal-industrial complex I made clear that I did not intend it to only refer to the agricultural sector because that has significant overlaps with others,[76]

---

75   Watson 2017, 181.

for example, the way the "animal health" sector connects agriculture to the pharmaceutical-industrial complex.[77] The animal-industrial complex then is significantly unbounded and any attempt to map it necessitates introducing some artificial boundaries unless one's aim is to map much of the entire global economy. Concurrently, though, it is important not to totally neglect such overlaps because they may represent points of intervention and be illustrative of apparent contradictions in human–animal relations; the immersion of the companion animal industry within agriculture via pet food, or the 2013 European horsemeat scandal which included meat from horses formerly used in racing sold to the public as "beef",[78] being two cases in point.

The idea of breaking, disrupting and dismantling the animal-industrial complex is a controversial idea. It could be seen as counter to democracy because it is a part of the food system that most people seem to prefer to perpetuate. Nevertheless, the argument in Chapter 5 was that it represents a practice conformity that is not a choice, but a practice-complex that most people are successfully enrolled into via several processes that comprise a generational universalism. The same of course can be said of car culture; a radical rethink of mobility that moves beyond the model of private car ownership would likewise be presently unpopular and yet have numerous co-benefits. It is not that such ideas are undemocratic but that there is a lack of democratic space in which to challenge highly normalised practice-complexes.

To recall, my 2012 definition of the animal-industrial complex referred to it as "a partly opaque and multiple set of networks and relationships between the corporate (agricultural) sector, governments, and public and private science. With economic, cultural, social, and affective dimensions it encompasses an extensive range of practices,

---

76 Twine 2012.
77 Twine 2013b.
78 This also meant that many omnivores unknowingly consumed horse passed off as cow (beef). This scandal and others such as the BSE or foot and mouth outbreaks brought usually hidden supply chains and animal death into public view, placing the animal-industrial complex uncomfortably on display, see "Horsemeat scandal: the essential guide", Lawrence 2013.

technologies, images, identities, and markets".[79] This began to pinpoint some important relations but did not yet reflect, at the time, my then early reading of practice theory. Nevertheless, there was already an attempt not to define the animal-industrial complex in terms of agency/ structure, or micro/macro distinctions. It alludes to elements of practice but could be reimagined to be more in line with a practice approach. As a very large-scale phenomenon which is able to shape the everyday practices of billions of humans, and to coordinate the lives and deaths of billions of nonhuman animals in a given year, the animal-industrial complex is a highly intricate system in which multiple practices, spaces and actors are tightly interlinked. The element of opacity can make the necessary research into the animal-industrial complex trickier to undertake but CAS and other scholarship have gone some way to providing better understandings of its practices and relations.

Succinct definitions may be useful but a more extensive attempt to map and represent the animal-industrial complex is necessary to both better grasp relations and practices and to imagine transitions. Figure 7.1 is such a visualisation which strives to map out the key practices and relations of the animal-industrial complex. It is presented as a web not to assume that any one being has created or is in control but because it fits well with the practice theory contention of the importance of breaking links between practices and practice-elements as a strategy for and within transition. Practices and their relations are organised around six key interlinked nodes: government, corporations and associations, farmed animals – management and slaughter, social institutions, consumers, and research and development. Each are further elaborated within the exterior ovals of the visualisation. UK examples are used as illustrative, such as the National Farmers' Union, the Department for Environment, Food and Rural Affairs (DEFRA) and the Biotechnology and Biological Sciences Research Council (BBSRC). The International Livestock Research Institute (ILRI) is given as an example of a research institute that encourages animal consumption in "developing" countries. Elaborations imply the unbounded nature of the animal-industrial complex pointing to

---

79   Twine 2012, 23.

connections with other practice-complexes and infrastructures around energy, transport; other domains that use nonhuman animals; social institutions such as the education system; and transnational corporations that are not *only* involved in the food system. Although the rationale of veganism is to exit from involvement in the animal-industrial complex, it is included under consumers due to interconnections such as the use of manure in arable farming, emphasising that it is truly difficult not to be caught in its web. The visualisation is also useful because it helps to pinpoint practices, relations and practitioners which the overall practice-complex depends upon. For example, how could the animal-industrial complex operate without a compliant veterinary profession, without Heavy Goods Vehicle (HGV) drivers and trucks, slaughterhouse workers and technologies, and of course, without consumers? Figure 7.1 is intended as a heuristic that can be developed through further research, contributing to broader work in the "counter-mapping" of the animal-industrial complex.[80]

This chapter concludes with a summary of interventions suggested by practice theory during this and the previous chapter (Table 7.2). These offer a potential toolkit for pathways to change, both for further research and for advocacy strategies for change. Some of these dovetail with pre-existing strategic practices that are familiar to animal advocates such as contesting many of the dominant meanings outlined in Table 7.1. The issue for this and other transition frameworks, however, is whether they can address the urgency of practice change required by the climate and related biodiversity crisis. Attempts to change societies are happening decades later than they should have, creating an anxious, narrowing window of opportunity. With this in mind, the final substantial chapter examines what more can be done in light of this urgency and situates the discussion around vegan transition and societal transformations in human–animal relations back within the question of the Capitalocene.

---

80   Barnes and White 2020.

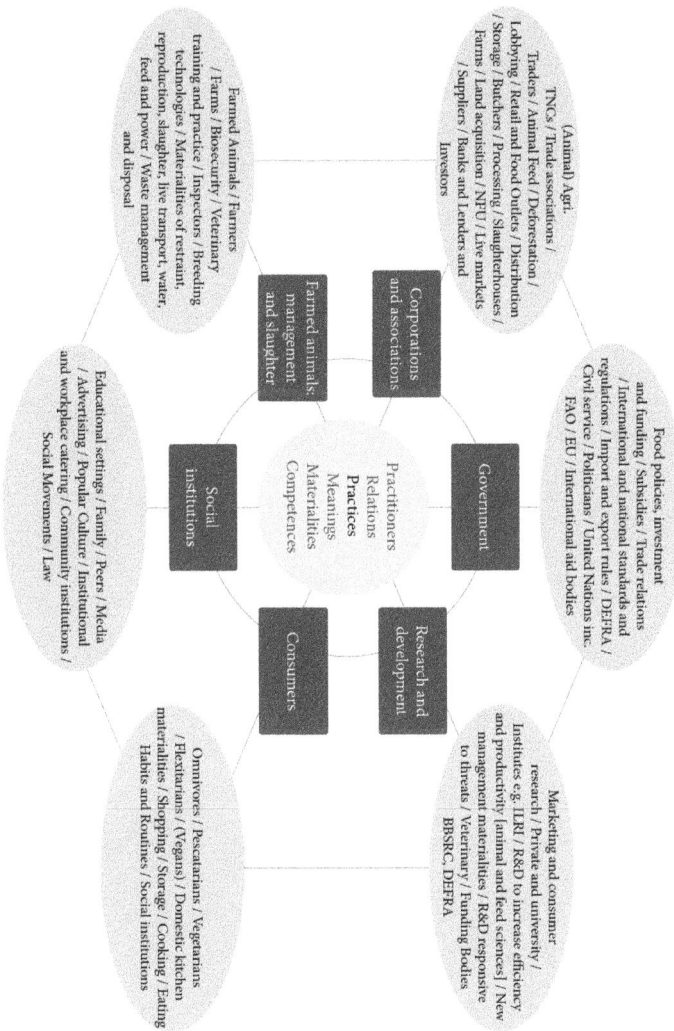

Figure 7.1 Picturing the animal-industrial complex through a practice theory lens.

Table 7.2 Interventions inspired by practice theory.

Type of intervention	Suggestions
Ease a competency. Interventions which save time or bypass preparation allowing a practice to become more easily integrated into routine. Make competences accessible and simplified.	Examples include batch cooking and recipe and cooking clubs. Labelling of vegan-friendly foods. Greenhouse gas labelling on all foods.
Contest meanings of a dominant practice.	Contest the conflation of meat with strength and virility. Demonstrate vegan strength and endurance. Question the cultural valorisation of these qualities.
Direct interventions at key practices, relations and practice connections considering those fundamental to the animal-industrial complex.	Could involve change from within key professions such as veterinary practice. Chart other social changes that may be weakening of the practice in question.
Promote meanings of practice.	Demonstrative veganism, food sharing, photo sharing. Recognise that elements of a practice extend beyond core practitioners. Extend bridges toward people and organisations that already share some elements of vegan practice. Consider meanings broadly. Health and taste meanings may initially be more attractive to non-practitioners.
Direct change at routines and habits. Aim at social collectives rather than only individuals.	Create new food routines. Consider the temporalities of everyday life and how to make sustainable food practices easier to become part of. Consider how the elements of mealtimes may be changing and how this is relevant to transition. For example, if lunch has become more snack-like then such plant-based options would make transition easier. Recognise that food-related routines are often formed collectively so designing transition pathways for diverse families, friendship groups and workplaces may be more effective. A more collective approach to

Type of intervention	Suggestions
	transition may also work with food producers. These include initiatives to transition "animal farms" to arable farms.
Create new spaces for a practice.	Spaces for knowledge sharing and support – websites, buddy schemes. Bundle with other established practices.
Organise events.	Events can be settings in which a novel practice and its elements are demonstrated: meanings, competences and materialities (including tastes and textures of food).
Bundle a niche sustainable practice with others (cooperative practices).	Vegan bundling such as "Vegan Runners" or "Vegan Organic". Bundle veganism into pre-existing practices and settings. Embed and extend vegan practice.
Substituting materials. Ask what materials from pre-existing practices are useful for vegan transition. Adaptation and experimentation.	Plant-based milks, plant-based meats, "veganisation" of culturally cherished meals. Vegan butchers.
Add materials. Adaptation and experimentation.	Foods that are novel to pre-existing routines.
Deconstruct vegan practice in terms of its three elements to better understand the cohering of elements together leading to a process of successful transition. Listen to the narratives of new practitioners.	This means recognising there are multiple pathways. Different meanings are salient for different people. The environmental, health and animal ethics meanings of veganism are part of the coherence of vegan practice but how the practice is communicated is important. The majority may be more open to health meanings. Exposure to practice elements can be amplified using events, relations and bundling.
Substituting practices and practice-bundles.	Substitute farms for sanctuaries but then re-craft the space. Crowd-sourced land acquisition initiatives to transition land away from animal agriculture to rewilding and arable use.

Type of intervention	Suggestions
Break and remake links between practices and seek to understand how a change in one practice can shape neighbouring practices.	A focus on how practices interlock and changing this is a further strategy. Social transitions to snacking shaped partly by eating while also doing something else could introduce opportunities for novel eating practices. Tying eating practices to workplace sustainability initiatives could also be another strategy.[81]
Build plant-based/vegan networks.	Form guilds, associations, groups and networks.
Invest/divest in a practice. Push for institutional patronage.	Contest the investments of institutions, banks and pension schemes in the animal-industrial complex. Subsidise vegan infrastructure.
Contest meanings that connect practices.	Intervene on dominant meanings of masculinity defined in terms of the disavowal of care and compassion toward other species. Contest regionalism and nationalism.
Contest broader food meanings.	Question entitlement to unbridled choice, autonomy and privacy. Question commodified food system and bundle veganism with examples of de-commodified food practice – such as home growing, community allotments, food sharing. Intersectional veganism should explore connections with the food justice movement.
Target sites and organisations which coordinate the food practices of many practitioners.	Intervene in catering policies in institutions such as schools, hospitals, prisons, workplaces.
Redefine the meal instead of being derivative.	Redefine breakfast from necessarily involving dairy milk, or any milk. Redefine main meals from being orientated around a large protein object (contest the conflation of animal "products" with protein and the overconsumption of protein).

81 Spurling, McMeekin et al. 2013, 37.

8
From plant-based capitalism to system change

Part I of this book established anthropocentric human–animal relations as partly constitutive of the climate crisis and documented the myriad ways in which climatic change is killing and otherwise impacting nonhuman animals (both "wild" and "domesticated") on a global scale. Part II has explored the potential for societies to enact deep reductions to animal consumption called for by a wide range of reports and peer-reviewed science. So far this has been examined via social science understandings of transition and three potential scenarios with differing commitments to tackling anthropocentrism and with divergent investments in the status quo capitalist economic system.

In distinguishing between plant-based transition, vegan transition and intersectional vegan transition, the previous chapter noted that only the latter two scenarios explicitly contest the anthropocentrism that has shaped the climate crisis, but only the final scenario of intersectional veganism is specifically further critical of capitalism as a social and economic formation that it sees as inescapably productive of the climate crisis. Part I also underlined how this places advocates of this scenario, such as some vegans and all practitioners of critical animal studies (CAS), in an overlapping critical space with a larger constituency who similarly see organising tenets of capitalism as producing the climate crisis – most notably advocates of the Capitalocene thesis, but also a wider group of environmentalists and climate scientists. It is unsurprising, however, that the first scenario of a plant-based transition, within the parameters of the pre-existing economic system, is that which presently demonstrates most traction.

This chapter examines in more detail the emergence of this "plant-based capitalism" and considers its ability to address the climate crisis. From the perspective of both vegans and those who see capitalism and the climate crisis as inextricably linked, a plant-based capitalism approach is limited as a critique of anthropocentrism as well as an effective strategy for addressing greenhouse gas emissions. Nevertheless, plant-based capitalism advocates would counter that it has the potential to radically reduce the dominance of animal agriculture in the global food system and so contribute to declines in emissions levels. This raises an important question for CAS researchers and animal advocates which guides the remainder of this chapter: to what extent should plant-based capitalism be embraced, if at all? Could it be counterproductive not to support it? An intersectional scenario, allied to a climate justice framing, implies an anti-capitalist veganism, but if plant-based capitalism is to be wholly rejected, is there a viable pathway which could effectively bundle together veganism and anti-capitalism? Such questions invite (intersectional) vegans to join a much larger constituency in prefiguring a world beyond capitalism, and all the accusations that attracts of idealism or utopianism.

Yet if the analysis of the ecosocialists invested in the Capitalocene framing discussed in Chapters 1 and 2 is correct, there is no alternative to adequately addressing the climate crisis than to transcend major fundamentals of capitalism. This presents both a challenge and an opportunity for coalition between intersectional vegans and potentially allied social movements, to construct an alternative social imaginary that better represents a significant critique of meat culture and anthropocentrism conjoined to an anti-capitalist climate justice framing, and that also delivers a better response to the climate and biodiversity crises.

This chapter begins with an exploration of the vision of plant-based capitalism, including its potential obstacles and limitations. It then moves on to explore and make the case for a vegan social imaginary that is intersectional, and so anti-capitalist. In doing so the discussion is intended to further illuminate the debate around transition within this second part of the book because it argues that the complexity of transition implies a certain amount of naivety in theories that imagine social change to be possible without attending to how pre-existing infrastructures of practice are kept in place by a nexus of classed, gendered, racialised and anthropocentric relations.

The dream of plant-based capitalism

Chapter 6 noted the proliferation of plant-based options in mainstream supermarkets, cafés and restaurants in recent years. To be middle class, urban and vegetarian or vegan, the facilitation of plant-based eating is better than ever. It may seem justifiable to claim that capitalistic profit-driven supermarkets are doing more than national governments to assist plant-based transition.[1] In this sense, it could seem that capitalism is delivering. Research in 2021 suggested meat consumption in the United Kingdom fell 17% between 2008 and 2019.[2] The rise of veganism is also geographically diverse. Even a meat-centred country like Argentina has at least 170 vegan or vegetarian restaurants or cafés (69 wholly vegan).[3] On the face of it, the growth and commercialisation of plant-based eating seems to have been a boon for the visibility and intelligibility of veganism across a wide variety of cultures. Cross-cultural institutional and government buy-in would embed such transitions further.

Alongside the introduction of plant-based ranges within major supermarkets, numerous companies specialising in plant-based foods have emerged and gained prominence. In May 2021 when Swedish oat milk company Oatly debuted on the Nasdaq stock exchange it was valued at US$13 billion, exceeding expectations. Although, for example, dairy milk consumption in the United States still outstrips plant-based milks tenfold, the former is decreasing and the latter

1 Such views need to be checked. An analysis of UK supermarkets found widespread targeting of consumers with cheap meat deals, see Campbell 2022.
2 Stewart, Piernas et al. 2021.
3 Data from https://tinyurl.com/42az95fm.

increasing.[4] Oatly are interesting also for their promotion of veganism in their public relations, even drawing upon activist framings in their advertising such as billboards proclaiming "It's like milk but made for humans". Their TV ads have also targeted the least likely vegan demographic, middle-aged men. In 2023 Oatly completed a new factory in England to supply the UK market with oat milk and other oat-based products, solely using UK–grown oats. As such companies seek expansion, they inevitably become embroiled in the financial aspects of global capitalism. In 2020 Oatly were subject to criticism for selling a minority stake to a consortium of celebrities and the US investment company Blackstone for US$200 million because of allegations that Blackstone had previously funded companies involved in Amazonian deforestation (which they denied) and given support to former US President Donald Trump. Oatly retorted:

> If we ever want to have a chance of reaching the global climate goals of cutting the greenhouse gas emissions by 50% before 2030 and reach net zero emissions by 2050, we need to speak a language that the capital markets can understand. Doing so can potentially trigger a massive flow of capital out of gas, oil, and soybean production in the Amazon and into greener projects and companies. With global private equity working toward a green future, we actually have a chance of saving the planet for future generations.[5]

In what is a familiar pathway for "ethical companies" which are then perceived to "sell-out" to big capital as they attempt growth, the approach of Oatly is to use the pre-existing mechanisms of capitalism to try to effect change. This vision of "change from the inside" perhaps makes Oatly the epitome of plant-based capitalism. Other companies have followed a similar path or have been completely bought by larger food companies. Increasingly, this is directly by meat or dairy companies. In 2021 the large Brazilian meat company JBS acquired Vivera, the third largest plant-based food company in Europe, valued

---

4   Burwood-Taylor 2021.
5   Oatly 2022.

at €341 million, and Canadian dairy company Saputo acquired Scottish vegan cheese maker Bute Island Foods. Plant-based meat company Beyond Meat had revenues in excess of US$400 million in 2020 and has entered into partnerships with mainstream brands such as McDonald's, Taco Bell and PepsiCo. It has production facilities in the United States, the Netherlands and China. Tyson Foods, the world's second largest meat processor after JBS, initially bought a 5% stake in Beyond Meat, later selling it in 2019, announcing plans to develop its own plant-based protein lines.

For the imaginary of plant-based capitalism to have credibility, such ventures from enormous global meat companies into plant-based food would need to be accompanied by concurrent downsizing of their core business. In the absence of this it appears more likely that investments are guided by the extra premium that may be yielded currently in the plant-based protein sector, or control of a rival food innovation. Derivative of the animal-based food system, the plant-based capitalism imaginary foregrounds material substitution as a way to usurp the dominance of the use of animals. Where this vision may potentially come unstuck is partly in its faith of using the pre-existing tools of the market and potentially in a simplistic misunderstanding of the economy as simply working along lines of supply and demand. In this way of thinking, producers and consumers rationally shift to become key actors that choose the lower carbon option and gradually shift the food system to one that is more or close to wholly plant-based. Superficially it might seem that this is happening but there is less evidence for downsizing or divestment.

Although not plant-based, the topic of cellular meat is included here within the imaginary of plant-based capitalism for two reasons. Firstly, both cellular meat companies and plant-based meat and milk companies tend to favour the mechanisms of the pre-existing global economy to achieve growth and transition and, secondly, both can be seen as examples of techno-optimist forms of "disruptive innovation" centred around a direct substitution of pre-existing animal "products". While plant-based meats and milks have the advantage of already being widely available, at least within richer nations, cellular meat remains a largely promissory materiality some way from widespread availability. Despite this, cellular meat research receives substantial funding (US$1 billion since 2013)[6] and may be seen by some as potentially the better long-term substitute as it could be intelligible as animal-sourced meat to consumers and conform to the pre-existing symbolic value that animal consumption appears to offer many. As Kivimaa et al. outline, the notion of disruptive innovation has emerged from overlaps between transition theory and the innovation management literature. Although their review of the concept uncovers a paucity of agreement over definition, a key aspect of the notion of disruption is to contrast it with incremental change, instead referring to the rapid disruption and transformation of a socio-technical system.[7] This chimes well with some proponents of cellular meat, forecasting the technology as a near-term systemic disrupter, such as the 2019 report by think tank RethinkX (literally called the disruption report) which imagines a highly technologised food system. The report envisions a suite of technologies, including precision fermentation and tissue engineered cellular meat, being so disruptive that "[b]y 2030, the number of cows in the U.S. will have fallen by 50% and the cattle farming industry will be all but bankrupt".[8] Disruptive innovation research is less beholden to unchecked techno-optimism and provides some sociological reasons for questioning such forecasts. One danger is that disruption is only imagined in a narrow technological sense, extracting technologies from their social, economic, political and cultural contexts, and succumbing to technological determinism, conflating technological change with wider socio-economic change. Kivimaa et al. exactly intend to underline a broader understanding of disruptive innovation as

> a high-intensity effect in the structure of the sociotechnical system(s), demonstrated as long-term change in more than one dimension or element, unlocking the stability and operation of incumbent technology and infrastructure, markets and business models, regulations and policy, actors, networks, and ownership structures, and/or practices, behaviour, and cultural models.[9]

---

6   Chemnitz and Becheva 2021, 61.
7   Kivimaa, Laakso et al. 2021.
8   Tubb and Seba 2019, 6.
9   Kivimaa, Laakso et al. 2021, 119.

This ought to counter naïve techno-optimism and complicate a potential pathway for cellular meat such that disruption in only one dimension may be insufficient to produce transformative change. In one sense the promissory hype is understandable; the notion of rapid disruption after all does entirely appeal to the urgency of change required to mitigate the climate and biodiversity crises. Nevertheless, translating novel food materialities like plant-based and cellular meat back into those transition theories previously explored is also further helpful in checking a simplistic technological determinism. As noted, proponents value the perceived simplicity of substitution as a disruptive innovation which has minimum disturbance at least for consumers. Yet even though cellular meat may come to be experienced as a seamless culinary change, from a practice theory perspective the materialities of cellular meat still may be compromised by tainted meanings related to a technologised food system. Viewed from the multi-level perspective, the pre-existing regime of the animal-industrial complex will not simply yield to a new foodscape, a point potentially underlined by the aforementioned corporate acquisitions.

Social scientists have tracked the trajectories of cellular meat for some time,[10] illustrating relatively slow progress on the technological obstacles related to its production. These obstacles mean that a sudden jump to large-scale commercialisation remains unlikely in the short term despite the well-publicised proto commercialisation event staged by the Californian company Eat Just (formerly Hampton Creek) at a private Singapore restaurant at the end of 2020.[11] There are also question marks over its energy inputs and emissions which have been raised by several studies and reports.[12] The imaginary of cellular meat is partly different from the rest of the plant-based capitalism vision in the sense that it taps into a broader autopoietic desire in biotechnology that marries research into the super growth potentials of biological materials with the neoliberal dream of continually circumventing limits

---

10   Chiles 2013, Stephens 2010; 2013; Stephens, Di Silvio et al. 2018; Stephens, Sexton and Driessen 2019.
11   McCormick 2021.
12   Chemnitz and Becheva 2021, 60; Mattick, Landis et al. 2015; Smetana, Mathys et al. 2015.

to growth.[13] This alchemic vision,[14] found also in the energy sector's hopes to one day substitute fossil fuels with nuclear fusion power, similarly pervades the technologised, molecular imaginary of a new food system where meat production is cleansed of its violence to animals, ecosystems and human labour. The material practices of cellular meat scientists in tandem with the entrepreneurial marketing work of well-funded start-up companies constitutes a significant attempt to produce novel food materialities intended to be intelligible as "ethical biocapital"[15] that might usurp not entirely dissimilar attempts by animal scientists to construct less ecologically impactful farmed animal bodies using genomics and gene editing. While the latter effort attempts to be protective of the animal-industrial complex, the former is committed to its disruption and replacement.

If not for the assumption that cellular meat could be the ultimate substitute (the need for which assumes a consumer unwilling to change), it would be hard to explain both the hype and investment in the technology. It is not hard to understand why proponents of the other two scenarios object to the plant-based capitalism scenario, inclusive of cellular meat. For advocates of vegan transition, it means being goaded into supporting corporate greenwashing; global firms that on the one hand may produce plant-based lines but on the other have their operations firmly within the animal-industrial complex. Unsurprisingly, vegan social media is full of debate over whether or not to support, for example, the McDonald's plant-based burger, and statements of preferential support for *wholly* vegan businesses. For intersectional vegans these are issues too, but there is a broader concern that the plant-based capitalism scenario is firmly embedded within capitalist practice and silent on broader social justice concerns. This is reflected in the work of CAS researchers generally taking a position against cellular meat (and the broader plant-based capitalism

---

13   Cooper 2008, 28; Twine 2010a, 101.
14   Thacker 2003.
15   Franklin 2003, 98. I previously used Franklin's concept of ethical biocapital (Twine 2010a) to understand how animal scientists produced new materialities (including lifeforms) with supposedly "built-in ethics" such as using genomics as a fix for sustainability. Cellular meat works similarly and further imagines a consumer relieved of any need for ethical reflection.

imaginary). Cole and Morgan[16] argued that "it reproduces the cultural visibility of meat, and thereby reproduces the already existing status hierarchy of food in Western diets". White has pointed to how the plant-based capitalism imaginary reduces vegans to a consumer status, more passive, less active(ist) and so less threatening.[17]

Nevertheless, some vegans have been vocally supportive of cellular meat; notably in 2008 People for the Ethical Treatment of Animals offered a US$1 million prize for its commercialisation by 2014. The organisation's former head of campaigns Bruce Friedrich helped form the Good Food Institute in 2015, a non-profit organisation which promotes both alternative plant-based proteins and cellular meat. CAS theorist Vasile Stănescu has directly critiqued Friedrich for his essentialist views that it is "human nature" to eat meat, and further problematised the view that cellular meat will simply substitute for traditional meat rather than merely add to consumption within a capitalist food system based around growth, profitability, and considerable forecasted increases in global animal consumption.[18] On this second point, Clay et al. have argued similarly in relation to plant-based milks, which "may afford at best an interruption to the challenges they claim to resolve. At worst, they could distract from the need for systemic changes by virtue of fitting so well within the contours of globalized industrial agri-food".[19] Evans and Johnson's analysis of framings presented at the 2018 US Department of Agriculture – Food and Drug Administration Joint Public Meeting on the Use of Animal Cell Culture Technology further demonstrated how cellular meat is being imagined as an addition rather than as a replacement to traditional meat by meat industry stakeholders.[20] Stănescu's point speaks to broader longstanding arguments in climate debates over renewable energy and fossil fuels which argue that in lieu of policies to suppress the infrastructure of the latter, the former is only able to modestly suppress fossil fuels.[21]

---

16   Cole and Morgan 2013, 212.
17   White 2018.
18   Stănescu 2021.
19   Clay, Sexton et al. 2020, 959.
20   Evans and Johnson 2021.

While a total rejection of the plant-based capitalism imaginary may be accused of unrealistically expecting a near-term decisive break from capitalism and of missing some of the potential opportunities opened up by the socio-technical deconstruction of "meat", the counterarguments expressed above call into question the potential naivety of assuming that it can be effective. From this perspective, putting energy into plant-based capitalism is potentially misguided not just because, like its kindred paradigm of ecological modernisation, it has nothing to say about issues of climate and social justice, but also, fundamentally, in the absence of the questioning of an economic model based around deregulation, commodification, consumerism and overconsumption, profit and growth, it is poorly equipped to deal with the climate and biodiversity crises. This is not the same as saying that innovation in plant-based foods, or even the more speculative cellular meat,[22] have zero chance of a role in a future just food system, but that the plant-based capitalism imaginary overstates their role, uncritically assumes a liberatory potential for (bio)technology,[23] and ignores the ways in which capitalism has worked, and continues to work, within an exploitative nexus of intra- and extra-human relations.

To guarantee that novel plant-based ways of eating actually do effectively displace animal consumption, real transformative cultural change and policies to begin dismantling the animal-industrial complex are required – something demonstrably unlikely under present-day capitalism. Here Fraser's critique of "progressive neoliberalism", wherein many countries in the Global North have attempted to combine a "progressive politics of recognition" in regard to issues such as gender, LGBTQI+ rights and even environmentalism within a capitalist political economy,[24] could be applicable to the plant-based capitalism imaginary. For accommodated, safe liberal

---

21 York 2012.
22 Dutkiewicz and Rosenberg (2021) have explored the possibility of disentangling cellular meat from its corporate trajectory, arguing for its public funding.
23 A special issue of the journal *Configurations* that I edited with Neil Stephens critically examined the liberatory potential of biotechnology for human–animal relations: https://muse.jhu.edu/issue/28727.
24 Fraser 2019.

feminism, read also safe liberal "veganism". Displacement and substitution of animal-based foods could occur but, for example, would need to transcend the anti-regulatory impulses of contemporary neoliberal capitalism. Here the limitations of the plant-based capitalism scenario overlap to an extent with those of the transition literature covered in the previous chapters. These can be explored further by identifying more specific obstacles to plant-based capitalism. It was argued above that plant-based capitalism manages to misunderstand capitalism, taking the notion of a "free market" as read. Yet one only needs to return to the obduracy of the animal-industrial complex to understand how animal consumption has grown and persisted within a decidedly *political* economy.[25] Political in the sense of both relationships and networks of influence between governments and corporations in a global economy, and in the broader sense of the ability of economic sectors (in this case the animal-industrial complex) to manipulate capital to ultimately shape and influence which practices attain cultural dominance. That under neoliberal globalisation the balance of power has in many instances shifted from the state to the private sphere is important here. The plant-based capitalism imaginary appears to demonstrate little awareness of the political economy in its assumption of a relatively simple transition of substitution. This criticism can also be directed toward some transition theories and briefly returns us to the discussion on practice theory and power from the previous chapter. A considerable obstacle to the plant-based capitalism scenario are the political and economic contexts of the animal-industrial complex which must be accounted for by a viable transition framework. In this vein sociologist Andrew Sayer has argued that unless practice theory "can be combined with political economic analysis, it is likely to be seen as safely depoliticizing".[26] By excluding such considerations, the plant-based capitalism imaginary maintains its techno-optimism.

---

25   See Neo and Emel 2017, 5–7.
26   Sayer 2013, 176.

Capitalist political economy and the maintenance of business-as-usual

In contrast, let us consider some political economic dimensions of the animal-industrial complex beyond the rudimentary claim that it is guided by profitability rather than public need. One of these noted earlier concerns the power of lobbying by the private sphere, in the case of meat lobbying organisations attempting to influence the Food and Agriculture Organization of the United Nations (FAO) noted in Chapter 4. More recently, bodies such as the International Meat Secretariat lobbied the UN to exclude more critical voices from the UN Food Systems Summit of September 2021. Industry interests such as the International Meat Secretariat aimed to use the summit as a platform to argue for increases in animal production.[27] Lobbying of governments and politicians by animal-industrial complex practitioners is a thoroughly normalised practice.

Further dimensions are subsidisation and protectionism. There are numerous examples of both the private and public financing of the meat and dairy industry. A 2019 report by Greenpeace, using data from the European Commission and EUROSTAT, found high levels of subsidisation in the EU, concluding that "between €28.5 billion and €32.6 billion go to livestock farms or farms producing fodder for livestock – between 18% and 20% of the EU's total annual budget".[28] The use of such high levels of public taxpayer money to prop up an industry with such significant environmental, animal welfare, human health and land-use consequences is astonishing. With such subsidisation it is not surprising that Europeans eat more than twice as much meat as recommended by national dietary guidelines.[29] However, even this subsidisation is dwarfed by private investment. In 2020 sustainable food campaigning group Feedback commissioned research by the Profundo organisation into the financing of large meat and dairy corporations. They found that "between 2015 and 2020, global meat and dairy companies received over US$478 billion in backing by over

---

27 Kevany 2021.
28 Greenpeace 2019, n.p.
29 Buckwell and Nadeu 2018, 7.

2,500 investment firms, banks, and pension funds".[30] Moreover, they found that "in April 2020, 3,000 investors backed the world's thirty-five largest meat and dairy corporations to the tune of US$228 billion",[31] leading the report's authors to conclude by arguing for defunding and divestment from large meat and dairy companies on environmental grounds. While such enormous levels of investment and subsidisation of the animal-industrial complex constitute a form of market and state protectionism, states may also intervene to protect the animal-industrial complex under situations of economic stress. For example, in late 2021 the UK government stepped in to allow six-month visas for EU-based slaughterhouse workers after labour shortages in the meat processing sector exacerbated after both Brexit and Covid-19 created a break in the supply chain.[32] This led to a backlog of 120,000 pigs on farms, many of which were slaughtered in situ. Sociologically, it is noteworthy that the animal "processing" and slaughter sector in the United Kingdom increasingly struggles to recruit from the home labour force. Analysis of the world's 35 largest meat and dairy companies has also found evidence of activities to downplay and undermine climate-related policies,[33] and the sector has acted similarly in relation to health messaging.[34] Wide-ranging attempts to prevent novel plant-based foods from using the language of "meat" and "milk" in their products also constitutes further market protectionism.

A third dimension pertinent to the political economy of the animal-industrial complex, though more abstract, is the politicised practices at play in state determinations of national dietary guidelines. In some respects, the UK pathway here does exhibit some liberalisation away from diets centred on animal consumption. The replacement of the Eatwell plate by the Eatwell Guide in 2016 demonstrated a move away from the conflation of meat with protein and incorporated more aspects of a plant-based diet such as dairy alternatives. However, this was on a human health rather than environmental basis. Broader

---

30   Feedback 2020, 5.
31   Feedback 2020, 6.
32   Partridge 2021.
33   Lazarus, McDermid and Jacquet 2021.
34   Clare, Maani and Milner 2022.

international politics have developed around the question of whether national dietary guidelines should incorporate sustainable diet or planetary health considerations into their recommendations. Numerous articles and reports have now looked at this issue and argued strongly for such an incorporation which tends to focus on the co-benefits of transitions to more plant-based diets, while acknowledging the complexity of differential access around the world.[35] Over 100 countries (mostly richer nations) now have national dietary guidelines.[36] A 2016 study found only four countries had incorporated sustainability into their guidelines, with only two of these – Sweden and Germany – advising moderating meat consumption on environmental grounds.[37] In the United States the Dietary Guidelines Advisory Committee had argued in 2015 for the inclusion of sustainability in that country's guidelines, which included advising reductions in meat consumption. However, after intense lobbying from the meat industry, the US Department of Agriculture and US Department of Health decided not to incorporate sustainability.[38] While politicians undoubtedly would deny this being a political decision, the optics read: protectionism for the animal-industrial complex over the health of both the human population and the planet. Other national dietary guidelines research has found unsurprisingly that "[f]ollowing the current national dietary guidelines of the G20 countries will not ensure global warming stays below 1.5°C"[39] and that global convergence upon the recommended US diet would increase emissions.[40]

Scientists connected to the EAT-*Lancet* report's "Planetary Health diet"[41] which, to recall, is a significantly lower meat diet, have argued that it would vastly improve sustainability measures vis-à-vis existing

35   Fischer and Garnett 2016; Loken, DeClerck et al. 2020; Lucas, Guo and Guillén-Gosálbez 2023; Merrigan, Griffin et al. 2015; Ritchie, Reay and Higgins 2018; Rose, Heller and Roberto 2019; Schwingshackl, Watzl and Meerpohl 2020; Springmann et al. 2020.
36   Catalogued here: https://tinyurl.com/2p8atfnm.
37   Fischer and Garnett 2016, 17.
38   Fischer and Garnett 2016, 37–38; Rose, Heller and Roberto 2019.
39   Loken, DeClerck et al. 2020.
40   Ritchie, Reay and Higgins 2018.
41   Willett, Rockström et al. 2019.

national dietary guidelines.[42] Other research has found evidence of orchestrated social media campaigns against the findings of EAT-*Lancet*, instead promoting meat consumption, re-emphasising the political context of struggles over the future of the global food system.[43] Furthermore, it is not just national dietary guidelines that are important here but pronouncements from larger scale tiers of governance or influence such as the UN. In 2021 the UN produced a discussion paper titled "Livestock-derived foods and sustainable healthy diets", which articulated its view that animal consumption should be increased in countries where rates are comparably low. This is part of the UN approach noted earlier which advocates meat, eggs and dairy for the global poor and is combined with assertions that "livestock-derived foods" are "essential at certain times of life (for small children and pregnant and lactating women)".[44] These are contestable nutritional claims and while the paper does stress the need for reductions in animal consumption in richer countries there are questions here over the political context of the uses of nutritional science and possible ways in which animal consumption becomes re-naturalised in discourses of global food in/security and hunger. In 2023 some of the global community of animal production-related scientists, many with direct ties to industry, signed the "Dublin Declaration" in favour of the "social role of livestock", indicating concern at mounting criticisms of their sector.

Further dimensions of the political economy of the animal-industrial complex are found in examples of corporate corruption and the repressive treatment of animal and environmental activists by several states. For example, in 2017 the chairman of JBS admitted paying US$123 million in bribes to Brazilian politicians to secure finance from state-owned banks.[45] In the same year JBS were fined US$8 million for buying cattle ungulates raised in illegally deforested areas.[46] In recent decades the United States and then Australia[47] have introduced so-called ag-gag laws which aim to

---

42 Springmann, Spajic et al. 2020; Loken, DeClerck et al. 2020.
43 Garcia, Galaz and Daume 2019.
44 UN Nutrition 2021, 3.
45 Godoy and Valle 2020.
46 Earthsight 2017.

criminalise the taking of video footage in factory farms by undercover investigators.[48] Activists in Latin America struggling against deforestation have been subject to lethal repression.[49]

All of these dimensions – lobbying, subsidisation, protectionism, attempts to control national dietary guidelines, obfuscation, corruption and repression – are relevant to sociological frameworks of transition – which, to be credible, need to account for the political nature of economic activity within contemporary capitalism.[50] These dimensions strengthen the status quo, thwart sustainable transition, and provide serious obstacles to the plant-based capitalism scenario, the proponents of which seem to mistakenly assume an apolitical economy. At the very least they show that attempting to instigate *transformative* change from within the norms of the current economic model, which fails to regulate such dimensions, is woefully naïve. Struggles for a plant-based food system need to contest these obstacles and strive for the divestment of these forms of subsidisation and protectionism without which the animal-industrial complex would likely fail to maintain hegemony.

Pointing out these limitations of the plant-based capitalism scenario during the first half of this chapter is not meant to wholly detract from the efforts of its advocates. It is well intentioned; yet imagining a set of food substitutions as an answer to the unsustainability of the food system lends itself to a simplified vision of social change shielded from the complexity of varied relations of power which turn out to be crucial to understanding transition and the roots of the climate crisis. Of course, it is easier to map out change within a pre-existing economic model, rather than beyond it.

---

47 The Farm Transparency Project lost a legal challenge against the *Surveillance Devices Act* in New South Wales, Whitbourn 2022.
48 Potter 2017.
49 Global Witness 2021.
50 Sayer 2013.

From capitalist realism to an intersectional veganism predicated on system change

The problem is that such a vision risks being inadequately transformative to effectively curtail the climate and biodiversity crises. The same charge may be levelled against the second scenario of vegan transition. This arguably has several advantages over the plant-based capitalism scenario because it recognises the value of questioning anthropocentrism as one exit strategy from these crises, and it foregrounds the potential of vegan values and practices in building new infrastructures and ultimately societies. This scenario is also understandably less accommodating to transitions which only pursue small reductions in animal consumption because it is fundamentally foregrounding a set of values and practices which contest the commodification of animals as food. Some versions of plant-based capitalism may similarly seek the replacement of the animal-based food system but tend to coyly background the history and meanings of veganism.

However, the limitations of the second scenario are also apparent. This can be seen in some animal advocacy organisations and activists who understand part of the connections between human–animal relations and the wider climate crisis but largely fail to stitch this to a broader comprehension of intersections with, for example, class inequalities, capitalism, the social construction of gender and "race", and the histories and perpetuation of colonialism. This is prone to construct a veganism insensitive to social class inequalities or cultural differences. The difference between scenarios 2 and 3 can be encapsulated as follows: yes, the climate crisis is a crisis of anthropocentrism (scenario 2), but not *just* a crisis of anthropocentrism (scenario 3).

The specific "wicked problem" for the climate crisis and human–animal relations is that the scenario which is most clearly directed at the roots of the crisis and is inclusive of other salient relations of power is also the most ambitious. That is the third scenario of intersectional veganism found within some instances of activism and promoted in fields such as CAS and ecofeminism. This problem is not unique to this domain and is seen also in mobility studies where one might contrast the "green capitalist" strategy of electric car transition with an approach that fundamentally questions mobility, its privatisation and its demand.[51] These approaches which question our dominant social system constitute the taboo areas of climate discourse mentioned during the Introduction, challenging hegemonic ways of life in the Global North and daring to contest a capitalist model with all its gendered, colonialist and anthropocentric biases.

Capitalist realism, the system's own ideological insistence that it is the only viable way to organise an economy,[52] continues to herd willing practitioners toward particular "market-friendly" climate policy, failing to see that profitability and short-term political and corporate interest are unlikely to deliver truly sustainable ways of living or biodiversity protections. Capitalist realism attempts to sustain itself by promoting the idea that economic growth can be decoupled from rising greenhouse gas emissions, as noted earlier in the sectoral example of an efficiency approach to animal agriculture,[53] but actual analyses around the potential of absolute decoupling contradict this[54] and attempts by politicians to promote this view are typically dependent on the creative accounting of national emissions.[55] As noted earlier, "green capitalist" innovation does not necessarily reduce emissions if it takes place within a context of growth and unregulated consumer choice – points which ultimately cast doubt upon the ability of capitalist decarbonisation. Absolute reductions in emissions would be achieved by vegan transition but the intersectional vegan approach rather than the plant-based capitalism scenario would better secure this because of its oppositional stance to capitalism and its support for a managed, regulated and egalitarian economy. The first two scenarios are silent on

---

51 Henderson 2020; Haas 2021.
52 Fisher 2009.
53 See also Twine 2010a, 165.
54 Jackson 2009; Parrique, Barth et al. 2019.
55 An example being the UK government's rhetoric of itself as a climate leader (especially heightened around the time of COP26 in late 2021) which used emissions reductions figures based on excluding outsourced imported emissions, emissions from aviation and shipping, and emissions promoted overseas by the financing practices of the City of London. Significantly, greenhouse gases do not comprehend the concept of national borders.

the systemic capitalist underpinnings of the climate and biodiversity crises.

The onus then falls upon this third imaginary of intersectional veganism, the focus of the remainder of this chapter, to better articulate how reimagined human–animal relations can be part of the solution to the climate crisis and, in particular, how veganism can be part of an overall "just transition", but one that does not accept the normative confines of capitalism. I begin by returning to the endeavour of the early part of this book in bringing CAS (and intersectional veganism) closer together with anti-capitalist critique. In a novel move, I brought CAS into conversation with the Capitalocene framework, inclusive of various examples of feminist, post- colonial and ecosocialist thought which view capitalism as structurally inclined to degrade ecosystems. Hierarchical social divisions such as those along lines of class, gender, "race" and species have been an integral part of those practices which have ultimately led to exponential increases in greenhouse gas emissions.

Farmed animals, for example, have provided sources of cheap labour, "cheap fodder" for the low-waged human working class,[56] and enormous profits for the owners of meat and dairy transnational corporations. It is consequently unsurprising that such companies will fight to hold onto their business model,[57] or that governments will support them and keep even the reduction of animal consumption off the agenda at crucial meetings, such as the United Nations Framework Convention on Climate Change's Conference of the Parties in 2021 (COP26). Yet from an emerging *critical* conversation, capitalism is framed as a self-consuming, auto-cannibalistic economic system that is not salvageable through a "greening", or a Green New Deal, because ecological, human and nonhuman animal flourishing are simply not the goals of capitalism.[58] Rather the

---

56 The animalisation of the poor and racialised is often compounded by a low-quality animal-based diet which maintains ill-health in these groups, which can be further capitalised upon by the pharmaceutical industry, see Twine 2012, 18–19.

57 Just as fossil fuel companies are fighting to extend their practices.

inner logic of such a system manifests itself in the form of an incessant drive for economic expansion for the sake of class-based profits and accumulation. Nature and human labor are exploited to the fullest to fuel this juggernaut, while the destruction wrought on each is externalized so as to not fall on the system's own accounts.[59]

The problem for capitalism is that the climate crisis is the return of this externalisation, the stark materialisation of its own denial. It is also potentially the dead end of deferment for attempts at spatial and temporal fixes that have aimed to address crises in capitalist accumulation.[60] At this point, prior to articulating how an intersectional vegan imaginary could develop and what it could include, the specifics of anti-capitalist critique from the perspective of a CAS-infused Capitalocene framework can be specified. This book, and many others, have articulated several elements which can comprise anti-capitalist critique. Here these have specifically included drawing attention to its violent systemic exploitation of other animals for profit and its disinterest in public health, planetary health and animal suffering because it is not oriented to address such issues. A framework for anti-capitalist critique is outlined in the work of Fraser and Jaeggi.[61] They articulate three modes – functionalist, moral and ethical – summarised in Table 8.1, which I have further elaborated through a CAS lens.

In their nuanced conversation Fraser and Jaeggi argue that, on their own, functionalist and moral critiques are insufficient, and that functionalist critiques often overlap with a normative component through which they assess whether capitalism can be said "to work". By insufficient, they argue that a critical theory of capitalism needs to go beyond these critiques to demonstrate social-theoretical analysis that

---

58   This does not mean that aspects of a Green New Deal are not very necessary, such as renewable energy transition, but that they are likely to fail ecology *within* a capitalist model premised upon growth and capital accumulation.

59   Foster, Clark and York 2008, 6.

60   Harvey 2002; 2015, 151–52.

61   Fraser and Jaeggi 2018, 116.

Table 8.1 A critical animal studies critique of capitalism using Fraser and Jaeggi's (2018) framework.

Modes of critique	Fraser and Jaeggi definitions (Fraser and Jaeggi 2018, 116–30)	A critical animal studies elaboration
Functionalist	Capitalism does not work as an economic and social system. Seen as inherently dysfunctional and prone to crisis. In time capitalism undermines itself through a declining rate of profit, high levels of poverty or through the destruction of the ecological conditions necessary for social and economic life.	The externalisations of the animal-industrial complex (greenhouse gas emissions, enormous land demands, deforestation, poor public health, zoonotic disease, antibiotic use, air pollution, and water overuse and pollution) make a substantial contribution to inequality and ecological destruction.
Moral	Capitalism is seen as morally indefensible in its destruction of people's lives and their means of subsistence. Unjust and inequitable. The social surplus is stolen from society and privately appropriated. Entrenches deep-seated unequal relations along lines of class, gender, "race" and empire.	Capitalism performs a morally indefensible, violent and repetitive war against nonhuman animals both on land and in the seas. The immorality of the animal-industrial complex is further elaborated in the privatisation of its profits and its role in exacerbating classed, gendered and racialised inequalities.
Ethical	Capitalism produces an impoverished and meaningless life. It pervades our everyday life and changes how we value things, our relations and ourselves. Promotes divided forms of life such as splitting production from social reproduction, economy from polity, and society from nature. Impoverished democracy.	Capitalism obfuscates the possibility of more meaningful human–animal relations. Critical animal studies argues that rich, lived relations with nonhuman animals provide greater meaning than constructing animals as food. Simultaneously, giving other animals their own space is valued.

understands the processes and practices of capitalism that exposes it to such criticism and can show them to be *specific* to capitalism rather than economic activity generally.[62] For our purposes this is the continual work of clarifying the animal-industrial complex as a substantial infrastructural part of capitalism. Certainly, one could have large-scale animal agriculture under a different economic system, but the logic of capitalism has taken this in specific directions of growth in pursuit of ever more capital accumulation via various temporal and spatial strategies right down to the genome. From a CAS perspective, this has given shape to specific capitalist moral failings around the instrumentalisation of nonhuman animal life. While advocates for the animal-industrial complex underline its contribution to human food security,[63] it could be argued that as a major part of the global food system it *has* succeeded in its capitalist goals of growth, capital accumulation and the transfer of wealth to a small minority. However, "success" here is revelatory of the irrationality of capitalism, seen alongside the moral and ethical critiques above, and the contribution of the animal-industrial complex to the dysfunction of global ecology and the climate crisis.

At this point it is important to recall that Capitalocene and kindred theorists operate an "extra-economic" understanding of capitalism. As noted in Chapter 1, this includes a re-historicisation of capitalist development as dependent upon cheapened lives caught up in both waged and unwaged relations along lines of class, gender, "race" and species,[64] and extending understanding of the crisis tendencies of capitalism into its political, ecological and social-reproductive contradictions.[65] Once wedded to this expanded definition, CAS implies an intersectional veganism capable of systemic critique in a way far beyond the capacities of both plant-based capitalism (scenario 1) and vegan transition (scenario 2). For this helps not only to situate

---

62 Fraser and Jaeggi 2018, 122–23.
63 Global food insecurity, human starvation and the non-communicable diseases of the animal product–centred nutrition transition undermine such a view.
64 Moore 2017; 2018.
65 Arruzza, Bhattacharya and Fraser 2019.

the intersectional vegan as a political subject ensconced in a project to contest the capitalist war of commodification against nonhuman animals but to locate and make sense of this struggle as part of anti-capitalist practice taking place within a broader crisis complex.[66]

Alliances and prefigurations toward post-capitalist society

This positioning is already seen in much work carried out by intersectional vegans. Some of this relates to necessary firefighting against the resurfacing of power relations such as reflexivity toward class privilege, whiteness, masculinity, able-bodiedness and heteronormativity in the popularisation and practice of veganism.[67] This work assists the other side of intersectional veganism, the more external focus concerned with alliances and connections. This internal and external reflexivity is largely lacking from scenarios 1 and 2, positions which sometimes instead harm the chances for external alliance and growth. For example, plant-based capitalism has little or nothing to say about accessibility to a healthy food system.

Situating intersectional vegans within this broader crisis complex is also useful for enhancing alliances. While the connections between veganism and the ecological contradiction of the climate crisis have been well covered here, it is worth briefly mentioning connections between veganism and the political and social-reproductive crisis tendencies. Arruzza et al. articulate the political crisis tendency as a false division between the political and economic. This simultaneously strives to present an economic sphere as apolitical but also marketises key decision making on the direction of societies[68] disenfranchising human citizens. Outlining some of the key political contexts of the animal-industrial complex earlier in this chapter and underlying the normativity of anthropocentrism in Chapters 4 and 5 illustrates the relevance of this contradiction to intersectional veganism. The failure of plant-based capitalism to adequately account for the political

---

66 Arruzza, Bhattacharya and Fraser 2019, 16; Fraser 2019, 7–8, 36.
67 On vegan masculinities, see Oliver 2023.
68 Arruzza, Bhattacharya and Fraser 2019, 50.

dimensions of the (food) economy occurs precisely because of the political contradiction and democratic deficit in contemporary capitalism. The social-reproductive contradiction pertaining to the disavowal of unwaged care work upon which capitalism depends also speaks to intersectional veganism, not only in capitalism's structural consolidation of class, gendered and racialised inequalities but also in its implicit devaluation of vegan values of care and compassion and its key casting of cheapened animals as food in the role of the maintenance and reproduction of waged labour.

The idea of a crisis complex also potentially germinates a "resistance complex", and these connections are suggestive of political alliances against the climate crisis and its roots in the Capitalocene. Much of this work is underway in both critical theory and the activism of civil society but lacks the coherence that could be provided by the Capitalocene framework. Intersectional veganism is the preferred scenario here because, like the edges of jigsaw pieces, it offers those crucial points of connection with which to align overlapping social movements. For example, this means taking seriously the ways in which capitalism threatens democracy and allying with movements for radical democracy (e.g., children's rights) which reject a complacency inherent to the conflation of democracy with an impoverished and corrupted representative form. Furthermore, the imaginary of intersectional veganism needs to go beyond resistance to the commodification of other animals contextualised within a broader intersectional framework, to a consideration of elements of societal social reproduction.[69]

For instance, the critique of the global food system (a major player in economic production and social reproduction) from an intersectional vegan perspective cannot be content to only resist the commodification of animals but must extend into problematising the

---

69 Understandably the focus here is on the place of the food system in social reproduction, but there is a broader point about how the privatisation of other human needs (housing, health, education and utilities) acts to constrain sustainable transition. For instance, given the impact of the climate crisis on human and nonhuman homelessness and refugees, this is a considerable issue for vegan-inflected understandings of climate justice.

commodification of food itself. This is essential for integrating intersectional veganism into (broadened) notions of both climate justice and just transition. Although de-commodification of the entire food system is a vast undertaking, and part of transitioning to a post-capitalist society, it could take food and agriculture away from being guided by profit accumulation, thus undermining the rationale of the animal-industrial complex, to a properly sustainable food system guided instead by planetary health, healthy low-carbon diets, biodiversity, and nonhuman animal flourishing. At the moment, in a poorly regulated food system, there is no real incentive for these to constitute its fundamental pillars. This call for a de-commodified food system mirrors similar explorations of how free public transport or community renewable energy generation could be significant greenhouse gas mitigation policies.[70] The state subsidisation of these sectors tends to currently work as corporate welfare rather than social welfare, which fails to satisfy any demands for climate justice.

From the perspective of intersectional veganism, concepts of just transition and climate justice are valuable but need to be shorn of their humanist presuppositions. To talk about "justice" at this point in history and to construct it as applicable only to the human is startling and merely echoes the same exceptionalism that has shaped the ecological and biodiversity crisis. In contrast, a vegan climate justice animalises the concept's vital focus on class, gendered and racialised inequities between and within countries to include impacts upon nonhuman animals wrought by these crises. As highlighted in the Introduction, it also works to contest the potential limitations of justice frameworks in, for example, being too abstract and disavowing of emotions as central to moral relationships. Vegan climate justice prioritises accessible vegan transition in the Global North[71] but refuses the FAO assumption and normalisation of "livestock" as the antidote to

---

70 A further option might be to suggest a version of universal basic income that is directed toward the provision of free, healthy, low-carbon vegan food. Though such policies would not obviously curb the high consumption practices of the wealthy. I have not discussed meat taxes because I think their danger is in acting as a tax on the poor.
71 Predicated upon the historical carbon debt of rich nations.

food insecurity in the Global South. The sorts of alliances this points to must include a forged common purpose between vegan climate justice and the food sovereignty movement with its vital emphasis on the human right to healthy food, its opposition to the unchecked power of food transnational corporations and its struggle for public and community ownership of food production.[72] Relatedly, Masefield has argued that there is a pressing need for CAS to engage with food justice discourse and the topic of human hunger and malnutrition.[73] Vegan just transition means considering both the consequences of sustainability transitions for nonhuman animals and the livelihoods of pre-existing famers of nonhuman animals,[74] imagining ways in which such actors can participate in vegan transition, while examining the diversity of agricultural contexts around the world. That Global South countries have been encouraged by transnational corporations and others to follow dietary meatification is a climate justice issue. Moving beyond animal-based food systems will require contextual strategies that understand regional differences and specificities in animal consumption and production.[75] There is an urgent need to empower and give voice to citizens in these countries on the interface between food, health, animal ethics and climate. This aspect is not radically different from what is required in the Global North, but as an issue of justice, Global South citizens must specifically be empowered to create food systems now that can assure food sovereignty, arrest their slide into high rates of diet-related illness and incorporate low emissions agriculture. Citizens in countries on the frontline of experiencing

---

72 Schanbacher 2019.
73 Masefield 2021. Relatedly, there is an ongoing debate over the use of dairy or non-dairy RUTFs (ready to use therapeutic foods) in the treatment of human malnutrition, see Welsh 2021. For an outline of the vegan food justice movement, see Murphy and Mook 2022.Global North
74 Understandably the temptation from a vegan perspective is to demonise such farmers. This is interesting especially in different visions of the downsizing and end of farmed animal agriculture whether that be in terms of quick sudden collapse or something more socially managed. An intersectional vegan perspective inevitably means acting with compassion to those in this sector and working toward just transition for all.
75 Morris, Kaljonen et al. 2021.

climate impacts would also be justified in calling for vegan transition in the Global North.

In the case of UK government policy, a proposed environmental land management scheme would be a small step in the right direction of changing the food system (albeit still within the constraints of the wrong economic model and shifting the identity of farmers because it would reward them for delivering environmental improvements. However, this is a missed opportunity to embed transitions from animal to vegan farming practice as argued for in a series of initiatives[76] and featured in the award-winning short film *73 Cows*,[77] which follows the sustainable transition of an English beef farmer to vegan farming. As highlighted in Chapter 4, the UK government has consistently ignored expert advice to instigate policy to even reduce meat/dairy production or consumption. Only a small number of countries so far have demonstrated leadership, with relevant policies including the Danish prioritisation of plant-based agriculture[78] and, relatedly, the Dutch decision to reduce its "livestock" numbers by a third.[79]

The prefigurative practices of intersectional veganism

Specific sorts of transitions are examples of prefiguration, experimental practices associated with, but not limited to, anarchism, socialism, feminism, veganism and queer politics, creating new social arrangements and institutions that attempt to model a future society.[80] Many attempts at sustainable eating (however defined are forms of prefiguration, but intersectional veganism has goals which aim for a

---

76 Examples include bundling veganism with agroecology (Morais, Teixeira et al. 2021), the New Economics Foundation *Grow Green* report commissioned by The Vegan Society (NEF 2017), the report *Planting Value in the Food System* by The Vegan Society (2021a) and the organisation Refarmd (https://en.refarmd.com/about).
77 https://vimeo.com/293352305.
78 Vegconomist 2021.
79 A decision based on the environmental problem of manure rather than greenhouse gas emissions, Levitt 2021.
80 Maeckelbergh 2011; Yates 2015; Törnberg 2021.

more socially transformative reinvention of social practices. This returns us to the focus of Chapters 6 and 7 on sociological theories of transition in which forms of prefiguration are those proto-practices, bundles and new complexes, or novel niches. In this light, prefigurative practices are specifically oppositional to incumbent practice orthodoxies. Just as practice theory acknowledges the contemporary visibility of the old in fossilised materialities, eroding competences and meanings losing their practitioners, everyday life also offers glimpses of the future (and certainly also experimental paths that ultimately will not be taken). Even though societies beyond capitalism seem difficult to imagine, let alone bring into being, their ingredients are, in this understanding, already partly present in the social milieu of today.

Yates identifies two main ways in which prefiguration has been used in the social movement literature both "to mean the building of movement 'alternatives' or institutions and those who take it to be a way in which protest is performed".[81] The latter understanding is important because the process of prefiguration is supposed to approximate the goal, for example, of new norms of inclusive participatory radical democracy, and this chimes well with the aforementioned "internal" role of movements (such as intersectional veganism) in disabling the resurfacing of pre-existing power norms in the course of politico-social-ethical experimentation. Following practice theory, the second sense of prefiguration as the building of alternatives could be framed as involving bundling together practices that embed new meanings, norms, identities, materialities and forms of know-how; and intersectional knowledge here is similarly vital. Vegan practice is an alternative experiment which does all these things, and intersectional veganism is its best hope for protection against capitalist and other co-option risks.

A further conceptual overlap for prefiguration and practice theory is Foucault's concept of heterotopia,[82] understood as "counter-sites, a

---

81 Yates 2015, 2.
82 A number of CAS researchers have started to use this concept. Most notably, Paula Arcari (2020, 2–8; 2023) whose practice theory inflected exploration of the cultural persistence of "meat" employs vegan heterotopia as a vantage point from which to understand how "normal spaces" comprise a network of

kind of effectively enacted utopia in which the real sites, all the other real sites that can be found within the culture, are simultaneously represented, contested, and inverted".[83] This opens the possibility to map out a heterotopology[84] of different spaces of resistance and inversion made possible by novel practice performances. Vegan heterotopia contests animal commodification, but other spaces further contest the commodification of food generally. It is within this convergence that the elements of a new food system that could speak to vegan climate justice and a union of intersectional veganism with food sovereignty may be glimpsed.

Although the scenario definitions I have used here preclude any supermarket space from being understood as vegan, from their commercial perspective supermarkets are involved in creating "vegan" spaces. These are constructed as those novel sections selling meat substitutes, rather than the fruit, vegetable or beans and pulses aisles, which if named as such, would force people to think of veganism as already familiar and partly practised. Earlier I noted the vegan café and restaurant, and now there are also vegan butchers. In my city a vegan café now occupies the site of a former hunting shop. Furthermore, the United Kingdom has a network of vegetarian and vegan food co-operatives, small businesses which sell sustainable food, including from local farms. Though such co-operatives strive to operate partly insulated from market norms, they are presently caught in a bind because constraints around pricing mean that their produce may be out of reach of disadvantaged communities. In the United Kingdom and Ireland there is also the vegan organic network of over 20 vegan farms.[85] Temporary event spaces also come and go such as vegan potlucks and larger vegan fairs.

In the aforementioned documentary *73 Cows*, the transition of a farmer to vegan agriculture involved all cows being taken in by an animal sanctuary. Sanctuaries too are vegan heterotopia, potentially

---

social practices that maintain the edibility of animals. I thank her for bringing the concept to my attention.

83   Foucault 1984, 3.
84   Foucault 1984, 4.
85   See https://veganorganic.net/uk-farms-directory/.

prefiguring future human–animal relations. Contrary to some assumptions,[86] vegan imaginaries are not necessarily "exterminatory" regarding farmed animals. While clear that further vegan normalisation is associated with vastly reduced national and global herd sizes and the important environmental, biodiversity and climate mitigation benefits that would bring, most vegans would arguably prefer a sustained small population of formerly farmed animal species, where appropriate.[87] Animal sanctuaries are potentially radical spaces in which practices of care, respect and communication can take place between humans and rescued animals. These novel affective relations demonstrate alternative human–animal relations to the commodified norm; nonhuman animals are afforded individual respect and care. Human volunteers or visitors experience the difference this makes to the well-being of the rescued and come to appreciate the value this interaction also has for their own mental health. That lived interaction with nonhuman animals, where appropriate, is mutually more rewarding than the violent commodified form is a hallmark of the hopeful CAS ethical critique of capitalism articulated earlier. Donaldson and Kymlicka have explored how sanctuaries might move beyond a refuge role toward a space for the exploration of novel relations and ways of being for the animals themselves. They envisage sanctuaries as intentional interspecies communities in which the agency of nonhuman animals participates in defining the space. Such a site can be normalised, they suggest,[88] via local partnerships with

---

86 Haraway (2008, 80) has opined on vegans that "their work to avoid eating or wearing any animal products would consign most domestic animals to the status of curated heritage collections or to just plain extermination as kinds and as individuals". At its worst this is an academic take on the much-lambasted omnivore question "what would we do with all the animals if nobody ate them?". It is also unclear why living as a "curated heritage collection" (as if that was the only option) would not be better experientially for animals compared to living short lives as soon to be killed commodities.

87 Where farmed animals have been bred beyond functional survival as in the case of broiler chickens this would arguably be inappropriate. I have explored such questions in more detail, for example in relation to ideas of rewilding (see Twine 2013a).

88 Donaldson and Kymlicka 2015, 68. For more on sanctuaries, see also Abrell 2021; Scotton 2017.

farmers, small business, education settings, special needs support structures or community gardens. Sanctuaries and micro-sanctuaries can also become spaces of learning, embedded and bundled into many other practices, more effectively escaping their resemblance to farms. The US micro-sanctuary movement is based around a principle of collective liberation, extending its focus to be inclusive of intra-human relations of power and demonstrating how it has been shaped by an iteration of intersectional veganism.[89]

In Chapter 6 the ability of practices to remake space was noted, with the social diffusion of practice seen as "their re-enactment in multiple sites".[90] Yet practice theorists also rightly recognise that practitioners differ in their access to space and in the degree to which space may act as a limitation.[91] This simple observation is a highly significant consideration for the potential of prefiguration, practice-bundling and embedding new ways of living, which gets to the heart of setting practice and prefiguration within their political economic context as well as understanding the ability of capitalism to protect the status quo. Staying with the example of sanctuaries, as one soon to be micro-sanctuary owner expressed it, "I had no plan to start a sanctuary; in my mind, that was something that people with access to a lot of money, land, and resources did. I barely had rent".[92] While part of the rationale of the *micro*-sanctuary movement is exactly to try to navigate these challenges, it is undoubtedly the case that access to land and money act to constrain the potential for sanctuaries and other vegan heterotopia to be "re-enacted in multiple sites". Their existence is a testimony to the determined political agency of practitioners rather than actual societal patronage.

Specific pro-vegan organisations campaigning to extend practice constitute a further example. In the United Kingdom this includes groups such as Viva! and The Vegan Society[93] and further afield

---

89 See https://microsanctuary.org/.
90 Shove, Pantzar and Watson 2012, 132.
91 Shove, Pantzar and Watson 2012, 131.
92 Moore 2021, 245.
93 Neither of these groups employ an explicit intersectional framing of veganism. This is regrettable because it excludes an analysis of the roots of animal exploitation and anthropocentrism as intersecting with classed,

organisations such as the Palestinian Animal League, an animal protection charity formed in 2011, or the California-based Food Empowerment Project, formed in 2007. The Palestinian Animal League focuses on community engagement, especially with young people in Palestine, specifically "seeing the goal of seeking justice for both people and animals as an interlinked challenge which can, and should, be tackled in tandem".[94] Its projects include providing an animal ambulance, helping stray animals, running a vegan not-for-profit café and vegan food tours. Meanwhile, the Food Empowerment Project is a vegan food justice organisation, and one of the few whose work meshes together shared interests of both intersectional veganism and food sovereignty. Their mission is to "encourage healthy food choices that reflect a more compassionate society by spotlighting the abuse of animals on farms, the depletion of natural resources, unfair working conditions for produce workers, and the unavailability of healthy foods in low-income areas".[95] They campaign around all these issues including, for example, a prominent food justice campaign on chocolate regarding the problem of child labour and slavery in its production. These sorts of campaigns speak to work on the decolonisation of dietary desire[96] wherein normalised Western diets bear the heavy imprint of colonialism be that in their meatification or in the human slavery-based embedding of high-processed sugar consumption. Just as land presently used for animal feed and pasture can be reclaimed, so can that used for sugar cultivation.[97] In this way, decolonising diet extends beyond animal consumption, a point that must be acknowledged by any alliance between intersectional veganism and food sovereignty. This underlines the aforementioned importance of intersectional knowledge to alternative practice prefigurations precisely as a counter to the role of class, gender and colonial histories in shaping diet. Thus, decolonising and degendering diet, and

---

gendered and racialised power relations. Yet both the Food Empowerment Project and the Palestinian Animal League are creative examples of intersectional veganism.

94   See https://pal.ps/en/about-us/.
95   See https://foodispower.org/mission-and-values/.
96   Jones 2010, 196; see Harper 2010 for a discussion about sugar.
97   Springmann and Freund 2022.

deconstructing social class habitus around eating practices, are indispensable strategies for both vegan climate justice and food transition broadly.

To vegan heterotopology, sites of food de-commodification may be added, as these can pivot veganism into possible post-capitalism. In among our heavily commodified food system exist spaces and practices where production and consumption attempt to exist outside the cash nexus. Examples, most familiar to the United Kingdom, include the practices of home-grown fruit and vegetables, food and seed sharing. CAS geographer Richard White has written of the radicality of the humble allotment as part of a potential post-capitalist foodscape[98] and community farms where one can volunteer in return for a share of the harvest are similar. Despite being a British institution, land devoted to allotments is surpassed tenfold by golf courses.[99] Far older practices such as gleaning, an entitlement of the poor to gather leftover food from farmers' fields, or contemporary practices such as dumpster diving,[100] highlight food waste and strive to resist food commodification. Sean Parson has analysed the latter in relation to the anarchist organisation Food not Bombs that emerged in the late 1980s in the United States, which reclaims supermarket food waste to feed those in need. Parson considers the possibility that groups like Food not Bombs are able to "undermine the commodity logic of food" and to move people's relationship with food "away from market relationships".[101] The recent emergence and normalisation of food banks in countries such as the United Kingdom as a response to rising levels of poverty, despite their often more formal relationship with supermarkets, also, in certain respects, constitute spaces of food de-commodification. A small number of vegan cafés combine a concern with food justice accessibility, employing a "pay what you want" policy. While not strict de-commodification, it is a creative way to introduce intersectional practice.

---

98   White 2018.
99   Shrubsole 2019.
100 In the United Kingdom this is termed "skipping".
101 Parson 2019, 409. Food not Bombs only serves vegetarian or vegan food.

These examples of vegan heterotopology that also can merge into practices of food de-commodification constitute acts of resistance and part prefigurations of an alternative food system. However, that the incumbent economic system acts as a check upon alternative practices is invitational toward intersectional veganism turning toward the politics of land ownership. This is not a big leap for veganism. For example, in the United Kingdom, land ownership is enmeshed in the history of aristocratic hunting estates. As Shrubsole underlines, "A staggering 550,000 acres [2225 square kilometres] of England is given over to grouse moors – an area of land the size of Greater London".[102] Private property is thus historically embedded in human exceptionalism and now prevents many niche-like innovations from becoming scaled up. It is telling, for example, that the large-scale rewilding plans for considerable swathes of Scotland[103] are only being made possible by a small number of wealthy landowners who see the ecological benefits of doing so. The politics of land ownership impinge on everything from the ability of people to grow their own food, the extension of animal sanctuaries, farm transition, reforestation and rewilding. It is little surprise that local food sustainability initiatives often involve attempts to create forms of community ownership so that land can be reclaimed for communal food production. Awareness of these issues is already present in organisations such as the Vegan Land Movement,[104] a UK community interest company which crowdfunds to buy land previously used for animal agriculture to convert for rewilding, community orchards or vegan organic growing. However, the familiar vegan argument that advocates land-use changes once land is no longer required to grow fodder crops or graze animals does tend to simplify this imagined transition without attending to the ways in which the political economy of land ownership may constrain it.

---

102 Shrubsole 2019, 57.
103 Carrell 2019. Rewilding raises a multitude of animal ethics issues which include the envisaged relationship to animal agriculture, the killing of deer and the orientation to "invasive" species. It is also a very contested term (see Carver, Convery et al. 2021) with regard to its degree of emphasis on the reintroduction of "lost" species. For a thoroughly capitalist attempt to extend rewilding, see https://www.realwildestates.com/.
104 https://globalvegancrowdfunder.org/vegan-land-movement-cic/.

These points on land and private property strengthen the case for intersectional veganism not just because the politics of land ownership speak thoroughly to the nexus of classed, gendered, racialised and anthropocentric relations, but because it appears unlikely that transitions to a low-carbon food system, and related conversions of land from animal agriculture to socially just forms of plant-based agriculture and types of rewilding, can be achieved without contesting the concentration of power in land ownership that has unfolded during the Capitalocene. Radical policy interventions are required here but again are rendered less likely due to the ideological climate of an anti-regulatory neoliberalism and because incumbent ruling governments tend to represent the interests of (and are comprised of) the property-owning classes. It is important to note not only that incumbent governments are unlikely to act fast enough to transform land ownership but that no government has shown the climate leadership to incentivise the diffusion of vegan practice in spite of its low-carbon credentials, noted in Chapter 4. Veganism finds itself caught up within a broader schism between just and real sustainable transitions and the reluctance of political leaders to advance particular solutions that do not fit with capitalistic frameworks of growth – but do actually entail changes to the way people live in the Global North.

This ideological reluctance was illustrated both internationally and in the United Kingdom in the run-up to and during the United Nations Framework Convention on Climate Change's Conference of the Parties in late 2021 (COP26). Just before the meeting, when the United Kingdom were presenting their "net-zero" plans, the government's own Behavioural Insights Unit[105] published a paper calling for a shift in dietary habits toward plant-based foods as well as other strategies such as reducing air travel via the promotion of domestic tourism. The paper had appeared on the website of the Department of Business but then had been "hastily deleted".[106] Thus the UK government has its own

---

105 Also known as the Nudge Unit due to being influenced by Thaler and Sunstein 2008. The paper was entitled "Net zero: principles for successful behaviour change initiatives" (Londakova, Park et al. 2021).

106 What was telling was the comment from the department's spokesperson, saying: "We have no plans whatsoever to dictate consumer behaviour in this

advisory research group that includes social scientists interested in policies that can shift practices and social norms, but it chose to ignore it.[107] Coming out of COP26 itself and speaking more to international policy, announcements on new voluntary agreements for the control of methane emissions and the cessation of deforestation managed to omit mention of the key role of the animal-industrial complex in both, thus avoiding decreased animal agriculture as a policy lever. In the United States, concerns were raised that then Agricultural Secretary Tom Vilsack, a former dairy industry lobbyist, had been enacting protectionist policies for animal agriculture.[108] Meanwhile, states with large animal agriculture interests such as Brazil and Argentina were accused of trying to water down the *Mitigation of Climate Change* section of the Intergovernmental Panel on Climate Change's Sixth Assessment Report (IPCC 2022) by seeking to have statements removed noting the climate benefits of promoting plant-based diets and of curbing meat and dairy consumption.[109] Governmental (and by implication corporate) influence on IPCC reports undermines their integrity. Given that COP26 failed in its goals to keep a 1.5°C target alive, it is astonishing that leaders ignored a mitigation strategy that would help to bend the emissions curve and also deliver human health benefits.

In this frustrating context of a backlash from interests of the animal-industrial complex it is not surprising, yet understandable, that intersectional vegans and the broader climate justice movement would refocus upon the need for a *system* change beyond capitalism. The vegan and de-commodified heterotopology discussed above is of course fragmentary, only a sample of practices for doing both the food system and the economy differently. They require anchoring within a

---

way. For that reason, our net zero strategy published yesterday contained no such plans", see Islam 2021.

107 Despite then UK Chief Scientist Patrick Vallance taking the view that behaviour change is needed to tackle the climate crisis, see Carrington 2021b.

108 Molidor 2021.

109 IPCC 2022. See Carter and Dowler 2021. The *Summary for Policymakers* section and later synthesis report of AR6 *were* watered down with prior recommendations for plant-based diets present in a leaked version of IPCC (2022) subsequently removed in the final version.

far broader set of post-capitalist theorising.[110] This can involve stronger alliances and identification with analyses such as the feminist and ecosocialist positions associated or allied to the Capitalocene framework and conversing with a broader range of contemporary work around alternative economic models. This encapsulates a set of approaches and ideas such as degrowth,[111] sufficiency transitions,[112] new ways of measuring quality of life, and the "diverse economies" paradigm.[113] It is within such alliances that a more compelling imaginary can be forged, one that could enrol growing numbers of practitioners, as people become increasingly aware of the failures of climate governance and market-led solutions.

This chapter has brought together and critically reflected upon three scenarios – plant-based capitalism, vegan transition and intersectional vegan transition. These have been presented as already identifiable scenarios that have been taken up by various groups of practitioners internationally, that have more or less linked degrees of practice change in human–animal relations to the task of addressing the climate and biodiversity crises. I have favoured the scenario of intersectional vegan for several reasons. The first two reasons it shares with scenario 2: its implied emissions cuts go further, and it partly understands these crises as arising from a contradiction in, and a crisis of, normative anthropocentrism. Thirdly, its theorisation of these crises, like the Capitalocene framework, embeds them within the unfolding of capitalism. Fourthly, in being faithful to historical understandings of veganism as never only about nonhuman animals, and in theorising the roots of animal exploitation within capitalism, colonialism and male dominance, it is uniquely placed to be in alliance with the climate justice movement. Finally, its intersectional focus is further important for situating the theorisation of transition and social change within overlapping entrenched meanings, materialities and infrastructures,

---

110 Although one would expect to see greater localisation of food production, I should not be read as implying a turn against scaled-up production or a partly globalised food system.
111 Amate and de Molina 2013; Hickel 2021.
112 Sandberg 2021.
113 Gibson-Graham 2008; Schmid and Smith 2021.

and so is preventative of a social-theoretical account that misunderstands "sustainable transition" as separable from relations of power.

In adding to an appreciation of the complexity of transition during the second part of this book, and in advocating for a scenario which may be theoretically accurate and politically inclusive but practically and politically ambitious, I have, some may contend, confirmed the prognosis of the climate crisis as not just a "wicked" problem but perhaps an intractable one. In the final concluding chapter, I reflect upon these prospects and summarise a potentially hopeful perspective on the climate crisis that has centred human–animal relations, and in so doing, has reflected sociologically on the creative possibilities for living both differently and better.

Conclusion: Unearthing hope from real pessimism in the alliances yet to be

To avoid a sense of nihilistic inertia, much writing on the climate crisis strives to conclude with an optimistic narrative. Yet false optimism in the face of evidence to the contrary is potentially as dangerous. Therefore, to be clear: there are unfortunately grounds for real pessimism. More positively, this book has highlighted a major opportunity for international policymakers to approach the emergency via transformative change to human–animal relations.

I started this book underlining how awareness of the climate crisis has accompanied its exacerbation. Along the way I have noted the antipathy of neoliberalism toward regulation and meaningful democracy. On a global level the geopolitical persistence of nationalisms and wars thwarts international cooperation. Our dominant economics can be judged by their performance and there is no evidence to suggest that capitalist organisation, which has remade ecologies in which the interconnected webs of life are being eroded, can urgently be adapted into a system that forsakes short-term accumulation in favour of climate and interspecies justice. Furthermore, the entrenched and interconnected relations of power within capitalism, operating along lines of gender, "race", class, generation and species, all mitigate against the transformational change required to address the climate and associated crises. The ability of capitalism to quickly instigate and normalise new practices is a key dimension of its world-making and a central contributor to the climate crisis. To think that this can simply be adapted to better greener practices and infrastructures is to assume they complement the accumulative rationale of the market or the social world it has helped to shape. Bringing the anthropocentrism of capitalism to the fore in this book and theorising the possibility of transition to alternatives has been an exercise in illustrating how human exceptionalism is embedded in the social and how the status quo is further protected by the politics of economic hegemony. When national leaders seem content for large swathes of their populations to live in poverty it could be fanciful to think that they would devise policy uniformly aimed at the welfare of future human populations let alone those of nonhuman animals. Given all this I would expect an intensifying need for mass mobilisation and civil disobedience due to the incumbency bias of nominally "democratic" societies.

Moreover, according to the Capitalocene framing there is a great cultural misunderstanding of the climate crisis at play in which it is barely historicised and the reasons for its emergences left nebulous. This book has added to critique of the Anthropocene narrative, which manages to obscure the relations of power inherent to the practices that have ultimately raised greenhouse gas emissions so sharply and panders to the capitalist realism of our times. Capitalism did not invent social inequality, racism, patriarchy or anthropocentrism, but it entrenched them and rendered them useful for capital accumulation. The first two chapters of this book drew out an affinity between critical animal studies (CAS) and the Capitalocene framework. This strengthens both and underlines a historicisation of the climate crisis as inextricably bound to intersecting unequal relations and practices, in which it can now be seen as a crisis of class relations, of racialised geopolitics, of hegemonic masculinity, and of anthropocentrism. Specifically, CAS situates the ecological crisis, in part, as a problem of "human" self-image in which the ongoing capitalist war against nonhuman animals and animality acts to perform and reiterate human exceptionalism while securing state and corporate hegemony. Retiring the Anthropocene concept and replacing it with the Capitalocene already provides intimations of hope in better understanding crises, but first it was necessary to detail, in the third chapter, the capitalogenic undermining of life.

This highlighted the numerous ways in which 1.2°C of warming is already catastrophic for animals (including homo sapiens) and is rightly framed as the beginning of a sixth mass extinction. Starkly confirming that narrations of the climate crisis that omit its experience for other animals are caught within the same cultural anthropocentrism at play within its emergence, the fourth chapter was centred on examining the emissions link between the animal-industrial complex and climate change. It should not be a great revelation that the global capitalist conversion of land over four times the size of the United States for use by animal agriculture has hugely destructive ecological consequences, of which the climate is but one. While the first part of the chapter had to contend with the anthropocentrism of the social sciences and humanities in their attempts to contribute to climate knowledge, the second half was concerned with the politics of scientific knowledge around the animal agriculture–climate change link. Important questions were raised over the work of the Food and Agriculture Organization of the United Nations on this topic, not least its protectionist stance toward the animal-industrial complex, given its work to also highlight its significant contribution to emissions. This protectionism was seen to be out of touch with a wide range of peer-reviewed science that has heralded a consensus over the need to reduce animal consumption for climate mitigation and other co-benefits. Despite there being a distance between this consensus and the focused intersectional critique of anthropocentrism found within ecofeminism and CAS, it at least signposts a more hopeful alternative future narrative based around changes to land use, biodiversity promotion and the transition to a low-carbon, and from this perspective, a plant-based or plant-centric food system. As far as largely anthropocentric social imaginaries go, the promise of a more sustainable food system that also would improve human health *ought* to capture the imagination of a broad range of "stakeholders".[1] This

---

1   One of the peculiarities of the present "developed" country food system is that governments seem broadly content to allow their populations to suffer the consequences of poor diet even if it affects national productivity and

describes what I named as the scenario 1 transition – a shift to more plant-based diets *within* the pre-existing rubric of capitalist organisation. This was contrasted with two further scenarios which foreground a critique of anthropocentrism – vegan transition and intersectional veganism – with only the latter explicitly oppositional to capitalism and primed for coalition with a broader climate justice politics.

After having demonstrated the significant role of human–animal relations in the emergences of the climate crisis, and how its effects are unequivocally already affecting nonhuman species, the shift into Part II of the book could be summarised as examining the *depth* of transformation required to properly address the climate crisis. Will an emphasis on reducing animal consumption (particularly in richer nations) satisfy the urgently required mitigations of greenhouse gas emissions? Or should the approach to the climate crisis be targeted at what this book understands as one of the key contributing dimensions, the embedded normalisation of cultural anthropocentrism and human sovereignty, including their broader imbrication within classed, racialised and gendered politics? This implies a considerably more radical critique of the animal-industrial complex and a goal of its abolition.

Certainly, this latter approach is that of the synergy outlined earlier between the Capitalocene framework and CAS. The orientation of Capitalocene (sympathetic) narratives has been to regard capitalism as interwoven with the web of life and dependent upon the cheapening of nonhuman animals and other species, and of dehumanised humans.[2] This perspective grounds sufficient scepticism toward something like scenario 1, that shifting to a more plant-based capitalism could be effective in tackling the climate crisis, especially when one aspect of the social reproduction of anthropocentric capitalism has been the broad cultural circulation of human exceptionalist meanings that make even scenario 1 challenging. Although it could be potentially self-defeating

---

competitiveness. Peculiar because a plant-based imaginary can conform with the biopolitical interests of the state.

2   Moore 2017, 600.

to abandon all developments taking place under its rubric, the danger is that if one scratches at the hope of scenario 1, it begins to look naïve.

In light of this, Part II was concerned with a detailed exploration of what transition to post-anthropocentric cultures would entail. Firstly, this turned to childhood studies to better understand how prevailing inconsistent meanings around human–animal relations are successfully reproduced between generations. That field's critique of childhood innocence mythologies was extended to highlight how they are protective of both anthropocentrism and, relatedly, the climate status quo. Chapter 5 argued that children should have the right to contest meat culture, contra the normalised imposition of a generational universalism of animal consumption. The climate crisis amplifies the "fleischgeist", centring the slaughterhouse as a site of violence, unwillingly disturbed from its space of sequestration. Moreover, it incites pre-existent discourses of children's rights and underlines the generational injustice at play as children are excluded from the political sphere that is presently curtailing their futures. Worse, they are still often excluded from adequate climate education and unlikely to be exposed to links between food and climate. Nevertheless, children resist, and vegan climate activist Greta Thunberg was identified as a pivotal killjoy figure, accentuating both the politicisation of childhood and the climate crisis. The school strike movement she initiated has also questioned the inadequacies of contemporary middle- to older-aged (male) adult-dominated "democracies". Critical animal pedagogy, which has emerged out of CAS, was presented as a vital educational philosophy for contesting cultural anthropocentrism, but one that remains challenging to establish in formal educational settings.

The focus then turned to social science theories of transition, especially practice theory as a promising approach because of its potential to transcend the agency/structure distinction that tends to limit other approaches. Theorising vegan transition using practice theory afforded a deconstructive approach to the practice which made clearer the elements of the practice that need to coexist for it to attract practitioners and become further societally embedded. That veganism involves a politics over the meanings of food is not surprising, but practice theory offers a systematic way in which to consider interventions on food meanings and how they connect different practices together. The remainder of Chapter 7 worked to reflect upon the ability of practice theory to conceptualise power and scale as a means to improve the theorisation of the animal-industrial complex, culminating in a suggested visualisation of the animal-industrial complex and a table of possible practice theory-inspired recommended interventions that could work toward its dismantling.

Transition theories however must be aware of the broader political economy invested in the status quo, be that fossil fuels or the animal-industrial complex. Therefore, the final chapter was also an exercise in outlining how various practices shaping the animal-industrial complex nationally and globally (e.g., lobbying, subsidising, financial investing) act to both scale up the practice theory understanding of social inertia and caution against overly techno-optimistic accounts of plant-based capitalism. Several sources of potential naivety in scenario 1 were identified such as the assumption that global meat corporations will voluntarily begin to downsize profitable operations and investments, or the lack of questioning within a plant-based capitalist paradigm of how the inequalities, deregulation, consumerism and growth imperatives of global capitalism all act to exacerbate a climate crisis it purports to help reverse.

For this reason, it was necessary to foreground intersectional vegan transition (scenario 3) as the imaginary that not only tackles anthropocentrism head on as an underlying reason for the climate crisis but understands it intersectionally, so to better be in alliance with a broader vision of climate justice. With regards to the global food system, it was argued that such an alliance necessitates a cessation to the commodifying violence perpetrated against nonhuman animals and a de-commodification of the food system itself. Here systemic change means that the goals of a food system must be planetary health, human health, and justice and care for nonhuman animal species. It is appropriate and unsurprising that the alliance of intersectional veganism and climate justice should, as argued in Chapter 8, return us to the politics of land, the accelerated expropriation of which partly constituted the " Age of Capital" after 1450.[3]

---

3   Moore 2017, 610.

That much transformative change seemed unlikely retrospectively is an important check on the sober recognition of pessimism becoming apathy. No strategy for tackling the climate crisis can avoid being seen as "hopelessly ambitious" and requiring profound, urgent change to globally interconnected societies. Such is the systemic nature of the problem. This is why the previous chapter culminated in discussion of alliances and prefiguration to demonstrate already existing hopeful practices that are both lived oppositions to anthropocentrism and novel ways of being human. Only after the "resolution" of the political struggles identified by CAS and the Capitalocene framework will it be known whether the seeds of the future are already being practised in the here and now, or whether insisting a critique of anthropocentrism as an unavoidable part of resolving the climate crisis is, instead, an exercise in writing epitaphs for paths that will never be fully taken.

Yet in arguing the need for post-anthropocentric transition, in which an element of the transcendence of capitalogenic climate change is found in a radical correction to the gendered fantasy of the human as existing above and apart from the rest of the bios, there is genuine hope and liberatory optimism. The imaginary of climate justice already understands that social inequalities are inseparable from an understanding of the emergences of the climate crisis and that its remedy cannot be reduced to a more efficiently managed capitalism but is dependent upon contesting the consumption of the rich, the self-entitlement of the Global North and the empathic atrophy of hegemonic masculinity. If this imaginary can find allyship with intersectional veganism then the potential for a significantly more powerful understanding of, and opposition to, the climate crisis comes into view.

It is beyond timely that kindred post-capitalist narratives, the progressive left (and beyond) take seriously both the questioning of human–animal relations and the systemic problem of the animal-industrial complex.[4] Put another way, the ethico-political questioning of human–animal relations insists itself, not as a panacea to what is a broad crisis complex, but nevertheless, as an obligatory component of contemporary liberatory and climate discourse. It is now

---

4   For such calls, see Dickstein, Dutkiewicz et al. 2022; Taylor and Taylor 2022.

untenable to claim that the cultural anthropocentrism entrenched during the Capitalocene is unrelated to the tragically increasing multispecies death toll as global heating escalates.

References

2 Degrees Institute (2022). Atmospheric CO_2 levels graph. 2 Degrees Institute. https://www.2degreesinstitute.org/.

Abdullah, S., C. Helps, S. Tasker, H. Newbury and R. Wall (2019). Pathogens in fleas collected from cats and dogs: distribution and prevalence in the UK. *Parasites and Vectors* 12(1): 71. https://doi.org/10.1186/s13071-019-3326-x.

Abdullah, S., C. Helps, S. Tasker, H. Newbury and R. Wall (2018). Prevalence and distribution of borrelia and babesia species in ticks feeding on dogs in the UK. *Medical and Veterinary Entomology* 32(1): 14–22. https://doi.org/10.1111/mve.12257.

Abdullah, S., C. Helps, S. Tasker, H. Newbury and R. Wall (2016). Ticks infesting domestic dogs in the UK: a large-scale surveillance programme. *Parasites and Vectors* 9(1): 391. https://doi.org/10.1186/s13071-016-1673-4.

Abrell, E. (2021). *Saving Animals: Multispecies Ecologies of Rescue and Care.* Minneapolis: University of Minnesota Press.

Acampora, R.R. (2006). *Corporal Compassion: Animal Ethics and Philosophy of Body.* Pittsburgh: University of Pittsburgh Press.

Acker, J. (1992). From sex roles to gendered institutions. *Contemporary Sociology* 21(5): 565–69. https://doi.org/10.2307/2075528.

Adams, C.J. (1994). *Neither Man nor Beast: Feminism and the Defense of Animals.* New York: Continuum.

Adams, C.J. (1990). *The Sexual Politics of Meat.* New York: Continuum.

Adams, M. (2016). *Ecological Crisis, Sustainability, and the Psychosocial Subject: Beyond Behaviour Change.* London: Palgrave.

Adelman, S. (2018). The Sustainable Development Goals, anthropocentrism and neoliberalism. In D. French and L.J. Kotzé, eds. *Sustainable Development Goals: Law, Theory and Implementation*, 15–40. Cheltenham: Edward Elgar.

Agamben, G. (1998). *Homo Sacer: Sovereign Power and Bare Life*. Stanford: Stanford University Press.

Ahmed, S. (2010a). Feminist killjoys (and other willful subjects). *Scholar and Feminist Online* 8(3). http://sfonline.barnard.edu/polyphonic/print_ahmed.htm.

Ahmed, S. (2010b). Killing joy: feminism and the history of happiness. *Signs: Journal of Women in Culture and Society* 35(3): 571–94. https://doi.org/10.1086/648513.

Ahmed, S. (2010c). *The Promise of Happiness*. Durham, NC: Duke University Press.

Ajzen, I. (1985). From intentions to actions: a theory of planned behavior. In J. Kuhl and J. Beckman, eds. *Action Control: From Cognition to Behavior*, 11–39. Heidelberg: Springer.

Alaimo, S. (2009). Insurgent vulnerability: masculinist consumerism, feminist activism, and the science of global climate change. *Women, Gender and Research* 3(3–4): 22–35.

Aleksandrowicz, L., R. Green, E.J.M. Joy, P. Smith and A. Haines (2016). The impacts of dietary change on greenhouse gas emissions, land use, water use, and health: a systematic review. *PLOS One* 11(11): e0165797. https://doi.org/10.1371/journal.pone.0165797.

Allister, M. (2004). *Eco-man: New Perspectives on Masculinity and Nature*. Charlottesville: University of Virginia Press.

Almiron, N. (2020a). Meat taboo: climate change and the EU meat lobby. In J. Hannan, ed. *Meatsplaining: The Animal Agriculture Industry and the Rhetoric of Denial*, 163–86. Sydney: Sydney University Press.

Almiron, N. (2020b). The "animal-based food taboo": climate change denial and deontological codes in journalism. *Frontiers in Communication* 5(96): 512956. https://doi.org/10.3389/fcomm.2020.512956.

Almiron, N., M. Rodrigo-Alsina and J.A. Moreno (2022). Manufacturing ignorance: think tanks, climate change and the animal-based diet. *Environmental Politics* 31(4): 576–97. https://doi.org/10.1080/09644016.2021.1933842.

Almiron, N. and M. Zoppeddu (2015). Eating meat and climate change: the media blind spot; a study of Spanish and Italian press coverage. *Environmental Communication* 9(3): 307–25. https://doi.org/10.1080/17524032.2014.953968.

Altizer, S., R.S. Ostfeld, P.T.J. Johnson, S. Kutz and C.D. Harvell (2013). Climate change and infectious diseases: from evidence to a predictive framework. *Science* 341(6145): 514–19. https://doi.org/10.1126/science.1239401.

References

Altvater, E. (2007). The social and natural environment of fossil capitalism. *Socialist Register* 43. https://socialistregister.com/index.php/srv/article/view/5857/2753.

Alvaro, C. (2020). Vegan parents and children: zero parental compromise. *Ethics and Education* 15(4): 476–98. https://tinyurl.com/2fjs2sk7.

Alvaro, C. (2019). Veganism and children: a response to Marcus William Hunt. *Journal of Agricultural and Environmental Ethics* 32(4): 647–61. https://doi.org/10.1007/s10806-019-09797-w.

Amate, J.I. and M.G. de Molina (2013). "Sustainable de-growth" in agriculture and food: an agro-ecological perspective on Spain's agri-food system (year 2000). *Journal of Cleaner Production* 38: 27–35. https://tinyurl.com/4sx9wtfn.

American Dietetic Association (2009). Position of the American Dietetic Association: vegetarian diets. *Journal of the American Dietetic Association* 109(7): 1266–82. https://doi.org/10.1016/j.jada.2009.05.027.

Anderson, M.J. and A.A. Thompson (2004). Multivariate control charts for ecological and environmental monitoring. *Ecological Applications* 14(6): 1921–35. https://doi.org/10.1890/03-5379.

Antal, M., G. Mattioli and I. Rattle (2020). Let's focus more on negative trends: a comment on the transitions research agenda. *Environmental Innovation and Societal Transitions* 34: 359–62. https://doi.org/10.1016/j.eist.2020.02.001.

Antonelli, A., C. Fry, R.J. Smith, M.S.J. Simmonds, P.J. Kersey, H.W. Pritchard et al. (2020). *State of the World's Plants and Fungi 2020*. London: Royal Botanic Gardens, Kew. https://doi.org/10.34885/172.

Anza, I., D. Vidal, J. Feliu, E. Crespo and R. Mateo (2016). Differences in the vulnerability of waterbird species to botulism outbreaks in Mediterranean wetlands: an assessment of ecological and physiological factors. *Applied and Environmental Microbiology* 82(10): 3092–99. https://doi.org/10.1128/AEM.00119-16.

Arcari, P. (2020). *Making Sense of "Food" Animals: A Critical Exploration of the Persistence of "Meat"*. London: Palgrave Macmillan.

Arcari, P. (2019). "Dynamic" non-human animals in theories of practice: views from the subaltern. In C. Maller and Y. Strengers, eds. *Social Practices and Dynamic Non-humans*, 63–86. London: Palgrave.

Arcari, P. (2017). Normalised, human-centric discourses of meat and animals in climate change, sustainability and food security literature. *Agriculture and Human Values* 34(1): 69–86. https://doi.org/10.1007/s10460-016-9697-0.

Ariès, P. (1962). *Centuries of Childhood*. London: Penguin.

Armitage, T.W.K., G.E. Manucharyan, A.A. Petty, R. Kwok and A.F. Thompson (2020). Enhanced eddy activity in the Beaufort Gyre in response to sea ice loss. *Nature Communications* 11(1): 761. https://tinyurl.com/54ex55zn.

Arnold, D.G., ed. (2011). *The Ethics of Global Climate Change*. Cambridge, UK: Cambridge University Press.

Aronoff, K. (2019). The Republican party is the political arm of the fossil fuel industry. *Guardian*, 27 March. https://tinyurl.com/kvcnadkv.

Arruzza, C., T. Bhattacharya and N. Fraser (2019). *Feminism for the 99%: A Manifesto*. London: Verso Books.

Atkins, L. and S. Michie (2013). Changing eating behaviour: what can we learn from behavioural science? *Nutrition Bulletin* 38(1): 30–35. https://doi.org/10.1111/nbu.12004.

Australian Associated Press (2018). Scott Morrison tells students striking over climate change to be "less activist". *Guardian*, 26 November. https://tinyurl.com/mr3pfa4k.

Australia's Natural History Museum Directors (2020). Joint Final Statement: Impact of fires on biodiversity on a scale not seen since species records were first kept – Loss is in the "trillions" of animals due to climate change crisis. 4 February. https://tinyurl.com/2tvvu23t.

Bailey, R., A. Froggatt and L. Wellesley (2014). *Livestock: Climate Change's Forgotten Sector*. London: Chatham House.

Baptista, R. (1999). The diffusion of process innovations: a selective review. *International Journal of the Economics of Business* 6(1): 107–29. https://doi.org/10.1080/13571519984359.

Bar-On, Y.M., R. Phillips and R. Milo (2018). The biomass distribution on Earth. *Proceedings of the National Academy of Sciences* 115(25): 6506–11. https://doi.org/10.1073/pnas.1711842115.

Barclay, E. and B. Resnick (2019). How big was the global climate strike? 4 million people, activists estimate. *Vox*, 22 September. https://tinyurl.com/5n7x242v.

Barnes, A. and R. White (2020). Mapping emotions: exploring the impact of the Aussie farms map. *Journal of Contemporary Criminal Justice* 36(3): 303–26. https://doi.org/10.1177/1043986220910306.

Baroni, L., M. Berati, M. Candilera and M. Tettamanti (2014). Total environmental impact of three main dietary patterns in relation to the content of animal and plant food. *Foods* 3(3): 443–60. https://doi.org/10.3390/foods3030443.

Bathiany, S., V. Dakos, M. Scheffer and T.M. Lenton (2018). Climate models predict increasing temperature variability in poor countries. *Science Advances* 4(5): eaar5809. https://doi.org/10.1126/sciadv.aar5809.

Bathiany, S., H. Dijkstra, M. Crucifix, V. Dakos, V. Brovkin, M.S. Williamson et al. (2016). Beyond bifurcation: using complex models to understand and predict abrupt climate change. *Dynamics and Statistics of the Climate System* 1(1): dzw004. https://doi.org/10.1093/climsys/dzw004.

Bauman, Z. (1989). *Modernity and the Holocaust*. Cambridge, UK: Polity Press.

References

Bavas, J. (2014). About 100,000 bats dead after heatwave in southern Queensland. *ABC News*, 8 January. https://tinyurl.com/4spfn95k.

BBC (2020). Australia's fires "killed or harmed three billion animals". *BBC News*, 28 July. https://tinyurl.com/2zbjjpy7.

Beardsell, A., G. Gauthier, D. Fortier, J.-F. Therrien and J. Bêty (2017). Vulnerability to geomorphological hazards of an Arctic cliff-nesting raptor, the rough-legged hawk. *Arctic Science* 3(2): 203–19. https://doi.org/10.1139/as-2016-0025.

Beck, U. (1992). *Risk Society: Towards a New Modernity*. London: Sage.

Bednaršek, N., G.A. Tarling, D.C.E. Bakker, S. Fielding, E.M. Jones, H.J. Venables et al. (2012). Extensive dissolution of live pteropods in the Southern Ocean. *Nature Geoscience* 5(12): 881–85. https://doi.org/10.1038/ngeo1635.

Beechey, V. (1977). Some notes on female wage labour in capitalist production. *Capital and Class* 3: 45–66.

Behrens, P., J.C. Kiefte-de Jong, T. Bosker, J.F.D. Rodrigues, A. de Koning and A. Tukker (2017). Evaluating the environmental impacts of dietary recommendations. *Proceedings of the National Academy of Sciences* 114(51): 13412–17. https://doi.org/10.1073/pnas.1711889114.

Benítez-López, A., R. Alkemade, A.M. Schipper, D.J. Ingram, P.A. Verweij, J.A.J. Eikelboom et al. (2017). The impact of hunting on tropical mammal and bird populations. *Science* 356(6334): 180–83. https://tinyurl.com/48tsja7a.

Bennett, C.E., R. Thomas, M. Williams, J. Zalasiewicz, M. Edgeworth, H. Miller et al. (2018). The broiler chicken as a signal of a human reconfigured biosphere. *Royal Society Open Science* 5(12): 180325. https://tinyurl.com/3xacc2f9.

Bennett, J. (2020). *Being Property Once Myself: Blackness and the End of Man*. Boston: Harvard University Press.

Bergmann, I.M. (2019). Interspecies sustainability to ensure animal protection: lessons from the thoroughbred racing industry. *Sustainability* 11(19): 5539. https://doi.org/10.3390/su11195539.

Berners-Lee, M., C. Hoolohan, H. Cammack and C.N. Hewitt (2012). The relative greenhouse gas impacts of realistic dietary choices. *Energy Policy* 43: 184–90. https://doi.org/10.1016/j.enpol.2011.12.054.

Berners-Lee, M., C. Kennelly, R. Watson and C.N. Hewitt (2018). Current global food production is sufficient to meet human nutritional needs in 2050 provided there is radical societal adaptation. *Elementa: Science of the Anthropocene* 6: 52. https://doi.org/10.1525/elementa.310.

Bernhardt, E.S., E.J. Rosi and M.O. Gessner (2017). Synthetic chemicals as agents of global change. *Frontiers in Ecology and the Environment* 15(2): 84–90. https://doi.org/10.1002/fee.1450.

Berteaux, D., G. Gauthier, F. Domine, R.A. Ims, S.F. Lamoureux, E. Lévesque et al. (2016). Effects of changing permafrost and snow conditions on tundra wildlife: critical places and times. *Arctic Science* 3(2): 65–90. https://doi.org/10.1139/as-2016-0023.

Best, S. (2014). *The Politics of Total Liberation: Revolution for the 21st Century.* London: Palgrave.

Best, S. (2009). The rise of critical animal studies: putting theory into action and animal liberation into higher education. *Journal for Critical Animal Studies* 7(1): 9–53.

Best, S., A. Nocella, R. Kahn, C. Gigliotti and L. Kemmerer (2007). Introducing critical animal studies. *Journal for Critical Animal Studies* 5(1): 4–5.

Bett, B., P. Kiunga, J. Gachohi, C. Sindato, D. Mbotha, T. Robinson et al. (2017). Effects of climate change on the occurrence and distribution of livestock diseases. *Preventive Veterinary Medicine* 137: 119–29. https://doi.org/10.1016/j.prevetmed.2016.11.019.

Bicchieri, C. (2017). *Norms in the Wild: How to Diagnose, Measure, and Change Social Norms.* Oxford: Oxford University Press.

Biermann, F., T. Hickmann, C.A. Sénit, M. Beisheim, S. Bernstein, P. Chasek et al. (2022). Scientific evidence on the political impact of the Sustainable Development Goals. *Nature Sustainability* 5: 795–800. https://doi.org/10.1038/s41893-022-00909-5.

Birtchnell, T. (2012). Elites, elements and events: practice theory and scale. *Journal of Transport Geography* 24: 497–502. https://tinyurl.com/mrrz7svf.

Biswas, T. and N. Mattheis (2022). Strikingly educational: a childist perspective on children's civil disobedience for climate justice. *Educational Philosophy and Theory* 54(2): 145–57. https://doi.org/10.1080/00131857.2021.1880390.

Bittel, J. (2020). "Extinct" toad rediscovery offers hope amid amphibian apocalypse. *National Geographic*, 28 April. https://tinyurl.com/56n9hn2h.

Blöschl, G., J. Hall, A. Viglione, R.A.P. Perdigão, J. Parajka, B. Merz et al. (2019). Changing climate both increases and decreases European river floods. *Nature* 573(7772): 108–11. https://doi.org/10.1038/s41586-019-1495-6.

Bodenstein, D. (2012). *Steven the Vegan.* Pennsauken, NJ: BookBaby.

Boisseron, B. (2018). *Afro-dog: Blackness and the Animal Question.* New York: Columbia University Press.

Bolam, F.C., L. Mair, M. Angelico, T.M. Brooks, M. Burgman, C. Hermes et al. (2021). How many bird and mammal extinctions has recent conservation action prevented? *Conservation Letters* 14(1): e12762. https://doi.org/10.1111/conl.12762.

Boonstra, R., K. Bodner, C. Bosson, B. Delehanty, E.S. Richardson, N.J. Lunn et al. (2020). The stress of Arctic warming on polar bears. *Global Change Biology* 26(8): 4197–214. https://doi.org/10.1111/gcb.15142.

Boscardin, L. (2018). Greenwashing the animal-industrial complex: sustainable intensification and the livestock revolution. In D.H. Constance, J.T. Konefal and M. Hatanaka, eds. *Contested Sustainability Discourses in the Agrifood System*, 111–26. London: Routledge.

Boscardin, L. (2017). Capitalizing on nature, naturalizing capitalism: an analysis of the "livestock revolution", planetary boundaries, and green tendencies in the animal-industrial complex. In D. Nibert, ed. *Animal Oppression and Capitalism*, vol. 1, 259–76. Santa Barbara: ABC-CLIO.

Boscardin, L. and L. Bossert (2015). Sustainable development and nonhuman animals: why anthropocentric concepts of sustainability are outdated and need to be extended. In S. Meisch, J. Lundershausen, L. Bossert and M. Rockoff, eds. *Ethics of Science in the Research for Sustainable Development*, 323–52. Baden-Baden: Nomos.

Both, C., S. Bouwhuis, C.M. Lessells and M.E. Visser (2006). Climate change and population declines in a long-distance migratory bird. *Nature* 441(7089): 81–83. https://doi.org/10.1038/nature04539.

Boutayeb, A. (2006). The double burden of communicable and non-communicable diseases in developing countries. *Transactions of the Royal Society of Tropical Medicine and Hygiene* 100(3): 191–99. https://tinyurl.com/54uec979.

Bowles, N., S. Alexander and M. Hadjikakou (2019). The livestock sector and planetary boundaries: a "limits to growth" perspective with dietary implications. *Ecological Economics* 160: 128–36. https://doi.org/10.1016/j.ecolecon.2019.01.033.

Braidotti, R. (2022). *Posthuman Feminism*. Cambridge, UK: Polity Press.

Brannen, P. (2019). The Anthropocene is a joke. *Atlantic*, 13 August. https://tinyurl.com/3e676znb.

Bratanova, B., S. Loughnan and B. Bastian (2011). The effect of categorization as food on the perceived moral standing of animals. *Appetite* 57(1): 193–96. https://doi.org/10.1016/j.appet.2011.04.020.

Braverman, I. (2015). En-listing life: red is the color of threatened species lists. In K. Gillespie and R.-C. Collard, eds. *Critical Animal Geographies: Politics, Intersections, and Hierarchies in a Multispecies World*, 184–202. London: Routledge.

Bray, H.J., S.C. Zambrano, A. Chur-Hansen and R.A. Ankeny (2016). Not appropriate dinner table conversation? Talking to children about meat production. *Appetite* 100: 1–9.

Bristow, E. and A. Fitzgerald (2011). Global climate change and the industrial animal agriculture link: the construction of risk. *Society and Animals* 19(3): 205–24. https://doi.org/10.1163/156853011X578893.

British Dietetic Association (2017). British Dietetic Association confirms well-planned vegan diets can support healthy living in people of all ages. 7 August. https://tinyurl.com/yn23d6ev.

Broome, J. (2012). *Climate Matters: Ethics in a Warming World*. New York: W.W. Norton & Company.

Brown, D. (2012). *Climate Change Ethics: Navigating the Perfect Moral Storm*. London: Routledge.

Bruckner, B., K. Hubacek, Y. Shan, H. Zhong and K. Feng (2022). Impacts of poverty alleviation on national and global carbon emissions. *Nature Sustainability* 5: 311–20. https://doi.org/10.1038/s41893-021-00842-z.

Bryngelsson, D., S. Wirsenius, F. Hedenus and U. Sonesson (2016). How can the EU climate targets be met? A combined analysis of technological and demand-side changes in food and agriculture. *Food Policy* 59: 152–64. https://doi.org/10.1016/j.foodpol.2015.12.012.

Buckwell, A. and E. Nadeu (2018). *What Is the Safe Operating Space for EU Livestock?* Brussels: RISE Foundation.

Bullard, R.D. (1990). *Dumping in Dixie: Race, Class, and Environmental Quality*. London: Routledge.

Büntgen, U., L. Greuter, K. Bollmann, H. Jenny, A. Liebhold, J.D. Galván et al. (2017). Elevational range shifts in four mountain ungulate species from the Swiss Alps. *Ecosphere* 8(4): e01761. https://doi.org/10.1002/ecs2.1761.

Burchell, K., R. Rettie and K. Patel (2013). Marketing social norms: social marketing and the "social norm approach". *Journal of Consumer Behaviour* 12(1): 1–9. https://doi.org/10.1002/cb.1395.

Burgen, S. (2022). "Poor meat and ill-treated animals": Spain in uproar over minister's remarks. *Guardian*, 7 January. https://tinyurl.com/2p8wc5y3.

Burgess, M.D., K.W. Smith, K.L. Evans, D. Leech, J.W. Pearce-Higgins, C.J. Branston et al. (2018). Tritrophic phenological match: mismatch in space and time. *Nature Ecology and Evolution* 2(6): 970–75. https://doi.org/10.1038/s41559-018-0543-1.

Burman, E. (2007). *Deconstructing Developmental Psychology*. London: Routledge.

Burwood-Taylor, L. (2021). ESG-stacked Oatly debuts at $13bn valuation, but execs say plant-based industry "still has work to do". *Agfunder News*, 20 May. https://tinyurl.com/yckhe43z.

Busby, M. (2019). Arron Banks jokes about Greta Thunberg and "freak yachting accidents". *Guardian*, 15 August. https://tinyurl.com/2p9xuznk.

References

Buttlar, B. and E. Walther (2018). Measuring the meat paradox: how ambivalence towards meat influences moral disengagement. *Appetite* 128: 152–58. https://doi.org/10.1016/j.appet.2018.06.011.

Caesar, L., S. Rahmstorf, A. Robinson, G. Feulner and V. Saba (2018). Observed fingerprint of a weakening Atlantic Ocean overturning circulation. *Nature* 556(7700): 191–96. https://doi.org/10.1038/s41586-018-0006-5.

Cairns, K.J. and J. Johnston (2018). On (not) knowing where your food comes from: meat, mothering and ethical eating. *Agriculture and Human Values* 35: 569–80. https://doi.org/10.1007/s10460-018-9849-5.

Cairns, V., C. Wallenhorst, S. Rietbrock and C. Martinez (2019). Incidence of Lyme disease in the UK: a population-based cohort study. *BMJ Open* 9(7): e025916. https://doi.org/10.1136/bmjopen-2018-025916.

Calarco, M.R. (2021). *Animal Studies: The Key Concepts*. London: Routledge.

Campbell, D. (2022). UK supermarkets accused of "bombarding" shoppers with cheap meat. *Guardian*, 26 March. https://tinyurl.com/yckzrvbd.

Campbell-Lendrum, D., L. Manga, M. Bagayoko and J. Sommerfeld (2015). Climate change and vector-borne diseases: what are the implications for public health research and policy? *Philosophical Transactions of the Royal Society B: Biological Sciences* 370(1665): 20130552. https://doi.org/10.1098/rstb.2013.0552.

Cardoso, P., P.S. Barton, K. Birkhofer, F. Chichorro, C. Deacon, T. Fartmann et al. (2020). Scientists' warning to humanity on insect extinctions. *Biological Conservation* 242: 108426. https://doi.org/10.1016/j.biocon.2020.108426.

Carleton, T.A., A. Jina, M.T. Delgado, M. Greenstone, T. Houser, S.M. Hsiang et al. (2020). *Valuing the Global Mortality Consequences of Climate Change Accounting for Adaptation Costs and Benefits*. Working paper 27599. Cambridge, MA: National Bureau of Economic Research. https://doi.org/10.3386/w27599.

Carlsson-Kanyama, A. and A.D. González (2009). Potential contributions of food consumption patterns to climate change. *American Journal of Clinical Nutrition* 89(5): 1704S–1709S. https://doi.org/10.3945/ajcn.2009.26736AA.

Carrell, S. (2019). Danish billionaires plan to rewild large swath of Scottish Highlands. *Guardian*, 21 March. https://tinyurl.com/yckkjrxw.

Carrington, D. (2022). Climate crisis: past eight years were the eight hottest ever, says UN. *Guardian*, 6 November. https://tinyurl.com/4vkvjr79.

Carrington, D. (2021a). Europe and US could reach "peak meat" in 2025. *Guardian*, 23 March. https://tinyurl.com/4ecsf8uf.

Carrington, D. (2021b). Changes in behaviour needed to tackle climate crisis, says UK chief scientist. *Guardian*, 9 November. https://tinyurl.com/83at2dmf.

Carrington, D. (2020). UK school and hospital caterers vow to cut meat served by 20%. *Guardian*, 16 April. https://tinyurl.com/5yasj69d.

Carrington, D. (2019). Why *The Guardian* is changing the language it uses about the environment. *Guardian*, 17 May. https://tinyurl.com/muyzhkrb.

Carrington, D. (2018). Avoiding meat and dairy is "single biggest way" to reduce your impact on earth. *Guardian*, 31 May. https://tinyurl.com/ycxvp74j.

Carter, L. and C. Dowler (2021). Leaked documents reveal the fossil fuel and meat producing countries lobbying against climate action. *Unearthed*, 21 October. https://tinyurl.com/2s3spvt9.

Carver, S., I. Convery, S. Hawkins, R. Beyers, A. Eagle, Z. Kun et al. (2021). Guiding principles for rewilding. *Conservation Biology* 35(6): 1882–93. https://doi.org/10.1111/cobi.13730.

Cass, N., T. Schwanen and E. Shove. (2018). Infrastructures, intersections and societal transformations. *Technological Forecasting and Social Change* 137: 160–67. https://doi.org/10.1016/j.techfore.2018.07.039.

Castañeda Camey, I., L. Sabater, C. Owren and A.E. Boyer (2020). *Gender-based Violence and Environment Linkages: The Violence of Inequality*. J. Wen, ed. Gland: International Union for Conservation of Nature.

Cavicchioli, R., W.J. Ripple, K.N. Timmis, F. Azam, L.R. Bakken, M. Baylis et al. (2019). Scientists' warning to humanity: microorganisms and climate change. *Nature Reviews Microbiology* 17(9): 569–86. https://doi.org/10.1038/s41579-019-0222-5.

CCC (Climate Change Committee) (2020a). *Reducing UK Emissions: 2020 Progress Report to Parliament*. London: Climate Change Committee.

CCC (Climate Change Committee) (2020b). *The Sixth Carbon Budget: The UK's Path to Net Zero*. London: Climate Change Committee.

CCC (Climate Change Committee) (2019). *Behaviour Change, Public Engagement and Net Zero: A Report for the UK Committee on Climate Change*. London: Climate Change Committee.

Ceballos, G., P.R. Ehrlich, A.D. Barnosky, A. García, R.M. Pringle and T.M. Palmer (2015). Accelerated modern human-induced species losses: entering the sixth mass extinction. *Science Advances* 1(5): e1400253. https://doi.org/10.1126/sciadv.1400253.

Ceballos, G., P.R. Ehrlich and R. Dirzo (2017). Biological annihilation via the ongoing sixth mass extinction signaled by vertebrate population losses and declines. *Proceedings of the National Academy of Sciences* 114(30): E6089–E6096. https://doi.org/10.1073/pnas.1704949114.

Ceballos, G., P.R. Ehrlich and P.H. Raven (2020). Vertebrates on the brink as indicators of biological annihilation and the sixth mass extinction.

Proceedings of the National Academy of Sciences 117(24): 13596–602. https://doi.org/10.1073/pnas.1922686117.

Celermajer, D., S. Chatterjee, A. Cochrane, S. Fishel, A. Neimanis, A. O'Brien, et al. (2020). Justice through a multispecies lens. *Contemporary Political Theory* 19: 475–512. https://doi.org/10.1057/s41296-020-00386-5.

Cenamor, R. and S. Brandt (2019). *Ecomasculinities: Negotiating Male Gender Identity in U.S. Fiction*. Lanham: Rowman & Littlefield.

Centola, D. (2011). An experimental study of homophily in the adoption of health behavior. *Science* 334(6060): 1269–72. https://tinyurl.com/5y3prs78.

Chai, B.C., J.R. van der Voort, K. Grofelnik, H.G. Eliasdottir, I. Klöss and F.J.A. Perez-Cueto (2019). Which diet has the least environmental impact on our planet? A systematic review of vegan, vegetarian and omnivorous diets. *Sustainability* 11(15): 4110. https://doi.org/10.3390/su11154110.

Chapman, D.J. (2011). Environmental education and the politics of curriculum: a national case study. *Journal of Environmental Education* 42(3): 193–202. https://doi.org/10.1080/00958964.2010.526153.

Chatham House (2021). *Food System Impacts on Biodiversity Loss: Three Levers for Food System Transformation in Support of Nature*. [Benton, T.G., C. Bieg, H. Harwatt, R. Pudasaini and L. Wellesley]. Research paper. London: Chatham House.

Chaves, L.F. (2017). Climate change and the biology of insect vectors of human pathogens. In S.N. Johnson and T.H. Jones, eds. *Global Climate Change and Terrestrial Invertebrates*, 126–47. New York: John Wiley & Sons. https://doi.org/10.1002/9781119070894.ch8.

Chemnitz, C. and S. Becheva, eds (2021). *Meat Atlas: Facts and Figures about the Animals We Eat*. Berlin: Heinrich Böll Foundation.

Chiles, R.M. (2013). If they come, we will build it: in vitro meat and the discursive struggle over future agrofood expectations. *Agriculture and Human Values* 30(4): 511–23. https://doi.org/10.1007/s10460-013-9427-9.

Chingono, N. and Z. Ndebele (2019). More than 200 elephants in Zimbabwe die as drought crisis deepens. *Guardian*, 12 November. https://tinyurl.com/vy662f8c.

Chivers, D. (2016). Cowspiracy: stampeding in the wrong direction? *New Internationalist*, 10 February. https://bit.ly/3SffuSd.

Chomsky, N. (2017). Neoliberalism is destroying our democracy. *Nation*, 2 June. https://bit.ly/4aQYbOx.

Christakis, N.A. and J.H. Fowler (2007). The spread of obesity in a large social network over 32 years. *New England Journal of Medicine* 357(4): 370–79. https://doi.org/10.1056/NEJMsa066082.

Clare, K., N. Maani and J. Milner (2022). Meat, money and messaging: how the environmental and health harms of red and processed meat consumption are

framed by the meat industry. *Food Policy* 109: 102234.
https://doi.org/10.1016/j.foodpol.2022.102234.

Clark, M.A., N.G.G. Domingo, K. Colgan, S.K. Thakrar, D. Tilman, J. Lynch et al. (2020). Global food system emissions could preclude achieving the 1.5° and 2°C climate change target. *Science* 370(6517): 705–8. https://doi.org/10.1126/science.aba7357.

Clay, N., A.E. Sexton, T. Garnett and J. Lorimer (2020). Palatable disruption: the politics of plant milk. *Agriculture and Human Values* 37(4): 945–62. https://doi.org/10.1007/s10460-020-10022-y.

Cohen, J.M., D.J. Civitello, M.D. Venesky, T.A. McMahon and J.R. Rohr (2019). An interaction between climate change and infectious disease drove widespread amphibian declines. *Global Change Biology* 25(3): 927–37. https://doi.org/10.1111/gcb.14489.

Cohen, M.J. (2011). Is the UK preparing for "war"? Military metaphors, personal carbon allowances, and consumption rationing in historical perspective. *Climatic Change* 104(2): 199–222. https://tinyurl.com/2xz97csd.

Cole, J.R. and S. McCoskey (2013). Does global meat consumption follow an environmental Kuznets curve? *Sustainability: Science, Practice and Policy* 9(2): 26–36. https://doi.org/10.1080/15487733.2013.11908112.

Cole, M. (2014). "The greatest cause on Earth": the historical formation of veganism as an ethical practice. In N. Taylor and R. Twine, eds. *The Rise of Critical Animal Studies: From the Margins to the Centre*, 223–44. London: Routledge.

Cole, M. and K. Morgan (2013). Engineering freedom? A critique of biotechnological routes to animal liberation. *Configurations* 21(2): 201–29. https://doi.org/10.1353/con.2013.0015.

Cole, M. and K. Morgan (2011). Vegaphobia: derogatory discourses of veganism and the reproduction of speciesism in UK national newspapers. *British Journal of Sociology* 62(1): 134–53. https://tinyurl.com/2scm4ea8.

Cole, M. and K. Stewart (2020). Socializing superiority: the cultural denaturalization of children's relations with animals. In A. Cutter-Mackenzie-Knowles, K. Malone and E.B. Hacking, eds. *Research Handbook on Childhoodnature: Assemblages of Childhood and Nature Research*, 1237–61. Heidelberg: Springer.

Cole, M. and K. Stewart (2017a). "A new life in the countryside awaits": interactive lessons in the rural utopia in "farming" simulation games. *Discourse: Studies in the Cultural Politics of Education* 38(3): 402–15. https://doi.org/10.1080/01596306.2017.1306985.

References

Cole, M. and K. Stewart (2017b). Speciesism party: a vegan critique of sausage party. *ISLE: Interdisciplinary Studies in Literature and Environment* 24(4): 767–86. https://doi.org/10.1093/isle/isx075.

Cole, M. and K. Stewart (2014). *Our Children and Other Animals: The Cultural Construction of Human-Animal Relations in Childhood.* London: Routledge.

Colling, S. (2020). *Animal Resistance in the Global Capitalist Era.* East Lansing: MSU Press.

Connell, R. (1995). *Masculinities.* Cambridge, UK: Polity Press.

Cooper, M.E. (2008). *Life as Surplus: Biotechnology and Capitalism in the Neoliberal Era.* Seattle: University of Washington Press.

Coppin, D. (2003). Foucauldian hog futures: the birth of mega-hog farms. *Sociological Quarterly* 44(4): 597–616. https://doi.org/10.1111/j.1533-8525.2003.tb00527.x.

Cox, D.T.C., I.M.D. Maclean, A.S. Gardner and K.J. Gaston (2020). Global variation in diurnal asymmetry in temperature, cloud cover, specific humidity and precipitation and its association with leaf area index. *Global Change Biology* 26(12): 7099–111. https://doi.org/10.1111/gcb.15336.

Cox, L. (2019). Beef industry linked to 94% of land clearing in Great Barrier Reef catchments. *Guardian*, 7 August. https://tinyurl.com/58rayuxf.

Crary, A. and L. Gruen (2022). *Animal Crisis: A New Critical Theory.* Cambridge, UK: Polity Press.

Crenshaw, K. (1991). Mapping the margins: intersectionality, identity politics, and violence against women of color. *Stanford Law Review* 43(6): 1241–99.

Crenshaw, K. (1989). Demarginalizing the intersection of race and sex: a black feminist critique of anti-discrimination doctrine, feminist theory and antiracist politics. *University of Chicago Legal Forum* 1(8): 139–67.

Crippa, M., E. Solazzo, D. Guizzardi, F. Monforti-Ferrario, F.N. Tubiello and A.J.N.F Leip (2021). Food systems are responsible for a third of global anthropogenic GHG emissions. *Nature Food* 2(3): 198–209. https://doi.org/10.1038/s43016-021-00225-9.

Crist, E. (2018). Reimagining the human. *Science* 362(6420): 1242–44. https://doi.org/10.1126/science.aau6026.

Crist, E. (2013). On the poverty of our nomenclature. *Environmental Humanities* 3(1): 129–47. https://doi.org/10.1215/22011919-3611266.

Cronon, W. (1991). *Nature's Metropolis: Chicago and the Great West.* New York: W.W. Norton & Company.

Crutzen, P. (2002). Geology of mankind. *Nature* 415: 23. https://doi.org/10.1038/415023a.

Crutzen, P. and E. Stoermer (2000). The "Anthropocene". *Global Change Newsletter* 41: 17–18.

Cudworth, E. (2015). Killing animals: sociology, species relations and institutionalized violence. *Sociological Review* 63(1): 1–18.

Cusack, T., ed. (2014). *Framing the Ocean, 1700 to the Present: Envisaging the Sea as Social Space*. London: Routledge.

Cutter-Mackenzie-Knowles, A., K. Malone and E.B. Hacking, eds (2020). *Research Handbook on Childhoodnature: Assemblages of Childhood and Nature Research*. Heidelberg: Springer.

Dankelman, I., ed. (2010). *Gender and Climate Change: An Introduction*. London: Routledge.

Dankelman, I. (2002). Climate change: learning from gender analysis and women's experiences of organising for sustainable development. *Gender and Development* 10(2): 21–29. https://doi.org/10.1080/13552070215899.

Darst, R.G. and J.I. Dawson (2019). Putting meat on the (classroom) table: problems of denial and communication. In T. Lloro-Bidart and V.S. Banschbach, eds. *Animals in Environmental Education: Interdisciplinary Approaches to Curriculum and Pedagogy*, 215–36. London: Palgrave.

Dasgupta, P. (2021). *The Economics of Biodiversity: The Dasgupta Review*. London: HM Treasury.

David, M. and N. Stephens Griffin (2021). Mediating a global capitalist, speciesist moral vacuum: how two escaped pigs disrupted Dyson Appliances' state of nature. *Journal for Critical Animal Studies* 18(1): 73–99.

Davidson, E.A. (2012). Representative concentration pathways and mitigation scenarios for nitrous oxide. *Environmental Research Letters* 7(2): 024005. https://doi.org/10.1088/1748-9326/7/2/024005.

Davies, C. (2021). Vegans "muddying" debate with bogus climate change arguments: "another agenda". *Express*, 30 December. https://tinyurl.com/2rr7xwkx.

Davis, J. (2005). Educating for sustainability in the early years: creating cultural change in a child-care setting. *Australian Journal of Environmental Education* 21: 47–55. https://doi.org/10.1017/S081406260000094X.

Davis, J. and S. Elliot, eds (2014). *Research in Early Childhood Education for Sustainability: International Perspectives and Provocations*. London: Routledge.

Davis, M., S. Faurby and J-C. Svenning (2018). Mammal diversity will take millions of years to recover from the current biodiversity crisis. *Proceedings of the National Academy of Sciences* 115(44): 11262–67. https://doi.org/10.1073/pnas.1804906115.

De Backer, C., S. Erreygers, C. De Cort, F. Vandermoere, A. Dhoest and J. Vrinten (2020). Meat and masculinities: can differences in masculinity predict meat consumption, intentions to reduce meat and attitudes towards vegetarians? *Appetite* 147: 104559. https://doi.org/10.1016/j.appet.2019.104559.

De Groot, R.S., P. Ketner and A.H. Ovaa (1995). Selection and use of bio-indicators to assess the possible effects of climate change in Europe. *Journal of Biogeography* 22(4–5): 935–43. https://doi.org/10.2307/2845994.

De-Shalit, A. (1994). *Why Posterity Matters: Environmental Policies and Future Generations*. London: Routledge.

De Vos, J.M., L.N. Joppa, J.L. Gittleman, P.R. Stephens and S.L. Pimm (2015). Estimating the normal background rate of species extinction. *Conservation Biology* 29(2): 452–62. https://doi.org/10.1111/cobi.12380.

Dedehayir, O., R. Ortt, C. Riverola and F. Miralles (2017). Innovators and early adopters in the diffusion of innovations: a literature review. *International Journal of Innovation Management* 21(8): 1–27.

Deeply Good Magazine (2018). George Monbiot proposes new language for environmental protection. *Deeply Good Magazine*, 6 November. https://tinyurl.com/mvdvh5db.

DEFRA (Department for Environment, Food and Rural Affairs, UK) (2013). *Sustainable Consumption Report: Follow-up to the Green Food Project*. London: DEFRA.

Denton, F. (2002). Climate change vulnerability, impacts, and adaptation: why does gender matter? *Gender and Development* 10(2): 10–20. https://doi.org/10.1080/13552070215903.

Derrida, J. (2008). *The Animal That Therefore I Am*. New York: Fordham University Press.

Dewey, J. (1916). *Democracy and Education: An Introduction to the Philosophy of Education*. London: MacMillan.

Dhont, K., G. Hodson, S. Loughnan and C.E. Amiot (2019). Rethinking human–animal relations: the critical role of social psychology. *Group Processes and Intergroup Relations* 22(6): 769–84. https://doi.org/10.1177/1368430219864455.

Di Chiro, G. (2017). Welcome to the white (m)Anthropocene? A feminist-environmentalist critique. In S. MacGregor, ed. *Routledge Handbook of Gender and Environment*, 487–504. London: Routledge.

Díaz, S., J. Settele, E.S. Brondízio, H.T. Ngo, J. Agard, A. Arneth et al. (2019). Pervasive human-driven decline of life on Earth points to the need for transformative change. *Science* 366(6471): eaax3100. https://doi.org/10.1126/science.aax3100.

Dickstein, J., J. Dutkiewicz, J. Guha-Majumdar and D.R. Winter (2022). Veganism as left praxis. *Capitalism Nature Socialism* 33(3): 56–75. https://doi.org/10.1080/10455752.2020.1837895.

Dimbleby, H. (2021). *The National Food Strategy: The Plan*. London: National Food Strategy.

Dinker, K.G. (2021). Critical creatures: children as pioneers of posthuman pedagogies. *International Journal of Sociology and Social Policy* 41(3–4): 391–406. https://doi.org/10.1108/IJSSP-10-2020-0464.

Dinker, K.G. and H. Pedersen (2019). Critical animal pedagogy: explorations toward reflective practice. In A. Nocella, C. Drew and A.E. George, eds. *Education for Total Liberation: Critical Animal Pedagogy and Teaching against Speciesism*, 47–64. New York: Peter Lang US.

Dinker, K.G. and H. Pedersen (2016). Critical animal pedagogies: re-learning our relations with animal others. In H.E. Lees and N. Noddings, eds. *The Palgrave International Handbook of Alternative Education*, 415–30. London: Palgrave Macmillan UK.

Donaldson, S. and W. Kymlicka (2015). Farmed animal sanctuaries: the heart of the movement. *Politics and Animals* 1(1): 50–74.

Donnelly, A., A. Caffarra and B.F. O'Neill (2011). A review of climate-driven mismatches between interdependent phenophases in terrestrial and aquatic ecosystems. *International Journal of Biometeorology* 55(6): 805–17. https://doi.org/10.1007/s00484-011-0426-5.

Donovan, J. and C.J. Adams, eds (2007). *The Feminist Care Tradition in Animal Ethics: A Reader*. New York: Columbia University Press.

Dow, K. and T.E. Downing (2011). *The Atlas of Climate Change: Mapping the World's Greatest Challenge*. Abingdon: Earthscan.

Doyle, J. (2011). *Mediating Climate Change*. Farnham: Ashgate Publishing.

Dryzek, J.S., R.B. Norgaard and D. Schlosberg (2011). *The Oxford Handbook of Climate Change and Society*. Oxford: Oxford University Press.

Duhn, I. (2012). Making "place" for ecological sustainability in early childhood education. *Environmental Education Research* 18(1): 19–29. https://doi.org/10.1080/13504622.2011.572162.

Duignan, P.J., N.S. Stephens and K. Robb (2020). Fresh water skin disease in dolphins: a case definition based on pathology and environmental factors in Australia. *Scientific Reports* 10(1): 21979. https://doi.org/10.1038/s41598-020-78858-2.

Dunayer, J. (2001). *Animal Equality: Language and Liberation*. Derwood: Ryce.

Dunlap, R.E. and R.J. Brulle, eds (2015). *Climate Change and Society: Sociological Perspectives*. Oxford: Oxford University Press.

DuPuis, E.M. (2002). *Nature's Perfect Food: How Milk Became America's Drink*. New York: NYU Press.

Dutkiewicz, J. and G. Rosenberg (2021). Labriculture now. *Logic* 13. 17 May. https://logicmag.io/distribution/labriculture-now/.

References

Dyke, J., R. Watson and W. Knorr (2021). Climate scientists: concept of net zero is a dangerous trap. *The Conversation*, 22 April. https://theconversation.com/climate-scientists-concept-of-net-zero-is-a-dangerous-trap-157368.

Earnshaw, G.I. (1999). Equity as a paradigm for sustainability: evolving the process toward interspecies equity. *Animal Law* 5: 113–46.

Earthsight (2017). World's largest beef producer fined after buying cattle raised on illegally deforested land in Brazil. 27 March. https://tinyurl.com/yrbav364.

Edelman, L. (2004). *No Future: Queer Theory and the Death Drive*. Durham, NC: Duke University Press.

Edenhofer O., R. Pichs-Madruga, Y. Sokona, S. Kadner, J.C. Minx, S. Brunner et al. (2014). *Technical Summary*. In: *Climate Change 2014: Mitigation of Climate Change. Contribution of Working Group III to the Fifth Assessment Report of the Intergovernmental Panel on Climate Change* [Edenhofer, O., R. Pichs-Madruga, Y. Sokona, E. Farahani, S. Kadner, K. Seyboth, et al. (eds)]. Cambridge, UK: Cambridge University Press.

Eisen, M.B. and P.O. Brown (2022). Rapid global phaseout of animal agriculture has the potential to stabilize greenhouse gas levels for 30 years and offset 68 percent of CO_2 emissions this century. *PLOS Climate* 1(2): e0000010. https://doi.org/10.1371/journal.pclm.0000010.

Eker, S., G. Reese and M. Obersteiner (2019). Modelling the drivers of a widespread shift to sustainable diets. *Nature Sustainability* 2(8): 725–35. https://doi.org/10.1038/s41893-019-0331-1.

El Bilali, H. (2019). The multi-level perspective in research on sustainability transitions in agriculture and food systems: a systematic review. *Agriculture* 9(4): 74. https://doi.org/10.3390/agriculture9040074.

Elgin, B. (2021). Beef industry tries to erase its emissions with fuzzy methane math. *Bloomberg*, 19 October. https://tinyurl.com/yck7z3hv.

Elias, N. (1969). *The Civilizing Process, vol. 1, The History of Manners*. Oxford: Blackwell.

Ellis, E., M. Maslin, N. Boivin and A. Bauer (2016). Involve social scientists in defining the Anthropocene. *Nature* 540(7632): 192–93. https://doi.org/10.1038/540192a.

Erb, K.-H., C. Lauk, T. Kastner, A. Mayer, M.C. Theurl and H. Haberl (2016). Exploring the biophysical option space for feeding the world without deforestation. *Nature Communications* 7(1): 11382. https://doi.org/10.1038/ncomms11382.

ETC Group (2017). *Who Will Feed Us? The Industrial Food Chain vs. the Peasant Food Web*. Val David, QC: ETC.

European Commission (2020). *Farm to Fork Strategy: For a Fair, Healthy and Environmentally Friendly Food System*. Brussels: EC Group.

Evans, B. and H. Johnson (2021). Contesting and reinforcing the future of "meat" through problematization: analyzing the discourses in regulatory debates around animal cell-cultured meat. *Geoforum* 127: 81–91. https://doi.org/10.1016/j.geoforum.2021.10.001.

FAO (Food and Agriculture Organization of the United Nations) (2022). *GLEAM 3.0: Model Description* Rome: FAO. https://tinyurl.com/2p9f6scv.

FAO (Food and Agriculture Organization of the United Nations) (2020). Forest loss slows in South America, protected areas rise. 7 May. https://tinyurl.com/38fxchz8.

FAO (Food and Agriculture Organization of the United Nations) (2017). *GLEAM 2.0: Assessment of Greenhouse Gas Emissions and Mitigation Potential*. Rome: FAO. https://tinyurl.com/yc3rtdy5.

FAO (Food and Agriculture Organization of the United Nations) (2016). World meat production is projected to double by 2050. 27 May. https://tinyurl.com/8959bbwr.

Fazioli, K. and V. Mintzer (2020). Short-term effects of Hurricane Harvey on bottlenose dolphins (*Tursiops truncatus*) in Upper Galveston Bay, TX. *Estuaries and Coasts* 43(5): 1013–31. https://doi.org/10.1007/s12237-020-00751-y.

FEC (Food Ethics Council) (2009). *Livestock Consumption and Climate Change*. London: FEC.

FEC and WWF (Food Ethics Council and World Wildlife Fund) (2013). *Prime Cuts: Valuing the Meat We Eat*. London: FEC and WWF-UK.

Federici, S. (2004). *Caliban and the Witch: Women, the Body and Primitive Accumulation*. London: Penguin.

Feedback (2020). *Butchering the Planet: The Big-name Financiers Bankrolling Livestock Corporations and Climate Change*. London: Feedback.

Fischer, C.G. and T. Garnett (2016). *Plates, Pyramids, and Planets: Developments in National Healthy and Sustainable Dietary Guidelines; A State of Play Assessment*. Oxford: Food and Agriculture Organization of the United Nations and Food Climate Research Network.

Fisher, M. (2009). *Capitalist Realism: Is There No Alternative?* Winchester, UK: John Hunt Publishing.

Fitzgerald, A.J. (2018). *Animal Advocacy and Environmentalism: Understanding and Bridging the Divide*. New York: John Wiley & Sons.

Fitzgerald A.J., L. Kalof and T. Dietz (2009). Slaughterhouses and increased crime rates: an empirical analysis of the spillover from "the jungle" into the surrounding community. *Organization and Environment* 22(2): 158–84. https://doi.org/10.1177/1086026609338164.

References

Foley, J.A., N. Ramankutty, K.A. Brauman, E.S. Cassidy, J.S. Gerber, M. Johnston et al. (2011). Solutions for a cultivated planet. *Nature* 478(7369): 337–42. https://doi.org/10.1038/nature10452.

Foresight (2011). *The Future of Food and Farming: Final Project Report*. London: Government Office for Science.

Foster, J.B. (2009). *The Ecological Revolution: Making Peace with the Planet*. New York: Monthly Review Press.

Foster, J.B. (2000). *Marx's Ecology: Materialism and Nature*. New York: Monthly Review Press.

Foster, J.B. and B. Clark (2018). Women, nature, and capital in the industrial revolution. *Monthly Review*, 1 January. https://bit.ly/3RYTCZZ.

Foster, J.B., B. Clark and R. York (2008). Ecology: the moment of truth; an introduction. *Monthly Review* 60(3): 1–11.

Foster, J.B., R. York and B. Clark (2010). *The Ecological Rift: Capitalism's War on the Earth*. New York: Monthly Review Press.

Foucault, M. (1991) [1978]. Governmentality. In G. Burchell, C. Gordon and P. Miller, eds. *The Foucault Effect: Studies in Governmentality*, 87–104. G. Burchell and C. Gordon, trans. London: Harvester Wheatsheaf.

Foucault, M. (1990) [1978]. *The History of Sexuality*, vol. 1, *The Will to Knowledge*. R. Hurley, trans. London: Penguin.

Foucault, M. (1984) [1967]. Of other spaces: utopias and heterotopias. J. Miskowiec, trans. *Architecture, Mouvement, Continuité* 5: 1–9.

Franklin, E. (2020). Acts of killing, acts of meaning: an application of corpus pattern analysis to language of animal-killing. Unpublished doctoral dissertation. Lancaster University.

Franklin, S. (2007). *Dolly Mixtures: The Remaking of Genealogy*. Durham, NC: Duke University Press.

Franklin, S. (2003). Ethical biocapital: new strategies of stem cell culture. In S. Franklin and M. Lock, eds. *Remaking Life and Death: Towards an Anthropology of Biomedicine*, 97–128. Santa Fe: SAR Press.

Fraser, N. (2019). *The Old Is Dying and the New Cannot Be Born: From Progressive Neoliberalism to Trump and Beyond*. London: Verso Books.

Fraser, N. and R. Jaeggi (2018). *Capitalism: A Conversation in Critical Theory*. New York: John Wiley & Sons.

Freeman, B.G., J.A. Lee-Yaw, J.M. Sunday and A.L. Hargreaves (2018). Expanding, shifting, and shrinking: the impact of global warming on species' elevational distributions. *Global Ecology and Biogeography* 27(11): 1268–76. https://doi.org/10.1111/geb.12774.

Freeman, C.P. (2010). Meat's place on the campaign menu: how US environmental discourse negotiates vegetarianism. *Environmental Communication* 4(3): 255–76. https://doi.org/10.1080/17524032.2010.501998.

Frehner, B. (2011). *Finding Oil: The Nature of Petroleum Geology, 1859–1920*. Lincoln: University of Nebraska Press.

Freire, P. (2017). *Pedagogy of the Oppressed*. London: Penguin Classics.

Friedlander, J., C. Riedy and C. Bonfiglioli (2014). A meaty discourse: what makes meat news? *Food Studies: An Interdisciplinary Journal* 3(3): 27–43.

Friel, S., A.D. Dangour, T. Garnett, K. Lock, Z. Chalabi, I. Roberts et al. (2009). Public health benefits of strategies to reduce greenhouse-gas emissions: food and agriculture. *Lancet* 374(9706): 2016–25. https://doi.org/10.1016/S0140-6736(09)61753-0.

Frommel, A.Y., R. Maneja, D. Lowe, A.M. Malzahn, A.J. Geffen, A. Folkvord et al. (2012). Severe tissue damage in Atlantic cod larvae under increasing ocean acidification. *Nature Climate Change* 2(1): 42–46. https://doi.org/10.1038/nclimate1324.

Fulton, G.R. (2017). The Bramble Cay melomys: the first mammalian extinction due to human-induced climate change. *Pacific Conservation Biology* 23(1): 1. https://doi.org/10.1071/PCv23n1_ED.

Fung, S.-C. (2017). Canine-assisted reading programs for children with special educational needs: rationale and recommendations for the use of dogs in assisting learning. *Educational Review* 69(4): 435–50. https://doi.org/10.1080/00131911.2016.1228611.

Gaard, G. (2015). Ecofeminism and climate change. *Women's Studies International Forum* 49: 20–33.

Gaard, G., ed. (1993). *Ecofeminism: Women, Animals, Nature*. Philadelphia: Temple University Press.

Gallardo, B. and D.C. Aldridge (2013). Evaluating the combined threat of climate change and biological invasions on endangered species. *Biological Conservation* 160: 225–33. https://doi.org/10.1016/j.biocon.2013.02.001.

Gander, K. (2017). The vegan parents who are bringing up meat and dairy-free children. *Independent*, 18 May. https://tinyurl.com/y8dxw9m4.

Garcia, D., V. Galaz and S. Daume (2019). EAT-*Lancet* vs Yes2meat: the digital backlash to the planetary health diet. *Lancet* 394(10215): 2153–54. https://doi.org/10.1016/S0140-6736(19)32526-7.

Gardiner, S.M., S. Caney, D. Jamieson and H. Shue, eds (2010). *Climate Ethics: Essential Readings*. Oxford: Oxford University Press.

Gardiner, S.M. and D.A. Weisbach (2016). *Debating Climate Ethics*. Oxford: Oxford University Press.

Garlen, J.C. (2019). Interrogating innocence: "childhood" as exclusionary social practice. *Childhood* 26(1): 54–67. https://doi.org/10.1177/0907568218811484.

Garner, R. (2013). *A Theory of Justice for Animals: Animal Rights in a Nonideal World*. Oxford: Oxford University Press.

Garnett, T. (2009). Livestock-related greenhouse gas emissions: impacts and options for policy makers. *Environmental Science and Policy* 12(4): 491–503. https://doi.org/10.1016/j.envsci.2009.01.006.

Garnett, T. (2008). *Cooking Up a Storm: Food, Greenhouse Gas Emissions and Our Changing Climate*. University of Surrey: Food Climate Research Network.

Garnett, T., M.C. Appleby, A. Balmford, I.J. Bateman, T.G. Benton, P. Bloomer et al. (2013). Sustainable intensification in agriculture: premises and policies. *Science* 341(6141): 33–34. https://doi.org/10.1126/science.1234485.

Garvey, J. (2008). *The Ethics of Climate Change: Right and Wrong in a Warming World*. New York: Continuum.

Geanous, J. (2019). Vegan parents "starved toddler to death after feeding him only raw fruit and vegetables". *Metro*, 14 November. https://tinyurl.com/565ujrnz.

Geels, F.W. (2011). The multi-level perspective on sustainability transitions: responses to seven criticisms. *Environmental Innovation and Societal Transitions* 1(1): 24–40. https://doi.org/10.1016/j.eist.2011.02.002.

Geels, F.W. and J. Schot (2010). The dynamics of transitions: a socio-technical perspective. In J. Grin, J. Rotmans and J. Schot, eds. *Transitions to Sustainable Development: New Directions in the Study of Long-term Transformative Change*, 9–101. London: Routledge.

Gerber, P.J., H. Steinfeld, B. Henderson, A. Mottet, C. Opio, J. Dijkman et al. (2013). *Tackling Climate Change through Livestock: A Global Assessment of Emissions and Mitigation Opportunities*. Rome: Food and Agriculture Organization of the United Nations.

Ghosh, A. (2016). *The Great Derangement: Climate Change and the Unthinkable*. Chicago: University of Chicago Press.

Gibson-Graham, J.K. (2008). Diverse economies: performative practices for "other worlds". *Progress in Human Geography* 32(5): 613–32. https://doi.org/10.1177/0309132508090821.

Giddens, A. (2009). *The Politics of Climate Change*. Cambridge, UK: Polity Press.

Giddens, A. (1984). *The Constitution of Society: Outline of the Theory of Structuration*. Cambridge, UK: Polity Press.

Giordano, C. (2019). Police investigate "homophobic graffiti" at Birmingham school amid row over LGBT+ sex education. *Independent*, 8 March. https://tinyurl.com/eupprmry.

Giraud, E.H. (2021). *Veganism: Politics, Practice, and Theory*. London: Bloomsbury.

Giroux, H.A. (2011). *On Critical Pedagogy*. New York: Continuum.

Global Witness (2021). *Last Line of Defence: The Industries Causing the Climate Crisis and Attacks against Land and Environmental Defenders*. London: Global Witness.

Godfray, H.C.J., P. Aveyard, T. Garnett, J.W. Hall, T.J. Key, J. Lorimer et al. (2018). Meat consumption, health, and the environment. *Science* 361(6399): eaam5324. https://doi.org/10.1126/science.aam5324.

Godoy, J. and S. Valle (2020). Parent of Brazil's JBS pleads guilty to U.S. foreign bribery charges. *Reuters*, 14 October. https://tinyurl.com/2p82nddr.

Goldstein, H. and B. Majsa (2019). Greta Thunberg: "We need to change the system". *Feministeerium*, 14 March. https://tinyurl.com/bd6pccus.

González, A.D., B. Frostell and A. Carlsson-Kanyama (2011). Protein efficiency per unit energy and per unit greenhouse gas emissions: potential contribution of diet choices to climate change mitigation. *Food Policy* 36(5): 562–70. https://doi.org/10.1016/j.foodpol.2011.07.003.

Goodfellow, M. (2019). Put our colonial history on the curriculum – then we'll understand who we really are. *Guardian*, 5 December. https://tinyurl.com/26f74xv2.

Goodland, R. and J. Anhang (2012). Response to "Livestock and greenhouse gas emissions: the importance of getting the numbers right". *Animal Feed Science and Technology* 3(172): 252–56. https://tinyurl.com/2hwjeea2.

Goodland, R. and J. Anhang (2009). *Livestock and Climate Change: What If the Key Actors in Climate Change Are Cows, Pigs, and Chickens?* Washington, DC: Worldwatch.

Goodman, D., E.M. DuPuis and M.K. Goodman (2012). *Alternative Food Networks: Knowledge, Practice, and Politics*. London: Routledge.

Gore, M., J. Ratsimbazafy, A. Rajaonson, A. Lewis and J. Kahler (2016). Public perceptions of poaching risks in a biodiversity hotspot: implications for wildlife trafficking interventions. *Journal of Trafficking, Organized Crime, and Security* 2: 1–20.

Gössling, S. and S. Cohen (2014). Why sustainable transport policies will fail: EU climate policy in the light of transport taboos. *Journal of Transport Geography* 39: 197–207. https://doi.org/10.1016/j.jtrangeo.2014.07.010.

Gough, I. (2017). *Heat, Greed and Human Need: Climate Change, Capitalism and Sustainable Wellbeing*. Cheltenham: Edward Elgar.

Gould, S.J. (1981). *The Mismeasure of Man*. Harmondsworth: Penguin.

GRAIN and IATP (Institute for Agriculture and Trade Policy) (2018). *Emissions Impossible: How Big Meat and Dairy Are Heating Up the Planet*. Barcelona: GRAIN.

References

Grassian, D.T. (2020). The dietary behaviors of participants in UK-based meat reduction and vegan campaigns: a longitudinal, mixed-methods study. *Appetite* 154: 104788. https://doi.org/10.1016/j.appet.2020.104788.

Grau-Sologestoa, I. and U. Albarella (2019). The "long" sixteenth century: a key period of animal husbandry change in England. *Archaeological and Anthropological Sciences* 11: 2781–803. https://doi.org/10.1007/s12520-018-0723-6.

Greenpeace (2020a). *Farming for Failure: How European Animal Farming Fuels the Climate Emergency*. Brussels: Greenpeace.

Greenpeace (2020b). *Winging It: How the UK's Chicken Habit Is Fuelling the Climate and Nature Emergency*. London: Greenpeace UK.

Greenpeace (2019). *Feeding the Problem: The Dangerous Intensification of Animal Farming in Europe*. Brussels: Greenpeace.

Greenspoon, L., E. Krieger, R. Sender and R. Milo (2023). The global biomass of wild mammals. *Proceedings of the National Academy of Sciences* 120(10): e2204892120. https://doi.org/10.1073/pnas.2204892120.

Grin, J., J. Rotmans and J. Schot, eds (2010). *Transitions to Sustainable Development: New Directions in the Study of Long-term Transformative Change*. London: Routledge.

Gruen, L. (2015). *Entangled Empathy: An Alternative Ethic for Our Relationships with Animals*. New York: Lantern Books.

Gruen, L. (2011). *Ethics and Animals: An Introduction*. Cambridge, UK: Cambridge University Press.

Grusin, R., ed. (2017). *Anthropocene Feminism*. Minneapolis: University of Minnesota Press.

Gunderman, H. and R. White (2020). Critical posthumanism for all: a call to reject insect speciesism. *International Journal of Sociology and Social Policy* 41(3–4): 489–505. https://doi.org/10.1108/IJSSP-09-2019-0196.

Gunderson, R. (2013). From cattle to capital: exchange value, animal commodification, and barbarism. *Critical Sociology* 39(2): 259–75. https://doi.org/10.1177/0896920511421031.

Gunderson, R. (2011). The metabolic rifts of livestock agribusiness. *Organization and Environment* 24(4): 404–22. https://doi.org/10.1177/1086026611424764.

Gunderson, R. and D. Stuart (2014). Industrial animal agribusiness and environmental sociological theory. *International Journal of Sociology* 44(1): 54–74. https://doi.org/10.2753/IJS0020-7659440104.

Haas, T. (2021). From green energy to the green car state? The political economy of ecological modernisation in Germany. *New Political Economy* 26(4): 660–73. https://doi.org/10.1080/13563467.2020.1816949.

Haines, A., A.J. McMichael, K.R. Smith, I. Roberts, J. Woodcock, A. Markandya et al. (2009). Public health benefits of strategies to reduce greenhouse-gas emissions: overview and implications for policy makers. *Lancet* 374(9707): 2104–14. https://doi.org/10.1016/S0140-6736(09)61759-1.

Hallmann, C.A., M. Sorg, E. Jongejans, H. Siepel, N. Hofland, H. Schwan et al. (2017). More than 75 percent decline over 27 years in total flying insect biomass in protected areas. *PLOS One* 12(10): e0185809. https://doi.org/10.1371/journal.pone.0185809.

Hallström, E., A. Carlsson-Kanyama and P. Börjesson (2015). Environmental impact of dietary change: a systematic review. *Journal of Cleaner Production* 91: 1–11. https://doi.org/10.1016/j.jclepro.2014.12.008.

Hamilton, C.D., J. Vacquié-Garcia, K.M. Kovacs, R.A. Ims, J. Kohler and C. Lydersen (2019). Contrasting changes in space use induced by climate change in two Arctic marine mammal species. *Biology Letters* 15(3): 20180834. https://doi.org/10.1098/rsbl.2018.0834.

Hamilton, I., H. Kennard, A. McGushin, L. Höglund-Isaksson, G. Kiesewetter, M. Lott et al. (2021). The public health implications of the Paris Agreement: a modelling study. *Lancet Planetary Health* 5(2): e74–e83. https://doi.org/10.1016/S2542-5196(20)30249-7.

Haraway, D. (2015). Anthropocene, Capitalocene, Plantationocene, Chthulucene: making kin. *Environmental Humanities* 6: 159–65.

Haraway, D. (2008). *When Species Meet*. Minneapolis: University of Minnesota Press.

Haraway, D., N. Ishikawa, S.F. Gilbert, K. Olwig, A.L. Tsing and N. Bubandt (2016). Anthropologists are talking about the Anthropocene. *Ethnos* 81(3): 535–64. https://doi.org/10.1080/00141844.2015.1105838.

Hards, S. (2011). Social practice and the evolution of personal environmental values. *Environmental Values* 20(1): 23–42. https://doi.org/10.3197/096327111X12922350165996.

Hargreaves, T., N. Longhurst and G. Seyfang (2012). *Understanding Sustainability Innovations: Points of Intersection between the Multi-level Perspective and Social Practice Theory*. 3S working paper. Norwich: University of East Anglia.

Harper, A.B. (2010). Social justice beliefs and addictions to uncompassionate consumption: food for thought. In A.B. Harper, ed. *Sistah Vegan: Black Female Vegans Speak on Food Identity, Health, and Society*, 20–41. New York: Lantern.

Hartley, D. (2016). Anthropocene, Capitalocene and the problem of culture. In J.W. Moore, ed. *Anthropocene or Capitalocene? Nature, History, and the Crisis of Capitalism*, 154–65. Oakland: PM Press.

Harvey, D. (2015). *Seventeen Contradictions and the End of Capitalism*. London: Profile Books.

References

Harvey, D. (2005). *A Brief History of Neoliberalism*. Oxford: Oxford University Press.

Harvey, D. (2002). *Spaces of Capital: Towards a Critical Geography*. London: Routledge.

Harvey, F. (2020). The national curriculum barely mentions the climate crisis. Children deserve better. *Guardian*, 11 February. https://tinyurl.com/yurn7a7v.

Harvey, F. (2018). "Tipping points" could exacerbate climate crisis, scientists fear. *Guardian*, 9 October. https://tinyurl.com/yjv3u4b4.

Harwatt, H. (2019). Including animal to plant protein shifts in climate change mitigation policy: a proposed three-step strategy. *Climate Policy* 19(5): 533–41. https://doi.org/10.1080/14693062.2018.1528965.

Harwatt, H., J. Sabaté, G. Eshel, S. Soret and W. Ripple (2017). Substituting beans for beef as a contribution toward US climate change targets. *Climatic Change* 143(1): 261–70. https://doi.org/10.1007/s10584-017-1969-1.

Hayek, M.N., H. Harwatt, W.J. Ripple and N.D. Mueller (2021). The carbon opportunity cost of animal-sourced food production on land. *Nature Sustainability* 4(1): 21–24. https://doi.org/10.1038/s41893-020-00603-4.

Hayhow, D.B., M.A. Eaton, A.J. Stanbury, F. Burns, W.B. Kirby, N. Bailey et al. (2019). *State of Nature 2019*. State of Nature Partnership.

He, F., C. Zarfl, V. Bremerich, J.N.W. David, Z. Hogan, G. Kalinkat et al. (2019). The global decline of freshwater megafauna. *Global Change Biology* 25(11): 3883–92. https://doi.org/10.1111/gcb.14753.

Headey, D., K. Hirvonen and J. Hoddinott (2018). Animal sourced foods and child stunting. *American Journal of Agricultural Economics* 100(5): 1302–19. https://doi.org/10.1093/ajae/aay053.

Hedenus, F., S. Wirsenius and D.J.A. Johansson (2014). The importance of reduced meat and dairy consumption for meeting stringent climate change targets. *Climatic Change* 124(1): 79–91. https://doi.org/10.1007/s10584-014-1104-5.

Henderson, J. (2020). EVs are not the answer: a mobility justice critique of electric vehicle transitions. *Annals of the American Association of Geographers* 110(6): 1993–2010. https://doi.org/10.1080/24694452.2020.1744422.

Henley, J. (2020). Climate crisis could displace 1.2bn people by 2050, report warns. *Guardian*, 9 September. https://tinyurl.com/mrjuk5mz.

Henley, J. (2001). Calls for legal child sex rebound on luminaries of May 68. *Guardian*, 24 February. https://tinyurl.com/4rf6ja2y.

Hennion, A. (2004). Pragmatics of taste. In M. Jacobs and N. Hanrahan, eds. *The Blackwell Companion to the Sociology of Culture*, 131–44. Oxford: Blackwell.

Henry, R.C., P. Alexander, S. Rabin, P. Anthoni, M.D.A. Rounsevell and A. Arneth (2019). The role of global dietary transitions for safeguarding biodiversity.

Global Environmental Change 58: 101956.
https://doi.org/10.1016/j.gloenvcha.2019.101956.

Herrero, M., P. Gerber, T. Vellinga, T. Garnett, A. Leip, C. Opio et al. (2011). Livestock and greenhouse gas emissions: the importance of getting the numbers right. *Animal Feed Science and Technology* 166–67: 779–82. https://doi.org/10.1016/j.anifeedsci.2011.04.083.

Herrero, M., B. Henderson, P. Havlík, P.K. Thornton, R.T. Conant, P. Smith et al. (2016). Greenhouse gas mitigation potentials in the livestock sector. *Nature Climate Change* 6(5): 452–61. https://doi.org/10.1038/nclimate2925.

Herrero, M., P.K. Thornton, P. Gerber and R.S. Reid (2009). Livestock, livelihoods and the environment: understanding the trade-offs. *Current Opinion in Environmental Sustainability* 1(2): 111–20. https://doi.org/10.1016/j.cosust.2009.10.003.

Heuer, R.M. and M. Grosell (2014). Physiological impacts of elevated carbon dioxide and ocean acidification on fish. *American Journal of Physiology-Regulatory, Integrative and Comparative Physiology* 307(9): R1061–R1084. https://doi.org/10.1152/ajpregu.00064.2014.

Hickel, J. (2021). What does degrowth mean? A few points of clarification. *Globalizations* 18(7): 1105–1111. https://tinyurl.com/yxv4kk6m.

Hickman, C. (2019). A psychotherapist explains why some adults are reacting badly to young climate strikers. *The Conversation*, 11 October. https://tinyurl.com/yc8eshf3.

Higgins, R. (1994). Race, pollution, and the mastery of nature. *Environmental Ethics* 16(3): 251–64. https://doi.org/10.5840/enviroethics199416315.

Higginson, S., E. McKenna, T. Hargreaves, J. Chilvers and M. Thomson (2015). Diagramming social practice theory: an interdisciplinary experiment exploring practices as networks. *Indoor and Built Environment* 24(7): 950–69. https://doi.org/10.1177/1420326X15603439.

Hitch, G. (2020a). Bushfire royal commission: "Black Summer" played out exactly as scientists predicted it would. *ABC News*, 25 May. https://tinyurl.com/2s3mkmmu.

Hitch, G. (2020b). Bushfire royal commission hears that Black Summer smoke killed nearly 450 people. *ABC News*, 26 May. https://tinyurl.com/3aexe2v2.

Holmberg, A. and A. Alvinius (2020). Children's protest in relation to the climate emergency: a qualitative study on a new form of resistance promoting political and social change. *Childhood* 27(1): 78–92. https://doi.org/10.1177/0907568219879970.

hooks, b. (1994). *Teaching to Transgress: Education as the Practice of Freedom*. London: Routledge.

Horrocks, R. (1994). Masculinity in crisis. *Self and Society* 22(4): 25–29.

References

Hribal, J. (2010). *Fear of the Animal Planet: The Hidden History of Animal Resistance*. Oakland: CounterPunch and AK Press.

Hsu, A., G. Sheriff, T. Chakraborty and D. Manya (2021). Disproportionate exposure to urban heat island intensity across major US cities. *Nature Communications* 12(1): 2721. https://doi.org/10.1038/s41467-021-22799-5.

Hubeau, M., F. Marchand, I. Coteur, K. Mondelaers, L. Debruyne and G. Van Huylenbroeck (2017). A new agri-food systems sustainability approach to identify shared transformation pathways towards sustainability. *Ecological Economics* 131: 52–63.

Hui, A., T. Schatzki and E. Shove, eds (2017). *The Nexus of Practices: Connections, Constellations, Practitioners*. London: Routledge.

Hultman, M. (2017). Exploring industrial, ecomodern, and ecological masculinities. In S. MacGregor, ed. *Routledge Handbook of Gender and Environment*, 239–52. London: Routledge.

Hultman, M. and P. Pulé (2019). Ecological masculinities: a response to the Manthropocene question? In L. Gottzén, U. Mellström, T. Shefer and T. Grimbeek, eds. *Routledge International Handbook of Masculinity Studies*, 477–87. London: Routledge.

Hultman, M. and P. Pulé (2018). *Ecological Masculinities: Theoretical Foundations and Practical Guidance*. London: Routledge.

Humphreys, A.M., R. Govaerts, S.Z. Ficinski, E.N. Lughadha and M.S. Vorontsova (2019). Global dataset shows geography and life form predict modern plant extinction and rediscovery. *Nature Ecology and Evolution* 3(7): 1043–47. https://doi.org/10.1038/s41559-019-0906-2.

Hunnicutt, G. (2020). *Gender Violence in Ecofeminist Perspective: Intersections of Animal Oppression, Patriarchy, and Domination of the Earth*. London: Routledge.

Hunt, M.W. (2019). Veganism and children: physical and social well-being. *Journal of Agricultural and Environmental Ethics* 32(2): 269–91. https://doi.org/10.1007/s10806-019-09773-4.

Hussar, K.M. and P.L. Harris (2010). Children who choose not to eat meat: a study of early moral decision-making. *Social Development* 19(3): 627–41. https://doi.org/10.1111/j.1467-9507.2009.00547.x.

IDC (International Development Committee) (2013). *Global Food Security Report*. London: IDC.

Iknayan, K.J. and S.R. Beissinger (2018). Collapse of a desert bird community over the past century driven by climate change. *Proceedings of the National Academy of Sciences* 115(34): 8597–602. https://doi.org/10.1073/pnas.1805123115.

Institute for Agriculture and Trade Policy (2020). *Milking the Planet: How Big Dairy Is Heating Up the Planet and Hollowing Rural Communities.* Minneapolis, MN: Institute for Agriculture and Trade Policy.

IPBES (Intergovernmental Science-Policy Platform on Biodiversity and Ecosystem Services) (2020). *Workshop Report on Biodiversity and Pandemics of the Intergovernmental Platform on Biodiversity and Ecosystem Services.* Daszak, P., das Neves, C., Amuasi, J., Hayman, D., Kuiken, T., Roche, B., et al. Bonn: IPBES secretariat. https://doi.org/10.5281/zenodo.4147317.

IPBES (Intergovernmental Science-Policy Platform on Biodiversity and Ecosystem Services) (2019). *Global Assessment Report on Biodiversity and Ecosystem Services of the Intergovernmental Science-Policy Platform on Biodiversity and Ecosystem Services.* Bonn: IPBES secretariat. https://doi.org/10.5281/zenodo.3831673.

IPCC (Intergovernmental Panel on Climate Change) (2022). Climate change 2022: mitigation of climate change. In P.R. Shukla, J. Skea, R. Slade, A. Al Khourdajie, R. van Diemen, D. McCollum, et al. eds. *Contribution of Working Group III to the Sixth Assessment Report of the Intergovernmental Panel on Climate Change.* Cambridge, UK: Cambridge University Press. https://doi.org/10.1017/9781009157926.

IPCC (Intergovernmental Panel on Climate Change) (2019). *Climate Change and Land: An IPCC Special Report on Climate Change, Desertification, Land Degradation, Sustainable Land Management, Food Security, and Greenhouse Gas Fluxes in Terrestrial Ecosystems.* Geneva, Switzerland: IPCC.

IPCC (Intergovernmental Panel on Climate Change) (2018). *Global Warming of 1.5°C: An IPCC Special Report on the Impacts of Global Warming of 1.5°C above Pre-industrial Levels and Related Global Greenhouse Gas Emission Pathways, in the Context of Strengthening the Global Response to the Threat of Climate Change, Sustainable Development, and Efforts to Eradicate Poverty.* Geneva, Switzerland: IPCC.

IPCC (Intergovernmental Panel on Climate Change) (2007). *Contribution of Working Group III to the Fourth Assessment Report of the Intergovernmental Panel on Climate Change.* Geneva, Switzerland: IPCC.

Ipsos MORI (2019). *Vegan Society Poll.* London: Ipsos MORI.

Ipsos MORI (2016). *Vegan Society Poll.* London: Ipsos MORI.

Islam, F. (2021). Climate plan urging plant-based diet shift deleted. *BBC News*, 20 October. https://tinyurl.com/mry5vd84.

Ivanovich, C.C., T. Sun, D.R. Gordon and I.B. Ocko. (2023). Future warming from global food consumption. *Nature Climate Change* 13: 297–302. https://doi.org/10.1038/s41558-023-01605-8.

References

Jackson, T. (2009). *Prosperity without Growth: Economics for a Finite Planet.* London: Routledge.

Jansen, E., J.H. Christensen, T. Dokken, K.H. Nisancioglu, B.M. Vinther, E. Capron et al. (2020). Past perspectives on the present era of abrupt Arctic climate change. *Nature Climate Change* 10(8): 714–21. https://doi.org/10.1038/s41558-020-0860-7.

Jenkins, S. and R. Twine (2014). On the limits of food autonomy: rethinking choice and privacy. In N. Taylor and R. Twine, eds. *The Rise of Critical Animal Studies: From the Margins to the Centre*, 225–40. London: Routledge.

Johnson, K. (2020). Vegan school dinners to be served at 180 Leeds schools to help tackle climate change. *LeedsLive*, 20 January. https://tinyurl.com/yc7ht4xw.

Johnson, S.N. and T.H. Jones, eds (2017). *Global Climate Change and Terrestrial Invertebrates.* New York: John Wiley & Sons.

Jonassen, W. (2012). 7 years after Katrina, New Orleans is overrun by wild dogs. *Atlantic*, 24 August. https://tinyurl.com/ykduckfn.

Jones, P. (2010). Afterword: liberation as connection and the decolonization of desire. In A.B. Harper, ed. *Sistah Vegan: Black Female Vegans Speak on Food Identity, Health, and Society*, 187–201. New York: Lantern.

Jones, T., L.M. Divine, H. Renner, S. Knowles, K.A. Lefebvre, H.K. Burgess et al. (2019). Unusual mortality of tufted puffins (*Fratercula cirrhata*) in the Eastern Bering Sea. *PLOS One* 14(5): e0216532. https://doi.org/10.1371/journal.pone.0216532.

Jowit, J. (2008). UN says eat less meat to curb global warming. *Observer*, 7 September. https://tinyurl.com/3dkm4fsc.

Joy, M. (2010). *Why We Love Dogs, Eat Pigs, and Wear Cows: An Introduction to Carnism.* Newburyport: Conari Press.

Joyce, A., J. Hallett, T. Hannelly and G. Carey (2014). The impact of nutritional choices on global warming and policy implications: examining the link between dietary choices and greenhouse gas emissions. *Energy and Emission Control Technologies* 2: 33–43. https://doi.org/10.2147/EECT.S58518.

Kagan, S. (2016). What's wrong with speciesism? *Journal of Applied Philosophy* 33(1): 1–21. https://doi.org/10.1111/japp.12164.

Kahn, R. (2008). From education for sustainable development to ecopedagogy: sustaining capitalism or sustaining life? *Green Theory and Praxis: The Journal of Ecopedagogy* 4(1): 1–14.

Kaijser, A. and A. Kronsell (2014). Climate change through the lens of intersectionality. *Environmental Politics* 23(3): 417–33. https://doi.org/10.1080/09644016.2013.835203.

Karera, A. (2019). Blackness and the pitfalls of Anthropocene ethics. *Critical Philosophy of Race* 7(1): 32–56.

Keeling, L., H. Tunón, G.O. Antillón, C. Berg, M. Jones, L. Stuardo et al. (2019). Animal welfare and the United Nations Sustainable Development Goals. *Frontiers in Veterinary Science* 6: 336. https://tinyurl.com/5z6d5rhr.

Kelbert, A.W. (2016). Climate change is a racist crisis: that's why Black Lives Matter closed an airport. *Guardian*, 6 September. https://tinyurl.com/333v8emc.

Kershaw, J.L., C.A. Ramp, R. Sears, S. Plourde, P. Brosset, P.J.O. Miller et al. (2021). Declining reproductive success in the Gulf of St. Lawrence's humpback whales (*Megaptera novaeangliae*) reflects ecosystem shifts on their feeding grounds. *Global Change Biology* 27(5): 1027–41. https://doi.org/10.1111/gcb.15466.

Kevany, S. (2021). Livestock industry lobbying UN to support more meat production. *Guardian*, 21 September. https://tinyurl.com/yzt9hft6.

Kevles, D.J. (1995). *In the Name of Eugenics: Genetics and the Uses of Human Heredity*. Cambridge, MA: Harvard University Press.

Kidner, D.W. (2014). Why "anthropocentrism" is not anthropocentric. *Dialectical Anthropology* 38(4): 465–80. https://doi.org/10.1007/s10624-014-9345-2.

Kim, B.F., R.E. Santo, A.P. Scatterday, J.P. Fry, C.M. Synk, S.R. Cebron et al. (2020). Country-specific dietary shifts to mitigate climate and water crises. *Global Environmental Change* 62: 101926. https://tinyurl.com/2rf2e57r.

Kitzinger, J. (1997). Who are you kidding? Children, power and the struggle against sexual abuse. In A. James and A. Prout, eds. *Constructing and Reconstructing Childhood: Contemporary Issues in the Sociological Study of Childhood*, 161–86. London: Routledge.

Kivimaa, P., S. Laakso, A. Lonkila and M. Kaljonen (2021). Moving beyond disruptive innovation: a review of disruption in sustainability transitions. *Environmental Innovation and Societal Transitions* 38: 110–26. https://doi.org/10.1016/j.eist.2020.12.001.

Klein, N. (2014). *This Changes Everything: Capitalism vs. the Climate*. London: Penguin.

Klein, N. (2007). *The Shock Doctrine: The Rise of Disaster Capitalism*. London: Penguin.

Kleinman, D.L. and S. Suryanarayanan (2013). Dying bees and the social production of ignorance. *Science, Technology and Human Values* 38(4): 492–517. https://doi.org/10.1177/0162243912442575.

Koch, M. (2012). *Capitalism and Climate Change: Theoretical Discussion, Historical Development and Policy Responses*. London: Palgrave Macmillan.

Kochnev, A.A. (2019). Phenology of the Pacific walrus (*Odobenus rosmarus divergens*) in coastal waters of Wrangel Island: the impact of the sea ice

dynamics. *Biology Bulletin* 46(9): 1156–64.
https://doi.org/10.1134/S1062359019090061.

Koerner, S. (2019). Fox News apologized to Greta Thunberg after a pundit called her "mentally ill". *BuzzFeed News*, 24 September.
https://tinyurl.com/yc4d3u55.

Kolbert, E. (2014). *The Sixth Extinction: An Unnatural History.* New York: Picador.

Koljonen, M. (2019). Thinking and caring boys go vegan: two European books that introduce vegan identity to children. *Bookbird: A Journal of International Children's Literature* 57(3): 13–22.

Kopnina, H. (2020). Education for the future? Critical evaluation of Education for Sustainable Development Goals. *Journal of Environmental Education* 51(4): 280–91. https://doi.org/10.1080/00958964.2019.1710444.

Kopnina, H. and B. Cherniak (2015). Cultivating a value for non-human interests through the convergence of animal welfare, animal rights, and deep ecology in environmental education. *Education Sciences* 5(4): 363–79.
https://doi.org/10.3390/educsci5040363.

Kopnina, H., M. Sitka-Sage, S. Blenkinsop and L. Piersol (2020). Moving beyond innocence: educating children in a post-nature world. In A. Cutter-Mackenzie-Knowles, K. Malone and E.B. Hacking, eds. *Research Handbook on Childhoodnature: Assemblages of Childhood and Nature Research*, 603–21. Heidelberg: Springer.

Kovach, R.P., A.J. Gharrett and D.A. Tallmon (2012). Genetic change for earlier migration timing in a pink salmon population. *Proceedings of the Royal Society B: Biological Sciences* 279(1743): 3870–78.
https://doi.org/10.1098/rspb.2012.1158.

Krange, O., B.P. Kaltenborn and M. Hultman (2019). Cool dudes in Norway: climate change denial among conservative Norwegian men. *Environmental Sociology* 5(1): 1–11. https://doi.org/10.1080/23251042.2018.1488516.

Kristiansen, S., J. Painter and M. Shea (2021). Animal agriculture and climate change in the US and UK elite media: volume, responsibilities, causes and solutions. *Environmental Communication* 15(2): 153–72.
https://doi.org/10.1080/17524032.2020.1805344.

Kustar, A. and D. Patino-Echeverri (2021). A review of environmental life cycle assessments of diets: plant-based solutions are truly sustainable, even in the form of fast foods. *Sustainability* 13(17): 9926.
https://doi.org/10.3390/su13179926.

Laestadius, L.I., R.A. Neff, C.L. Barry and S. Frattaroli (2016). No meat, less meat, or better meat: understanding NGO messaging choices intended to alter meat consumption in light of climate change. *Environmental Communication* 10(1): 84–103. https://doi.org/10.1080/17524032.2014.981561.

Laestadius, L.I., R.A. Neff, C.L. Barry and S. Frattaroli (2014). "We don't tell people what to do": an examination of the factors influencing NGO decisions to campaign for reduced meat consumption in light of climate change. *Global Environmental Change* 29: 32–40. https://tinyurl.com/69auk7h8.

Laestadius, L.I., R.A. Neff, C.L. Barry and S. Frattaroli (2013). Meat consumption and climate change: the role of non-governmental organizations. *Climatic Change* 120(1): 25–38. https://doi.org/10.1007/s10584-013-0807-3.

Laffoley, D. and J.M. Baxter, eds (2019). *Ocean Deoxygenation: Everyone's Problem; Causes, Impacts, Consequences and Solutions*. Gland: International Union for Conservation of Nature.

Lan, X., K.W. Thoning and E.J. Dlugokencky (2022). *Trends in Globally-averaged CH4, N2O, and SF6 Determined from NOAA Global Monitoring Laboratory Measurements*. Version 2022-11. Global Monitoring Laboratory. https://doi.org/10.15138/P8XG-AA10.

Landale, J. (2009). Whitehall turf war saves cows' hides. *BBC News*, 25 November. https://tinyurl.com/2p9bzbt7.

Larson, C.L., S.E. Reed, A.M. Merenlender and K.R. Crooks (2016). Effects of recreation on animals revealed as widespread through a global systematic review. *PLOS One* 11(12): e0167259. https://tinyurl.com/2ev866ht.

Latour, B. (2017). *Facing Gaia: Eight Lectures on the New Climatic Regime*. New York: John Wiley & Sons.

Lavelle, K. and J. Feagin (2006). Hurricane Katrina: the race and class debate. *Monthly Review*, 1 July. https://tinyurl.com/yr7kme7s.

Lawo, D., M. Esau, P. Engelbutzeder and G. Stevens (2020). Going vegan: the role(s) of ICT in vegan practice transformation. *Sustainability* 12(12): 5184. https://doi.org/10.3390/su12125184.

Lawrence, F. (2013). Horsemeat scandal: the essential guide. *Guardian*, 15 February. https://tinyurl.com/yhjsbhr3.

Lazarus, O., S. McDermid and J. Jacquet (2021). The climate responsibilities of industrial meat and dairy producers. *Climatic Change* 165(1): 30. https://doi.org/10.1007/s10584-021-03047-7.

Le Roy, E. (1928). The origins of humanity and the evolution of mind. In P.R. Samson and D. Pitt, eds. *The Biosphere and Noosphere Reader: Global Environment Society and Change*, 60–69. London: Routledge.

Leather, S.R. (2013). Institutional vertebratism hampers insect conservation generally; not just saproxylic beetle conservation. *Animal Conservation* 16(4): 379–80. https://doi.org/10.1111/acv.12068.

Lee, N. (2013). *Childhood and Biopolitics: Climate Change, Life Processes and Human Futures*. London: Palgrave.

References

Lélé, S.M. (1991). Sustainable development: a critical review. *World Development* 19(6): 607–21. https://doi.org/10.1016/0305-750X(91)90197-P.

Lenton, T.M., J. Rockström, O. Gaffney, S. Rahmstorf, K. Richardson, W. Steffen et al. (2019). Climate tipping points: too risky to bet against. *Nature* 575(7784): 592–95. https://doi.org/10.1038/d41586-019-03595-0.

Lesk, C., E. Coffel, A.W. D'Amato, K. Dodds and R. Horton (2017). Threats to North American forests from southern pine beetle with warming winters. *Nature Climate Change* 7(10): 713–17. https://doi.org/10.1038/nclimate3375.

Levant, R.F. (1997). The masculinity crisis. *Journal of Men's Studies* 5(3): 221–31. https://doi.org/10.1177/106082659700500302.

Levitt, T. (2021). Netherlands announces €25bn plan to radically reduce livestock numbers. *Guardian*, 15 December. https://tinyurl.com/y2ej94np.

Lewis, A. (2020). Infantilizing companion animals through attachment theory: why shift to behavioral ecology-based paradigms for welfare? *Society and Animals* Published ahead of print [18 September published online]. https://doi.org/10.1163/15685306-bja10011.

Lewis, S.L. and M.A. Maslin (2015). Defining the Anthropocene. *Nature* 519(7542): 171–80. https://doi.org/10.1038/nature14258.

Lin, N., K.S. Cook and R.S. Burt, eds (2001). *Social Capital: Theory and Research*. London: Routledge.

Live Well for Life and WWF (World Wildlife Fund) (2013). *Adopting Healthy, Sustainable Diets: Key Opportunities and Barriers*. Gland: WWF.

Live Well for Life and WWF (World Wildlife Fund) (2012). *Food Patterns and Dietary Recommendations in Spain, France, and Sweden*. Gland: WWF.

Lloro-Bidart, T. and V.S. Banschbach, eds (2019). *Animals in Environmental Education: Interdisciplinary Approaches to Curriculum and Pedagogy*. London: Palgrave.

Logan, M. (2020). Challenging the anthropocentric approach of science curricula: ecological systems approaches to enabling the convergence of sustainability, science, and STEM education. In A. Cutter-Mackenzie-Knowles, K. Malone and E.B. Hacking, eds. *Research Handbook on Childhoodnature: Assemblages of Childhood and Nature Research*, 1181–208. Heidelberg: Springer.

Loken, B., F. DeClerck, A. Bhowmik, W. Willett, B. Griscom, M. Springmann et al. (2020). *Diets for a Better Future: Rebooting and Reimagining Healthy and Sustainable Food Systems in the G20*. Oslo: World Wildlife Fund and EAT.

Londakova, K., T. Park, J. Reynolds and S. Wodak (2021). *Net Zero: Principles for Successful Behaviour Change Initiatives; Key Principles from Past Government-led Behaviour Change and Public Engagement Initiatives*. BEIS research paper 2021/063. London: Department of Business, Energy and Industrial Strategy.

Longobardi, P. (2014). Plastic as shadow: the toxicity of objects in the Anthropocene. In T. Cusack, ed. *Framing the Ocean, 1700 to the Present: Envisaging the Sea as Social Space*, 181–91. London: Routledge.

Lonkila, A. and M. Kaljonen (2021). Promises of meat and milk alternatives: an integrative literature review on emergent research themes. *Agriculture and Human Values* 38(3): 625–39. https://doi.org/10.1007/s10460-020-10184-9.

Loughnan, S., N. Haslam and B. Bastian (2010). The role of meat consumption in the denial of moral status and mind to meat animals. *Appetite* 55(1): 156–59. https://doi.org/10.1016/j.appet.2010.05.043.

Lovich, J.E., C.B. Yackulic, J. Freilich, M. Agha, M. Austin, K.P. Meyer et al. (2014). Climatic variation and tortoise survival: has a desert species met its match? *Biological Conservation* 169: 214–24. https://tinyurl.com/2jmbt2pk.

Lucas, E., M. Guo and G. Guillén-Gosálbez (2023). Low-carbon diets can reduce global ecological and health costs. *Nature Food*. https://doi.org/10.1038/s43016-023-00749-2.

Luscombe, R. (2020). "Chill!": Greta Thunberg recycles Trump's mockery of her as he tries to stop votes. *Guardian*, 5 November. https://tinyurl.com/3v9hhkpf.

MacCormack, P. (2013). Gracious pedagogy. *Journal of Curriculum and Pedagogy* 10(1): 13–17. https://doi.org/10.1080/15505170.2013.789994.

MacGregor, S., ed. (2017). *Routledge Handbook of Gender and Environment*. London: Routledge.

MacGregor, S. (2009). A stranger silence still: the need for feminist social research on climate change. *Sociological Review* 57(2): 124–40.

MacGregor, S. and N. Seymour, eds (2017). *Men and Nature: Hegemonic Masculinities and Environmental Change*. Munich: RCC Perspectives.

Machovina, B., K.J. Feeley and W.J. Ripple (2015). Biodiversity conservation: the key is reducing meat consumption. *Science of the Total Environment* 536: 419–31. https://doi.org/10.1016/j.scitotenv.2015.07.022.

Maeckelbergh, M. (2011). Doing is believing: prefiguration as strategic practice in the alterglobalization movement. *Social Movement Studies* 10(1): 1–20. https://doi.org/10.1080/14742837.2011.545223.

Mallory, C.D. and M.S. Boyce (2018). Observed and predicted effects of climate change on Arctic caribou and reindeer. *Environmental Reviews* 26(1): 13–25. https://doi.org/10.1139/er-2017-0032.

Malm, A. (2016). *Fossil Capital*. London: Verso.

Malm, A. and A. Hornborg (2014). The geology of mankind? A critique of the Anthropocene narrative. *Anthropocene Review* 1(1): 62–69. https://doi.org/10.1177/2053019613516291.

Malone, K., M. Tesar and S. Arndt (2020). *Theorising Posthuman Childhood Studies*. Heidelberg: Springer.

References

Manes, S., M.J. Costello, H. Beckett, A. Debnath, E. Devenish-Nelson, K.-A. Grey et al. (2021). Endemism increases species' climate change risk in areas of global biodiversity importance. *Biological Conservation* 257: 109070. https://doi.org/10.1016/j.biocon.2021.109070.

Marlow, H.J., W.K. Hayes, S. Soret, R.L. Carter, E.R. Schwab and J. Sabaté (2009). Diet and the environment: does what you eat matter? *American Journal of Clinical Nutrition* 89(5): 1699S–1703S. https://tinyurl.com/42a9m29z.

Marsh, D.R., D.G. Schroeder, K.A. Dearden, J. Sternin and M. Sternin (2004). The power of positive deviance. *BMJ* 329(7475): 1177–79. https://doi.org/10.1136/bmj.329.7475.1177.

Marx, K. (1981) [1883]. *Capital: A Critique of Political Economy*, vol. 3. S. Moore, trans. London: Penguin.

Marx, K. (1976) [1867]. *Capital: A Critique of Political Economy*, vol. 1. S. Moore and E. Aveling, trans. London: Penguin.

Marsh, S. (2016). Meet the parents raising vegan babies. *Guardian*, 19 July. https://tinyurl.com/bdcvmcjj.

Masefield, A. (2021). Where the animal is loud, but CAS is silent: a critical analysis of entrenched anthropocentrism across contemporary food justice discourse. Paper presented at the 7th Biennial Conference of the European Association for Critical Animal Studies. Barcelona, Spain, 24 June.

Mattick, C.S., A.E. Landis, B.R. Allenby and N.J. Genovese (2015). Anticipatory life cycle analysis of in vitro biomass cultivation for cultured meat production in the United States. *Environmental Science and Technology* 49(19): 11941–49. https://doi.org/10.1021/acs.est.5b01614.

McBrien, J. (2016). Accumulating extinction: planetary catastrophism in the necrocene. In J.W. Moore, ed. *Anthropocene or Capitalocene? Nature, History, and the Crisis of Capitalism*, 116–37. Oakland: PM Press.

McCormick, E. (2021). Eat Just is racing to put "no-kill meat" on your plate. Is it too good to be true? *Guardian*, 16 June. https://tinyurl.com/2p8ksx3j.

McCright, A.M. (2010). The effects of gender on climate change knowledge and concern in the American public. *Population and Environment* 32: 66–87. https://doi.org/10.1007/s11111-010-0113-1.

McCright, A.M. and R.E. Dunlap (2011). Cool dudes: the denial of climate change among conservative white males in the United States. *Global Environmental Change* 21(4): 1163–72. https://doi.org/10.1016/j.gloenvcha.2011.06.003.

McDonald, S.E., A.M. Cody, L.J. Booth, J.R. Peers, C. O'Connor Luce, J.H. Williams et al. (2018). Animal cruelty among children in violent households: children's explanations of their behavior. *Journal of Family Violence* 33(7): 469–80. https://doi.org/10.1007/s10896-018-9970-7.

McGuire, L., S.B. Palmer and N.S. Faber (2023). The development of speciesism: age-related differences in the moral view of animals. *Social Psychological and Personality Science*. 14(2): 228–37. https://tinyurl.com/2p9epcd5.

McMichael, A.J., J.W. Powles, C.D. Butler and R. Uauy (2007). Food, livestock production, energy, climate change, and health. *Lancet* 370(9594): 1253–63. https://doi.org/10.1016/S0140-6736(07)61256-2.

McMichael, A.J., R.E. Woodruff and S. Hales (2006). Climate change and human health: present and future risks. *Lancet* 367(9513): 859–69. https://doi.org/10.1016/S0140-6736(06)68079-3.

McSweeney, E. and H. Young (2021). "The whole system is rotten": Life inside Europe's meat industry. *Guardian*, 28 September. https://tinyurl.com/43rs75tn.

McWilliams, J. (2010). Carnivorous climate skeptics in the media. *Atlantic*, 22 April. https://tinyurl.com/3jcuy75m.

Medlock, J.M. and S.A. Leach (2015). Effect of climate change on vector-borne disease risk in the UK. *Lancet Infectious Diseases* 15(6): 721–30. https://doi.org/10.1016/S1473-3099(15)70091-5.

Meier, T. and O. Christen (2013). Environmental impacts of dietary recommendations and dietary styles: Germany as an example. *Environmental Science and Technology* 47(2): 877–88. https://doi.org/10.1021/es302152v.

Meijer, E. (2019). *When Animals Speak: Toward an Interspecies Democracy*. New York: NYU Press.

Mendes, E. (2013). An application of the transtheoretical model to becoming vegan. *Social Work in Public Health* 28(2): 142–49. https://doi.org/10.1080/19371918.2011.561119.

Merrigan, K., T. Griffin, P. Wilde, K. Robien, J. Goldberg and W. Dietz (2015). Designing a sustainable diet. *Science* 350(6257): 165–66. https://doi.org/10.1126/science.aab2031.

Mertens, D. (2018). Why I am staying optimistic about the world and its wicked problems! *Discover Society*, 3 July. https://tinyurl.com/yckrmhx7.

Micheelsen, A., L. Havn, S.K. Poulsen, T.M. Larsen and L. Holm (2014). The acceptability of the new Nordic diet by participants in a controlled six-month dietary intervention. *Food Quality and Preference* 36: 20–26. https://doi.org/10.1016/j.foodqual.2014.02.003.

Mies, M. (1986). *Patriarchy and Accumulation on a World Scale: Women in the International Division of Labour*. London: Zed Books.

Milburn, J. (2022). Should vegans compromise? *Critical Review of International Social and Political Philosophy* 25(2): 281–93. https://doi.org/10.1080/13698230.2020.1737477.

References

Milman, O. (2022). "A barbaric federal program": US killed 1.75m animals last year – or 200 per hour. *Guardian*, 25 March. https://tinyurl.com/2bp8pm9j.

Milman, O. (2020). Teaching climate crisis in classrooms critical for children, top educators say. *Guardian*, 3 December. https://tinyurl.com/4ekme3x5.

Milman, O. (2017). Meat industry blamed for largest-ever "dead zone" in Gulf of Mexico. *Guardian*, 1 August. https://tinyurl.com/52j8uejr.

Molidor, J. (2021). USDA needs to put climate solutions before industry profits. *The Hill*, 11 October. https://tinyurl.com/bdh594wu.

Molloy, C. (2011). *Popular Media and Animals*. London: Palgrave.

Molnár, P.K., C.M. Bitz, M.M. Holland, J.E. Kay, S.R. Penk and S.C. Amstrup (2020). Fasting season length sets temporal limits for global polar bear persistence. *Nature Climate Change* 10(8): 732–38. https://doi.org/10.1038/s41558-020-0818-9.

Montford, K.S. and C. Taylor, eds (2020). *Colonialism and Animality: Anti-colonial Perspectives in Critical Animal Studies*. London: Routledge.

Moore, J.W. (2018). The Capitalocene part II: accumulation by appropriation and the centrality of unpaid work/energy. *Journal of Peasant Studies* 45(2): 237–79. https://doi.org/10.1080/03066150.2016.1272587.

Moore, J.W. (2017). The Capitalocene, part I: on the nature and origins of our ecological crisis. *Journal of Peasant Studies* 44(3): 594–630. https://doi.org/10.1080/03066150.2016.1235036.

Moore, J.W., ed. (2016a). *Anthropocene or Capitalocene? Nature, History, and the Crisis of Capitalism*. Oakland, CA: PM Press.

Moore, J.W. (2016b). The rise of cheap nature. In J.W. Moore, eds. *Anthropocene or Capitalocene? Nature, History, and the Crisis of Capitalism*. Oakland, CA: PM Press.

Moore, J.W. (2015). *Capitalism in the Web of Life: Ecology and the Accumulation of Capital*. London: Verso Books.

Moore, J.W. (2014). The end of cheap nature; or, How I learned to stop worrying about "the" environment and love the crisis of capitalism. In C. Suter and C. Chase-Dunn, eds. *Structures of the World Political Economy and the Future Global Conflict and Cooperation*, 285–314. New York: LIT Verlag.

Moore, J.W. (2011). Transcending the metabolic rift: a theory of crises in the capitalist world-ecology. *Journal of Peasant Studies* 38(1): 1–46. https://doi.org/10.1080/03066150.2010.538579.

Moore, L.J. and M. Kosut (2013). *Buzz: Urban Beekeeping and the Power of the Bee*. New York: NYU Press.

Moore, R. (2021). Evolutions and revolutions in caregiving. In J. Kong, ed. *Vegan Voices: Essays by Inspiring Changemakers*, 242–48. Brooklyn: Lantern.

Morais, T.G., R.F.M. Teixeira, C. Lauk, M.C. Theurl, W. Winiwarter, A. Mayer et al. (2021). Agroecological measures and circular economy strategies to ensure sufficient nitrogen for sustainable farming. *Global Environmental Change* 69: 102313. https://doi.org/10.1016/j.gloenvcha.2021.102313.

Morris, C., M. Kaljonen, K. Aavik, B. Balázs, M. Cole, B. Coles et al. (2021). Priorities for social science and humanities research on the challenges of moving beyond animal-based food systems. *Humanities and Social Sciences Communications* 8(1): 1–12. https://doi.org/10.1057/s41599-021-00714-z.

Moss, P. (2014). *Transformative Change and Real Utopias in Early Childhood Education: A Story of Democracy, Experimentation and Potentiality*. London: Routledge.

Mottet, A. and H. Steinfeld (2018). Cars or livestock: which contribute more to climate change? 18 September. https://tinyurl.com/sxbuun69.

Murphy, T. and A. Mook (2022). The vegan food justice movement. In R. Brears, ed. *The Palgrave Encyclopedia of Urban and Regional Futures*, 1–7. London: Palgrave. https://doi.org/10.1007/978-3-030-51812-7_219-1.

Murris, K. (2016). *The Posthuman Child: Educational Transformation through Philosophy with Picture Books*. London: Routledge.

Myers, G. (2007). *The Significance of Children and Animals: Social Development and Our Connections to Other Species*. West Lafayette: Purdue University Press.

Mylan, J., C. Morris, E. Beech and F.W. Geels (2019). Rage against the regime: niche-regime interactions in the societal embedding of plant-based milk. *Environmental Innovation and Societal Transitions* 31: 233–47. https://doi.org/10.1016/j.eist.2018.11.001.

National Health and Medical Research Council (2013). *Australian Dietary Guidelines*. Canberra: National Health and Medical Research Council. https://tinyurl.com/5n77pbry.

NEF (New Economics Foundation) (2017). Grow Green: Sustainable Solutions for the Farm of the Future. Birmingham: The Vegan Society. https://tinyurl.com/3bh2zmsh.

Nelson, M.E., M.W. Hamm, F.B. Hu, S.A. Abrams and T.S. Griffin (2016). Alignment of healthy dietary patterns and environmental sustainability: a systematic review. *Advances in Nutrition* 7(6): 1005–25. https://doi.org/10.3945/an.116.012567.

Neo, H. and J. Emel (2017). *Geographies of Meat: Politics, Economy and Culture*. London: Routledge.

Neslen, A. (2023). "The anti-livestock people are a pest": how UN food body played down role of farming in climate change. *Guardian*, 20 October. https://tinyurl.com/bdhn4t3k.

376

References

Nguyen, A. and M.J. Platow (2021). "I'll eat meat because that's what we do": the role of national norms and national social identification on meat eating. *Appetite* 164: 105287. https://doi.org/10.1016/j.appet.2021.105287.

Nibert, D. (2017a). *Animal Oppression and Capitalism*, vol. 1. Santa Barbara: ABC-CLIO.

Nibert, D. (2017b). *Animal Oppression and Capitalism*, vol. 2. Santa Barbara: ABC-CLIO.

Nicolini, D. (2017). Is small the only beautiful? Making sense of "large phenomena" from a practice-based perspective. In A. Hui, T. Schatzki and E. Shove, eds. *The Nexus of Practices: Connections, Constellations, Practitioners*, 98–113. London: Routledge.

Nijdam, D., T. Rood and H. Westhoek (2012). The price of protein: review of land use and carbon footprints from life cycle assessments of animal food products and their substitutes. *Food Policy* 37(6): 760–70. https://doi.org/10.1016/j.foodpol.2012.08.002.

Nimmo, R. (2018). Enfolding the biosocial collective: ontological politics in the evolution of social insects. *Humanimalia* 9(2): 28–46. https://doi.org/10.52537/humanimalia.9541.

Nixon, R. (2011). *Slow Violence and the Environmentalism of the Poor*. Cambridge, MA: Harvard University Press.

Nocella, A., C. Drew and A.E. George, eds (2019). *Education for Total Liberation: Critical Animal Pedagogy and Teaching against Speciesism*. New York: Peter Lang US.

Nolt, J. (2011). Nonanthropocentric climate ethics. *WIRES Climate Change* 2(5): 701–11. https://doi.org/10.1002/wcc.131.

Norgaard, K.M. (2011). *Living in Denial: Climate Change, Emotions, and Everyday Life*. Cambridge, MA: MIT Press.

Noske, B. (1989). *Humans and Other Animals: Beyond the Boundaries of Anthropology*. London: Pluto Press.

Oakley, J. (2019). What can an animal liberation perspective contribute to environmental education? In T. Lloro-Bidart and V.S. Banschbach, eds. *Animals in Environmental Education: Interdisciplinary Approaches to Curriculum and Pedagogy*, 19–34. London: Palgrave.

Oakley, J., G.P.L. Watson, C.L. Russell, A. Cutter-Mackenzie, L. Fawcett, G. Kuhl et al. (2010). Animal encounters in environmental education research: responding to the "question of the animal". *Canadian Journal of Environmental Education* 15: 86–102.

Oatly (2022). Change isn't easy. 10 August. https://www.oatly.com/int/climate-and-capital.

O'Connor, J. (1998). *Natural Causes: Essays in Ecological Marxism*. New York: Guilford Press.

Oksala, J. (2018). Feminism, capitalism, and ecology. *Hypatia* 33(2): 216–34. https://doi.org/10.1111/hypa.12395.

Oliver, C. (2023). Mock meat, masculinity, and redemption narratives: vegan men's negotiations and performances of gender and eating. *Social Movement Studies* 22(1): 62–79. https://doi.org/10.1080/14742837.2021.1989293.

Orange, D.M. (2016). *Climate Crisis, Psychoanalysis, and Radical Ethics*. London: Routledge.

O'Rourke, D. and N. Lollo (2015). Transforming consumption: from decoupling to behavior change, to system changes for sustainable consumption. *Annual Review of Environment and Resources* 40(1): 233–59. https://doi.org/10.1146/annurev-environ-102014-021224.

Orzechowski, K. (2022). Global animal slaughter statistics & charts: 2022 update. *Faunalytics*, 13 July. https://tinyurl.com/mr4dr3hh.

Ostfeld, R.S. and J.L. Brunner (2015). Climate change and ixodes tick-borne diseases of humans. *Philosophical Transactions of the Royal Society B: Biological Sciences* 370(1665): 20140051. https://doi.org/10.1098/rstb.2014.0051.

Oswald, Y., A. Owen and J.K. Steinberger (2020). Large inequality in international and intranational energy footprints between income groups and across consumption categories. *Nature Energy* 5(3): 231–39. https://doi.org/10.1038/s41560-020-0579-8.

Our World in Data (2023). Animals slaughtered to produce all meat, 1961 to 2021. https://tinyurl.com/szsbrkbk.

Pachirat, T. (2011). *Every Twelve Seconds: Industrialized Slaughter and the Politics of Sight*. New Haven: Yale University Press.

Pacifici, M., P. Visconti and C. Rondinini (2018). A framework for the identification of hotspots of climate change risk for mammals. *Global Change Biology* 24(4): 1626–36. https://doi.org/10.1111/gcb.13942.

Pagano, A.M. and T.M. Williams (2021). Physiological consequences of Arctic sea ice loss on large marine carnivores: unique responses by polar bears and narwhals. *Journal of Experimental Biology* 224(Suppl_1): jeb228049. https://doi.org/10.1242/jeb.228049.

Palmer, C. (2021). Assisting wild animals vulnerable to climate change: why ethical strategies diverge. *Journal of Applied Philosophy* 38(2): 179–95. https://doi.org/10.1111/japp.12358.

Palmer, C. (2018). Should we help wild animals suffering negative impacts from climate change? In S. Springer and H. Grimm, eds. *Professionals in Food Chains*, 2694–706. Wageningen: Wageningen Academic Publishers.

References

Palmer, C. (2011). Does nature matter? The place of the nonhuman in the ethics of climate change. In D.G. Arnold, ed. *The Ethics of Global Climate Change*, 272–91. Cambridge, UK: Cambridge University Press.

Palmer, G. (2009). *The Politics of Breastfeeding*. London: Pinter and Martin.

Parham, P.E., J. Waldock, G.K. Christophides and E. Michael (2015). Climate change and vector-borne diseases of humans. *Philosophical Transactions of the Royal Society B: Biological Sciences* 370(1665): 20140377. https://doi.org/10.1098/rstb.2014.0377.

Parker, R.W.R., J.L. Blanchard, C. Gardner, B.S. Green, K. Hartmann, P.H. Tyedmers et al. (2018). Fuel use and greenhouse gas emissions of world fisheries. *Nature Climate Change* 8(4): 333–37. https://doi.org/10.1038/s41558-018-0117-x.

Parkinson, C. (2019). *Animals, Anthropomorphism and Mediated Encounters*. London: Routledge.

Parkinson, C. (2018). Animal bodies and embodied visuality. *Antennae: The Journal of Nature in Visual Culture* 46: 51–64.

Parkinson, C., R. Twine and N. Griffin (2019). *Pathways to Veganism: Exploring Effective Messages in Vegan Transition*. Ormskirk: Centre for Human-Animal Studies, Edge Hill University.

Parrique, T., J. Barth, F. Briens, C. Kerschner, A. Kraus-Polk, A. Kuokkanen et al. (2019). *Decoupling Debunked: Evidence and Arguments against Green Growth as a Sole Strategy for Sustainability*. Brussels: European Environmental Bureau.

Parry, J. (2010). Gender and slaughter in popular gastronomy. *Feminism and Psychology* 20(3): 381–96. https://doi.org/10.1177/0959353510368129.

Parson, S. (2019). The politics of dumpstered soup: Food Not Bombs and the limits of decommodifying food. In R. Kinna and U. Gordon, eds. *Routledge Handbook of Radical Politics*, 405–16. London: Routledge.

Parson, S. and E. Ray (2020). Drill baby drill: labor, accumulation, and the sexualization of resource extraction. *Theory and Event* 23(1): 248–70.

Partridge, J. (2021). UK to allow temporary visas for butchers in latest post-Brexit U-turn. *Guardian*, 14 October. https://tinyurl.com/5fr3h98w.

Patel, R. and J.W. Moore (2018). *A History of the World in Seven Cheap Things: A Guide to Capitalism, Nature, and the Future of the Planet*. London: Verso Books.

Payne, J.L., A.M. Bush, N.A. Heim, M.L. Knope and D.J. McCauley (2016). Ecological selectivity of the emerging mass extinction in the oceans. *Science* 353(6305): 1284–86. https://doi.org/10.1126/science.aaf2416.

Pearse, R. (2017). Gender and climate change. *Wiley Interdisciplinary Reviews: Climate Change* 8(2): e451.

Pease, B. (2019). Recreating men's relationship with nature: toward a profeminist environmentalism. *Men and Masculinities* 22(1): 113–23.

Pecl, G.T., M.B. Araújo, J.D. Bell, J. Blanchard, T.C. Bonebrake, I.-C. Chen et al. (2017). Biodiversity redistribution under climate change: impacts on ecosystems and human well-being. *Science* 355(6332): eaai9214. https://doi.org/10.1126/science.aai9214.

Pedersen, H. (2021). Education, anthropocentrism, and interspecies sustainability: confronting institutional anxieties in omnicidal times. *Ethics and Education* 16(2): 164–77. https://doi.org/10.1080/17449642.2021.1896639.

Pedersen, H. (2019). The contested space of animals in education: a response to the "animal turn" in education for sustainable development. *Education Sciences* 9(3): 211. https://doi.org/10.3390/educsci9030211.

Pedersen, H. (2010). Is "the posthuman" educable? On the convergence of educational philosophy, animal studies, and posthumanist theory. *Discourse: Studies in the Cultural Politics of Education* 31(2): 237–50. https://doi.org/10.1080/01596301003679750.

Pelletier, N. and P. Tyedmers (2010). Forecasting potential global environmental costs of livestock production 2000–2050. *Proceedings of the National Academy of Sciences* 107(43): 18371–74. https://doi.org/10.1073/pnas.1004659107.

Pepper, A. (2019). Adapting to climate change: what we owe to other animals. *Journal of Applied Philosophy* 36(4): 592–607. https://doi.org/10.1111/japp.12337.

Phillips, A. (2014). Coyote booms, bear attacks and how climate change is wreaking havoc on the animal kingdom. *Think Progress*, 5 December. https://tinyurl.com/4sscwa8h.

Piatt, J.F., J.K. Parrish, H.M. Renner, S.K. Schoen, T.T. Jones, M.L. Arimitsu et al. (2020). Extreme mortality and reproductive failure of common murres resulting from the northeast Pacific marine heatwave of 2014–2016. *PLOS One* 15(1): e0226087. https://doi.org/10.1371/journal.pone.0226087.

Piazza, J., M.B. Ruby, S. Loughnan, M. Luong, J. Kulik, H.M. Watkins et al. (2015). Rationalizing meat consumption: the 4Ns. *Appetite* 91: 114–28. https://doi.org/10.1016/j.appet.2015.04.011.

Pieper, M., A. Michalke and T. Gaugler (2020). Calculation of external climate costs for food highlights inadequate pricing of animal products. *Nature Communications* 11(1): 6117. https://doi.org/10.1038/s41467-020-19474-6.

Pimentel, D. and M. Pimentel (2003). Sustainability of meat-based and plant-based diets and the environment. *American Journal of Clinical Nutrition* 78(3): 660S–663S. https://doi.org/10.1093/ajcn/78.3.660S.

References

Pitesky, M.E., K.R. Stackhouse and F.M. Mitloehner (2009). Clearing the air: livestock's contribution to climate change. *Advances in Agronomy* 103: 1–40. https://doi.org/10.1016/S0065-2113(09)03001-6.

Ploll, U., H. Petritz and T. Stern (2020). A social innovation perspective on dietary transitions: diffusion of vegetarianism and veganism in Austria. *Environmental Innovation and Societal Transitions* 36: 164–76. https://doi.org/10.1016/j.eist.2020.07.001.

Plumwood, V. (1997). Babe: the tale of the speaking meat. *Animal Issues* 1(1): 21; 1(2): 20–39.

Plumwood, V. (1996). Androcentrism and anthrocentrism: parallels and politics. *Ethics and the Environment* 1(2): 119–52.

Plumwood, V. (1993). *Feminism and the Mastery of Nature*. London: Routledge.

Plumwood, V. (1992). Feminism and ecofeminism: beyond the dualistic assumptions of women, men and nature. *Ecologist* 22(1): 8–13.

Polk, M. (2009). Gendering climate change through the transport sector. *Women, Gender and Research* (3–4): 73–78.

Poore, J. (2019). Addressing climate through food. Cambridge Climate Lecture Series. 28 February. https://tinyurl.com/2fka48s7.

Poore, J. and T. Nemecek (2019). Erratum for the research article "Reducing food's environmental impacts through producers and consumers". *Science* 363(6429). https://doi.org/10.1126/science.aaw9908.

Poore, J. and T. Nemecek (2018). Reducing food's environmental impacts through producers and consumers. *Science* 360(6392): 987–92. https://doi.org/10.1126/science.aaq0216.

Popkin, B.M. (2009). Reducing meat consumption has multiple benefits for the world's health. *Archives of Internal Medicine* 169(6): 543–45. https://doi.org/10.1001/archinternmed.2009.2.

Popkin, B.M. (2006). Global nutrition dynamics: the world is shifting rapidly toward a diet linked with noncommunicable diseases. *American Journal of Clinical Nutrition* 84(2): 289–98. https://doi.org/10.1093/ajcn/84.2.289.

Popkin, B.M. (1998). The nutrition transition and its health implications in lower-income countries. *Public Health Nutrition* 1(1): 5–21. https://doi.org/10.1079/PHN19980004.

Poppy, G.M. and J. Baverstock (2019). Rethinking the food system for human health in the Anthropocene. *Current Biology* 29(19): R972–R977. https://doi.org/10.1016/j.cub.2019.07.050.

Post, E., R.B. Alley, T.R. Christensen, M. Macias-Fauria, B.C. Forbes, M.N. Gooseff et al. (2019). The polar regions in a 2°C warmer world. *Science Advances* 5(12): eaaw9883. https://doi.org/10.1126/sciadv.aaw9883.

Postman, N. (1994). *The Disappearance of Childhood*. New York: Vintage Books.

Potter, W. (2017). Ag-gag laws: corporate attempts to keep consumers in the dark. *Griffith Journal of Law and Human Dignity* 5(1): 1–31.

Potts, A., ed. (2016). *Meat Culture*. Leiden: Brill.

Potts, A. and J. Parry (2010). Vegan sexuality: challenging heteronormative masculinity through meat-free sex. *Feminism and Psychology* 20(1):53–72. https://doi.org/10.1177/0959353509351181.

Pounds, A.J., M.R. Bustamante, L.A. Coloma, J.A. Consuegra, M.P.L. Fogden, P.N. Foster et al. (2006). Widespread amphibian extinctions from epidemic disease driven by global warming. *Nature* 439(7073): 161–67. https://doi.org/10.1038/nature04246.

Povey, R., B. Wellens and M. Conner (2001). Attitudes towards following meat, vegetarian and vegan diets: an examination of the role of ambivalence. *Appetite* 37(1): 15–26. https://doi.org/10.1006/appe.2001.0406.

Powles, J. (2009). Commentary: why diets need to change to avert harm from global warming. *International Journal of Epidemiology* 38(4): 1141–42. https://doi.org/10.1093/ije/dyp247.

Probyn-Rapsey, F. (2018). Anthropocentrism. In L. Gruen, ed. *Critical Terms for Animal Studies*, 47–63. Chicago: University of Chicago Press.

Probyn-Rapsey, F., S. Donaldson, G. Ioannides, T. Lea, K. Marsh, A. Neimanis et al. (2016). A sustainable campus: the Sydney declaration on interspecies sustainability. *Animal Studies Journal* 5(1): 110–51.

Prochaska, J.O. and C.C. DiClemente (1982). Transtheoretical therapy: toward a more integrative model of change. *Psychotherapy: Theory, Research and Practice* 19(3): 276–88. https://doi.org/10.1037/h0088437.

Prokosch, J., Z. Bernitz, H. Bernitz, B. Erni and R. Altwegg (2019). Are animals shrinking due to climate change? Temperature-mediated selection on body mass in mountain wagtails. *Oecologia* 189(3): 841–49. https://doi.org/10.1007/s00442-019-04368-2.

Prout, A. and A. James (1997). A new paradigm for the sociology of childhood? Provenance, promise and problems. In A. James and A. Prout, eds. *Constructing and Reconstructing Childhood: Contemporary Issues in the Sociological Study of Childhood*, 7–32. London: Routledge.

Pulé, P. and M. Hultman, eds (2021). *Men, Masculinities and Earth: Contending with the (m)Anthropocene*. London: Palgrave.

Quintana, I., E.F. Cifuentes, J.A. Dunnink, M. Ariza, D. Martínez-Medina, F.M. Fantacini et al. (2022). Severe conservation risks of roads on apex predators. *Scientific Reports* 12(1): 2902. https://doi.org/10.1038/s41598-022-05294-9.

Rabès, A., L. Seconda, B. Langevin, B. Allès, M. Touvier, S. Hercberg et al. (2020). Greenhouse gas emissions, energy demand and land use associated with omnivorous, pesco-vegetarian, vegetarian, and vegan diets accounting for farming practices. *Sustainable Production and Consumption* 22: 138–46. https://doi.org/10.1016/j.spc.2020.02.010.

Raffa, K.F., B.H. Aukema, B.J. Bentz, A.L. Carroll, J.A. Hicke, M.G. Turner et al. (2008). Cross-scale drivers of natural disturbances prone to anthropogenic amplification: the dynamics of bark beetle eruptions. *BioScience* 58(6): 501–17. https://doi.org/10.1641/B580607.

Ranganathan, J., D. Vennard, R. Waite, P. Dumas, B. Lipinski, T. Searchinger et al. (2016). *Shifting Diets for a Sustainable Food Future*. Washington, DC: World Resources Institute.

Räty, R. and A. Carlsson-Kanyama (2010). Energy consumption by gender in some European countries. *Energy Policy* 38(1): 646–49. https://doi.org/10.1016/j.enpol.2009.08.010.

Rawles, K. (2006). Sustainable development and animal welfare: the neglected dimension. In J. Turner and J. D'Silva, eds. *Animals, Ethics, and Trade: The Challenge of Animal Sentience*, 208–16. London: Earthscan.

Rawls, J. (1971). *A Theory of Justice*. Cambridge, MA: Harvard University Press.

Reckwitz, A. (2002). Toward a theory of social practices: a development in culturalist theorizing. *European Journal of Social Theory* 5(2): 243–63.

Redlener, C., C. Jenkins and I. Redlener (2019). Our planet is in crisis. But until we call it a crisis, no one will listen. *Guardian*, 31 July. https://tinyurl.com/2p9dhyd9.

Rijnsdorp, A.D., S.G. Bolam, C. Garcia, J.G. Hiddink, N.T. Hintzen, P.D. van Denderen et al. (2018). Estimating sensitivity of seabed habitats to disturbance by bottom trawling based on the longevity of benthic fauna. *Ecological Applications* 28(5): 1302–12. https://doi.org/10.1002/eap.1731.

Ripple, W.J., K. Abernethy, M.G. Betts, G. Chapron, R. Dirzo, M. Galetti et al. (2016). Bushmeat hunting and extinction risk to the world's mammals. *Royal Society Open Science* 3(10): 160498. https://doi.org/10.1098/rsos.160498.

Ripple, W.J., P. Smith, H. Haberl, S.A. Montzka, C. McAlpine and D.H. Boucher (2014). Ruminants, climate change and climate policy. *Nature Climate Change* 4(1): 2–5. https://doi.org/10.1038/nclimate2081.

Ripple, W.J., C. Wolf, T.M. Newsome, P. Barnard, W.R. Moomaw and P. Grandcolas (2020). World scientists' warning of a climate emergency. *Bioscience* 70(1): 8–12. https://doi.org/10.1093/biosci/biz152.

Risku-Norja, H., S. Kurppa and J. Helenius (2009). Dietary choices and greenhouse gas emissions: assessment of impact of vegetarian and organic options at national scale. *Progress in Industrial Ecology* 6(4): 340–54. https://doi.org/10.1504/PIE.2009.032323.

Risman, B.J. (2004). Gender as a social structure: theory wrestling with activism. *Gender and Society* 18: 429–51.

Ritchie, H. (2019). Half of the world's habitable land is used for agriculture. *Our World in Data*, 11 November. https://tinyurl.com/3wjkya8m.

Ritchie, H., D.S. Reay and P. Higgins (2018). The impact of global dietary guidelines on climate change. *Global Environmental Change* 49: 46–55. https://doi.org/10.1016/j.gloenvcha.2018.02.005.

Ritchie, H., P. Rosado and M. Roser (2019). Meat and Dairy Production. *Our World in Data*, November. https://tinyurl.com/jensjh5d.

Ritchie, H., M. Roser and P. Rosado (2020). CO_2 and Greenhouse Gas Emissions. *Our World in Data*, August. https://tinyurl.com/2p8dbcbc.

Ritvo, H. (1995). Possessing Mother Nature: genetic capital in eighteenth-century Britain. In J. Brewer and S. Staves, eds. *Early Modern Conceptions of Property*, 413–26. London: Routledge.

Rivera-Ferre, M.G. (2009). Supply vs. demand of agri-industrial meat and fish products: a chicken and egg paradigm? *International Journal of Sociology of Agriculture and Food* 16(2): 90–105.

Riverola, C., R. Ortt, F. Miralles and O. Dedehayir (2017). When do early adopters share or scare? A conceptual model. *ISPIM Conference Proceedings*, 1–12.

Rogers, R.A. (2008). Beasts, burgers, and hummers: meat and the crisis of masculinity in contemporary television advertisements. *Environmental Communication* 2(3): 281–301. https://doi.org/10.1080/17524030802390250.

Rojas-Downing, M.M., A.P. Nejadhashemi, T. Harrigan and S.A. Woznicki (2017). Climate change and livestock: impacts, adaptation, and mitigation. *Climate Risk Management* 16: 145–63. https://doi.org/10.1016/j.crm.2017.02.001.

Rosa, R., J.L. Rummer and P.L. Munday (2017). Biological responses of sharks to ocean acidification. *Biology Letters* 13(3): 20160796. https://doi.org/10.1098/rsbl.2016.0796.

Rose, D., M.C. Heller and C.A. Roberto (2019). Position of the Society for Nutrition Education and Behavior: the importance of including environmental sustainability in dietary guidance. *Journal of Nutrition Education and Behavior* 51(1): 3–15.e1. https://doi.org/10.1016/j.jneb.2018.07.006.

Rose, D.B., T. van Dooren and M. Chrulew, eds (2017). *Extinction Studies: Stories of Time, Death and Generations*. New York: Columbia University Press.

Rosenberg, G. (2016). *The 4-H Harvest: Sexuality and the State in Rural America*. Philadelphia: University of Pennsylvania Press.

Rosenberg, G. and J. Dutkiewicz (2023). The viral story of a girl and her goat explains how the meat industry indoctrinates children. *Vox*, 5 April. https://tinyurl.com/4ybcjz9c.

Roth, R. (2013). *V Is for Vegan: The ABCs of Being Kind*. Berkeley: North Atlantic Books.

Rothgerber, H. and D.L. Rosenfeld (2021). Meat-related cognitive dissonance: the social psychology of eating animals. *Social and Personality Psychology Compass* 15(5): e12592. https://doi.org/10.1111/spc3.12592.

Rowlatt, J. (2020). Humans waging "suicidal war" on nature – UN chief Antonio Guterres. *BBC News*, 2 December. https://tinyurl.com/4476bdsa.

Roy-Dufresne, E., T. Logan, J.A. Simon, G.L. Chmura and V. Millien (2013). Poleward expansion of the white-footed mouse (*Peromyscus leucopus*) under climate change: implications for the spread of Lyme disease. *PLOS One* 8(11): e80724. https://doi.org/10.1371/journal.pone.0080724.

Royle, C. (2016). Marxism and the Anthropocene. *International Socialism*. 151: 63–84.

Ruby, M.B. and S.J. Heine (2012). Too close to home: factors predicting meat avoidance. *Appetite* 59(1): 47–52. https://doi.org/10.1016/j.appet.2012.03.020.

Ruby, M.B. and S.J. Heine (2011). Meat, morals, and masculinity. *Appetite* 56(2): 447–50. https://doi.org/10.1016/j.appet.2011.01.018.

Rudy, A.P. (2019). On misunderstanding the second contradiction thesis. *Capitalism Nature Socialism* 30(4): 17–35. https://doi.org/10.1080/10455752.2019.1652663.

Russell, C. (2019). An intersectional approach to teaching and learning about humans and other animals in educational contexts. In T. Lloro-Bidart and V.S. Banschbach, eds. *Animals in Environmental Education: Interdisciplinary Approaches to Curriculum and Pedagogy*, 35–52. London: Palgrave.

Ryder, R.D. (1975). *Victims of Science: The Use of Animals in Research*. London: Davis-Poynter.

Saari, M.H. (2020). Re-examining the human-nonhuman animal relationship through humane education. In A. Cutter-Mackenzie-Knowles, K. Malone and E.B. Hacking, eds. *Research Handbook on Childhoodnature: Assemblages of Childhood and Nature Research*, 1263–73. Heidelberg: Springer.

Salleh, A. (2010). From metabolic rift to "metabolic value": reflections on environmental sociology and the alternative globalization movement. *Organization and Environment* 23(2): 205–19. https://doi.org/10.1177/1086026610372134.

Sanbonmatsu, J. (2005). Listen, ecological Marxist! (Yes, I said animals!). *Capitalism Nature Socialism* 16(2): 107–14. https://doi.org/10.1080/10455750500108385.

Sánchez-Bayo, F. and K.A.G. Wyckhuys (2019). Worldwide decline of the entomofauna: a review of its drivers. *Biological Conservation* 232: 8–27. https://doi.org/10.1016/j.biocon.2019.01.020.

Sandberg, M. (2021). Sufficiency transitions: a review of consumption changes for environmental sustainability. *Journal of Cleaner Production* 293: 126097. https://doi.org/10.1016/j.jclepro.2021.126097.

Sandel, M. (2012). *What Money Can't Buy.* London: Penguin.

Sans, P. and P. Combris (2015). World meat consumption patterns: an overview of the last fifty years (1961–2011). *Meat Science* 109: 106–11. https://doi.org/10.1016/j.meatsci.2015.05.012.

Sarmento, M.J., R.D.C. Marchi and G.D.P. Trevisan (2018). Beyond the modern "norm" of childhood: children at the margins as a challenge for the sociology of childhood. In C. Baraldi and T. Cockburn, eds. *Theorising Childhood,* 135–57. London: Palgrave.

Sasgen, I., B. Wouters, A.S. Gardner, M.D. King, M. Tedesco, F.W. Landerer et al. (2020). Return to rapid ice loss in Greenland and record loss in 2019 detected by the GRACE-FO satellites. *Communications Earth and Environment* 1(8). https://doi.org/10.1038/s43247-020-0010-1.

Sayer, A. (2013). Power, sustainability and well-being. In E. Shove and N. Spurling, eds. *Sustainable Practices: Social Theory and Climate Change,* 167–80. London: Routledge.

Scarborough, P., S. Allender, D. Clarke, K. Wickramasinghe and M. Rayner (2012). Modelling the health impact of environmentally sustainable dietary scenarios in the UK. *European Journal of Clinical Nutrition* 66(6): 710–15. https://doi.org/10.1038/ejcn.2012.34.

Scarborough, P., P.N. Appleby, A. Mizdrak, A.D.M. Briggs, R.C. Travis, K.E. Bradbury and T.J. Key (2014). Dietary greenhouse gas emissions of meat-eaters, fish-eaters, vegetarians and vegans in the UK. *Climatic Change* 125(2): 179–92. https://doi.org/10.1007/s10584-014-1169-1.

Schanbacher, W.D. (2019). *Food as a Human Right: Combatting Global Hunger and Forging a Path to Food Sovereignty.* Santa Barbara: ABC-CLIO.

Schatzki, T. (2018). On practice theory; or, What's practices got to do (got to do) with it? In C. Edwards-Groves, P. Grootenboer and J. Wilkinson, eds. *Education in an Era of Schooling,* 151–65. Heidelberg: Springer.

Schatzki, T. (2002). *The Site of the Social: A Philosophical Account of the Constitution of Social Life and Change.* University Park: Pennsylvania State University Press.

Schatzki, T. (1996). *Social Practices: A Wittgensteinian Approach to Human Activity and the Social.* Cambridge, UK: Cambridge University Press.

Schlottmann, C. and J. Sebo (2019). *Food, Animals, and the Environment: An Ethical Approach.* London: Routledge.

Schmid, B. and T.S.J. Smith (2021). Social transformation and postcapitalist possibility: emerging dialogues between practice theory and diverse economies. *Progress in Human Geography* 45(2): 253–75. https://doi.org/10.1177/0309132520905642.

Schoonebeek, M.V. (2015). What are children's views on speciesism? A literature review and personal journey. *Early Education* 58: 24–26.

Schulz, K.A. (2017). Decolonising the Anthropocene: the mytho-politics of human mastery. *E-International Relations*, 1 July. https://www.e-ir.info/2017/07/01/decolonising-the-anthropocene-the-mytho-politics-of-human-mastery/.

Schwabe, C.W. (1994). Animals in the ancient world. In A. Manning and J. Serpell eds. *Animals and Human Society: Changing Perspectives*, 36–58. London: Routledge.

Schwingshackl, L., B. Watzl and J.J. Meerpohl (2020). The healthiness and sustainability of food based dietary guidelines. *BMJ* 370: m2417. https://doi.org/10.1136/bmj.m2417.

Scott, K., C. Bakker and J. Quist (2011). Designing change by living change. *Design Studies* 33: 279–97.

Scotton, G. (2017). Duties to socialise with domesticated animals: farmed animal sanctuaries as frontiers of friendship. *Animal Studies Journal* 6(2): 86–108.

SDC (Sustainable Development Commission) (2009). *Setting the Table: Advice to Government on Priority Elements of Sustainable Diets*. London: SDC.

Searchinger, T., S. Wirsenius, T. Beringer and P. Dumas (2018). Assessing the efficiency of changes in land use for mitigating climate change. *Nature* 564(7735): 249–53. https://doi.org/10.1038/s41586-018-0757-z.

Sebo, J. (2022). *Saving Animals, Saving Ourselves: Why Animals Matter for Pandemics, Climate Change, and Other Catastrophes*. Oxford: Oxford University Press.

Secretariat of the Convention on Biological Diversity (2020). *Global Biodiversity Outlook 5*. Montreal: CBD and United Nations Environment Programme.

Shaw, C. (2015). *The Two Degrees Dangerous Limit for Climate Change: Public Understanding and Decision Making*. London: Routledge.

Shepon, A., G. Eshel, E. Noor and R. Milo (2018). The opportunity cost of animal-based diets exceeds all food losses. *Proceedings of the National Academy of Sciences* 115(15): 3804–9. https://tinyurl.com/yck8b25r.

Shove, E. (2010). Beyond the ABC: climate change policy and theories of social change. *Environment and Planning A: Economy and Space* 42(6): 1273–85. https://doi.org/10.1068/a42282.

Shove, E. (2003). *Comfort, Cleanliness and Convenience: The Social Organization of Normality*. London: Berg.

Shove, E., M. Pantzar and M. Watson (2012). *The Dynamics of Social Practice: Everyday Life and How It Changes*. London: Sage.

Shove, E. and N. Spurling (2013). *Sustainable Practices: Social Theory and Climate Change*. London: Routledge.

Shove, E. and F. Trentmann (2018). *Infrastructures in Practice: The Dynamics of Demand in Networked Societies*. London: Routledge.

Shove, E., M. Watson, M. Hand and J. Ingram (2007). *The Design of Everyday Life*. London: Berg Publishers.

Shrubsole, G. (2019). *Who Owns England? How We Lost Our Green and Pleasant Land, and How to Take It Back*. London: HarperCollins UK.

Signal, T., N. Taylor, K. Prentice, M. McDade and K.J. Burke (2017). Going to the dogs: a quasi-experimental assessment of animal assisted therapy for children who have experienced abuse. *Applied Developmental Science* 21(2): 81–93. https://doi.org/10.1080/10888691.2016.1165098.

Simpson, M. (2020). The Anthropocene as colonial discourse. *Environment and Planning D: Society and Space* 38(1): 53–71. https://doi.org/10.1177/0263775818764679.

Sinervo, B., F. Méndez-de-la-Cruz, D.B. Miles, B. Heulin, E. Bastiaans, M. Villagrán-Santa Cruz et al. (2010). Erosion of lizard diversity by climate change and altered thermal niches. *Science* 328(5980): 894–99. https://doi.org/10.1126/science.1184695.

Singer, H. (2016). Writing the fleischgeist. *Animal Studies Journal* 5(2): 183–201.

Singer, P. (2016). Why speciesism is wrong: a response to Kagan. *Journal of Applied Philosophy* 33(1): 31–35. https://doi.org/10.1111/japp.12165.

Siraj-Blatchford, J., C. Mogharreban and E. Park, eds (2016). *International Research on Education for Sustainable Development in Early Childhood*. Heidelberg: Springer.

Slater, T., A.E. Hogg and R. Mottram (2020). Ice-sheet losses track high-end sea-level rise projections. *Nature Climate Change* 10(10): 879–81. https://doi.org/10.1038/s41558-020-0893-y.

Slocum, R. (2004). Polar bears and energy-efficient lightbulbs: strategies to bring climate change home. *Environment and Planning D: Society and Space* 22(3): 413–38. https://doi.org/10.1068/d378.

Smetana, S., A. Mathys, A. Knoch and V. Heinz (2015). Meat alternatives: life cycle assessment of most known meat substitutes. *International Journal of Life Cycle Assessment* 20(9): 1254–67. https://doi.org/10.1007/s11367-015-0931-6.

Smil, V. (2011). Harvesting the biosphere: the human impact. *Population and Development Review* 37(4): 613–36. https://tinyurl.com/ysmsdk4y.

Smith, A. and R. Raven (2012). What is protective space? Reconsidering niches in transitions to sustainability. *Research Policy* 41(6): 1025–36. https://doi.org/10.1016/j.respol.2011.12.012.

Sobal, J. (2005). Men, meat, and marriage: models of masculinity. *Food and Foodways* 13(1–2): 135–58. https://doi.org/10.1080/07409710590915409.

Sobel, D. (1996). *Beyond Ecophobia*. Great Barrington: Orion Society.

Sollund, R. (2022). Wildlife trade and law enforcement: a proposal for a remodeling of CITES incorporating species justice, ecojustice, and environmental justice. *International Journal of Offender Therapy and Comparative Criminology* 66(9): 1017–35. https://doi.org/10.1177/0306624X221099492.

Soper, K. (2008). Alternative hedonism, cultural theory and the role of aesthetic revisioning. *Cultural Studies* 22(5): 567–87.

Southerton, D. (2013). Habits, routines and temporalities of consumption: from individual behaviours to the reproduction of everyday practices. *Time and Society* 22(3): 335–55. https://doi.org/10.1177/0961463X12464228.

Southerton, D. (2006). Analysing the temporal organisation of daily life: social constraints, practices and their allocation. *Sociology* 40(3): 435–54.

Specht, J. (2019). *Red Meat Republic: A Hoof-to-Table History of How Beef Changed America*. Princeton: Princeton University Press.

Springmann, M., M. Clark, D. Mason-D'Croz, K. Wiebe, B.L. Bodirsky, L. Lassaletta et al. (2018). Options for keeping the food system within environmental limits. *Nature* 562(7728): 519–25. https://doi.org/10.1038/s41586-018-0594-0.

Springmann, M. and F. Freund (2022). Options for reforming agricultural subsidies from health, climate, and economic perspectives. *Nature Communications* 13(1): 82. https://doi.org/10.1038/s41467-021-27645-2.

Springmann, M., C.J. Godfray, M. Rayner and P. Scarborough (2016). Analysis and valuation of the health and climate change co-benefits of dietary change. *Proceedings of the National Academy of Sciences* 113(15): 4146–51. https://doi.org/10.1073/pnas.1523119113.

Springmann, M., L. Spajic, M.A. Clark, J. Poore, A. Herforth, P. Webb et al. (2020). The healthiness and sustainability of national and global food based dietary guidelines: modelling study. *BMJ* 370: m2322. https://doi.org/10.1136/bmj.m2322.

Spurling, N., A. McMeekin, E. Shove, D. Southerton and D. Welch (2013). *Interventions in Practice: Re-framing Policy Approaches to Consumer Behaviour*. Manchester: Sustainable Practices Research Group, University of Manchester.

Staal, A., I. Fetzer, L. Wang-Erlandsson, J.H.C. Bosmans, S.C. Dekker, E.H. van Nes et al. (2020). Hysteresis of tropical forests in the 21st century. *Nature Communications* 11(1): 4978. https://doi.org/10.1038/s41467-020-18728-7.

Stainforth, T. and B. Brzezinski (2020). More than half of all CO_2 emissions since 1751 emitted in the last 30 years. Institute for European Environmental Policy, 29 April. https://tinyurl.com/au9a64fv.

Standen, A. and S. Wizansky (2007). Hello from *Meatpaper*. *Meatpaper* 1. http://meatpaper.com/articles/2007/1217_edletter.html.

Stănescu, V. (2021). Birds of a feather: animal liberation, veganism and social justice. Presentation to the 6th international animal rights conference "Animal Futures. Animal rights in activism and academia", organized by Loomus and Eesti Vegan Selts. Tallinn, Estonia. 5–6 May. https://tinyurl.com/yckvzyuz.

Stănescu, V. (2019). "Cowgate": meat eating and climate change denial. In N. Almiron and J. Xifra, eds. *Climate Change Denial and Public Relations: Strategic Communication and Interest Groups in Climate Inaction*, 178–94. London: Routledge.

Stănescu, V. (2018). "White power milk": milk, dietary racism, and the "alt-right". *Animal Studies Journal* 7(2): 103–28.

Stănescu, V. (2010). "Green" eggs and ham? The myth of sustainable meat and the danger of the local. *Journal for Critical Animal Studies* 8(1–2): 8–32.

Steffen, W., P. Crutzen and J.R. McNeill (2007). The Anthropocene: are humans now overwhelming the great forces of nature? *Ambio* 36: 614–21.

Steffen, W., J. Grinevald, P. Crutzen and J. McNeill (2011). The Anthropocene: conceptual and historical perspectives. *Philosophical Transactions of the Royal Society A: Mathematical, Physical and Engineering Sciences* 369(1938): 842–67. https://doi.org/10.1098/rsta.2010.0327.

Stehfest, E., L. Bouwman, D.P. van Vuuren, M.G.J. den Elzen, B. Eickhout and P. Kabat (2009). Climate benefits of changing diet. *Climatic Change* 95(1): 83–102. https://doi.org/10.1007/s10584-008-9534-6.

Steinfeld, H. and P. Gerber (2010). Livestock production and the global environment: consume less or produce better? *Proceedings of the National Academy of Sciences* 107(43): 18237–38. https://doi.org/10.1073/pnas.1012541107.

Steinfeld, H., P. Gerber, T.D. Wassenaar, V. Castel, M. Rosales and C. de Haan (2006). *Livestock's Long Shadow: Environmental Issues and Options*. Rome: Food and Agriculture Organization of the United Nations.

Stephens, N. (2013). Growing meat in laboratories: the promise, ontology, and ethical boundary-work of using muscle cells to make food. *Configurations* 21(2): 159–81.

Stephens, N. (2010). In vitro meat: zombies on the menu? *SCRIPTed* 7: 394–401.

Stephens, N., L. Di Silvio, I. Dunsford, M. Ellis, A. Glencross and A. Sexton (2018). Bringing cultured meat to market: technical, socio-political, and regulatory challenges in cellular agriculture. *Trends in Food Science and Technology* 78: 155–66. https://doi.org/10.1016/j.tifs.2018.04.010.

Stephens, N., A.E. Sexton and C. Driessen (2019). Making sense of making meat: key moments in the first 20 years of tissue engineering muscle to make food. *Frontiers in Sustainable Food Systems* 3. https://tinyurl.com/2tm4h8dh.

Stewart, C., C. Piernas, B. Cook and S.A. Jebb (2021). Trends in UK meat consumption: analysis of data from years 1–11 (2008–09 to 2018–19) of the National Diet and Nutrition Survey rolling programme. *Lancet Planetary Health* 5(10): e699–e708. https://doi.org/10.1016/S2542-5196(21)00228-X.

Stewart, K. and M. Cole (2020). Veganism has always been more about living an ethical life than just avoiding meat and dairy. *The Conversation*, 7 January. https://tinyurl.com/mf6axya6.

Stiasny, M.H., M. Sswat, F.H. Mittermayer, I.-B. Falk-Petersen, N.K. Schnell, V. Puvanendran et al. (2019). Divergent responses of Atlantic cod to ocean acidification and food limitation. *Global Change Biology* 25(3): 839–49. https://doi.org/10.1111/gcb.14554.

Stoddard, I., K. Anderson, S. Capstick, W. Carton, J. Depledge, K. Facer et al. (2021). Three decades of climate mitigation: why haven't we bent the global emissions curve? *Annual Review of Environment and Resources* 46(1): 653–89. https://doi.org/10.1146/annurev-environ-012220-011104.

Stoll-Kleemann, S. and T. O'Riordan (2015). The sustainability challenges of our meat and dairy diets. *Environment: Science and Policy for Sustainable Development* 57(3): 34–48. https://doi.org/10.1080/00139157.2015.1025644.

Stoppani, A. (1873). Corso di Geologia, Milan: G. Bernardoni and E.G. Brigola. Excerpt titled "First period of the Anthropozoic era", V. Federighi, trans., in E. Turpin and V. Federighi, eds (2012). *A new element, a new force, a new input: Antonio Stoppani's Anthropozoic*, 36–41. In E. Ellsworth and J. Kruse, eds. *Making the Geologic Now: Responses to Material Conditions of Contemporary Life*, 34–41. Brooklyn: Punctum Books.

Stuart, D. and R. Gunderson (2020). Human-animal relations in the Capitalocene: environmental impacts and alternatives. *Environmental Sociology* 6(1): 68–81. https://doi.org/10.1080/23251042.2019.1666784.

Suliman, S., C. Farbotko, H. Ransan-Cooper, K.E. McNamara, F. Thornton, C. McMichael et al. (2019). Indigenous (im)mobilities in the Anthropocene. *Mobilities* 14(3): 298–318. https://doi.org/10.1080/17450101.2019.1601828.

Sully, S., D.E. Burkepile, M.K. Donovan, G. Hodgson and R. van Woesik (2019). A global analysis of coral bleaching over the past two decades. *Nature Communications* 10: 1264. https://doi.org/10.1038/s41467-019-09238-2.

Sumpter, K.C. (2015). Masculinity and meat consumption: an analysis through the theoretical lens of hegemonic masculinity and alternative masculinity theories. *Sociology Compass* 9(2): 104–14.

Sun, Z., L. Scherer, A. Tukker, S.A. Spawn-Lee, M. Bruckner, H.K. Gibbs et al. (2022). Dietary change in high-income nations alone can lead to substantial double climate dividend. *Nature Food* 3(1): 29–37. https://doi.org/10.1038/s43016-021-00431-5.

Sushma, M. (2019). Avian botulism killed 18,000 birds at Sambhar: Govt report. *Down to Earth*, 21 November. https://tinyurl.com/539y7djs.

Swain, M., L. Blomqvist, J. McNamara and W.J. Ripple (2018). Reducing the environmental impact of global diets. *Science of the Total Environment* 610–11: 1207–9. https://doi.org/10.1016/j.scitotenv.2017.08.125.

Swyngedouw, E. (2013). Apocalypse now! Fear and doomsday pleasures. *Capitalism Nature Socialism* 24(1): 9–18. https://doi.org/10.1080/10455752.2012.759252.

Swyngedouw, E. (2010). Apocalypse forever? *Theory, Culture and Society* 27(2–3): 213–32. https://doi.org/10.1177/0263276409358728.

Symes, W.S., D.P. Edwards, J. Miettinen, F.E. Rheindt and L.R. Carrasco (2018). Combined impacts of deforestation and wildlife trade on tropical biodiversity are severely underestimated. *Nature Communications* 9(1): 4052. https://doi.org/10.1038/s41467-018-06579-2.

Tanaka, K.R., K.S. Van Houtan, E. Mailander, B.S. Dias, C. Galginaitis, J. O'Sullivan et al. (2021). North Pacific warming shifts the juvenile range of a marine apex predator. *Scientific Reports* 11(1): 3373. https://doi.org/10.1038/s41598-021-82424-9.

Taylor, A. and S. Taylor (2022). Our animals, ourselves: the socialist feminist case for animal liberation. *Lux* 3. https://tinyurl.com/2yrz5875.

Taylor, N. and R. Twine eds (2014). *The Rise of Critical Animal Studies: From the Margins to the Centre*. London: Routledge.

Thacker, E. (2003). Black magic, biotech and dark markets. In Sarai Collective, ed. *Sarai Reader 03: Shaping Technologies*, 134–43. Delhi: Centre for the Study of Developing Societies.

Thaler, R.H. and C.R. Sunstein (2008). *Nudge: Improving Decisions about Health, Wealth and Happiness*. New Haven: Yale University Press.

The Times (2019). The Times view on the Anthropocene epoch: our man-made world. *Times*, 27 May. https://tinyurl.com/yswe8nba.

The Vegan Society (2021a). *Planting Value in the Food System*. Birmingham: The Vegan Society. https://www.plantingvalueinfood.org.

References

The Vegan Society (2021b). Scottish Government agrees to add non-dairy milk to free nursery milk scheme. 24 February. https://tinyurl.com/yckrf8tv.

Theurl, M.C., C. Lauk, G. Kalt, A. Mayer, K. Kaltenegger, T.G. Morais et al. (2020). Food systems in a zero-deforestation world: dietary change is more important than intensification for climate targets in 2050. *Science of the Total Environment* 735: 139353. https://doi.org/10.1016/j.scitotenv.2020.139353.

Thomas, C.D., T.H. Jones and S.E. Hartley (2019). "Insectageddon": a call for more robust data and rigorous analyses. *Global Change Biology* 25(6): 1891–92. https://doi.org/10.1111/gcb.14608.

Thomas, T. (2020). Arctic wildfires emit 35% more CO_2 so far in 2020 than for whole of 2019. *Guardian*, 31 August. https://tinyurl.com/2p93w7fa.

Thornalley, D.J.R., D.W. Oppo, P. Ortega, J.I. Robson, C.M. Brierley, R. Davis et al. (2018). Anomalously weak Labrador Sea convection and Atlantic overturning during the past 150 years. *Nature* 556(7700): 227–30. https://doi.org/10.1038/s41586-018-0007-4.

Thurston, A.M., H. Stöckl and M. Ranganathan (2021). Natural hazards, disasters and violence against women and girls: a global mixed-methods systematic review. *BMJ Global Health* 6(4): e004377. https://tinyurl.com/59r4m24s.

Tilman, D. and M. Clark (2014). Global diets link environmental sustainability and human health. *Nature* 515(7528): 518–22. https://doi.org/10.1038/nature13959.

Tomas, W.M., C.N. Berlinck, R.M. Chiaravalloti, G.P. Faggioni, C. Strüssmann, R. Libonati et al. (2021). Distance sampling surveys reveal 17 million vertebrates directly killed by the 2020's wildfires in the Pantanal, Brazil. *Scientific Reports* 11(1): 23547. https://doi.org/10.1038/s41598-021-02844-5.

Tomlinson, I. (2013). Doubling food production to feed the 9 billion: a critical perspective on a key discourse of food security in the UK. *Journal of Rural Studies* 29: 81–90. https://doi.org/10.1016/j.jrurstud.2011.09.001.

Torjesen, I. (2019). WHO pulls support from initiative promoting global move to plant-based foods. *BMJ* 365: l1700. https://doi.org/10.1136/bmj.l1700.

Törnberg, A. (2021). Prefigurative politics and social change: a typology drawing on transition studies. *Distinktion: Journal of Social Theory* 22(1): 83–107. https://doi.org/10.1080/1600910X.2020.1856161.

Treich, N. (2022). The Dasgupta Review and the problem of anthropocentrism. *Environmental and Resource Economics* 83: 973–97. https://doi.org/10.1007/s10640-022-00663-4.

Tremmel, J.C. and K. Robinson (2014). *Climate Ethics: Environmental Justice and Climate Change*. London: Bloomsbury Publishing.

Trisos, C.H., C. Merow and A.L. Pigot (2020). The projected timing of abrupt ecological disruption from climate change. *Nature* 580(7804): 496–501. https://doi.org/10.1038/s41586-020-2189-9.

Tubb, C. and T. Seba (2019). *The Second Domestication of Plants and Animals, the Disruption of the Cow, and the Collapse of Industrial Livestock Farming.* San Francisco: Rethinkx.

Tucker, C.A. (2019). Food practices of environmentally conscientious New Zealanders. *Environmental Sociology* 5(1): 82–92. https://doi.org/10.1080/23251042.2018.1495038.

Turner-McGrievy, G.M., A.M. Leach, S. Wilcox and E.A. Frongillo (2016). Differences in environmental impact and food expenditures of four different plant-based diets and an omnivorous diet: results of a randomized, controlled intervention. *Journal of Hunger and Environmental Nutrition* 11(3): 382–95. https://doi.org/10.1080/19320248.2015.1066734.

Twine, R. (2023). Where are the nonhuman animals in the sociology of climate change? *Society and Animals* 31: 105–30. https://doi:10.1163/15685306-bja10025.

Twine, R. (2021a). Masculinity, nature, ecofeminism, and the "Anthropo"cene. In P. Pulé and M. Hultman, eds. *Men, Masculinities, and Earth: Contending with the (m)Anthropocene*, 117–33. London: Palgrave Macmillan.

Twine, R. (2021b). Emissions from animal agriculture: 16.5% is the new minimum figure. *Sustainability* 13(11): 6276. https://doi.org/10.3390/su13116276.

Twine, R. (2018). Materially constituting a sustainable food transition: the case of vegan eating practice. *Sociology* 52(1): 166–81. https://doi.org/10.1177/0038038517726647.

Twine, R. (2017). A practice theory framework for understanding vegan transition. *Animal Studies Journal* 6(2): 192–224.

Twine, R. (2015). Understanding snacking through a practice theory lens. *Sociology of Health and Illness* 37(8): 1270–84. https://bit.ly/48NE2al.

Twine, R. (2014a). Ecofeminism and veganism: revisiting the question of universalism. In L. Gruen and C.J. Adams, eds. *Ecofeminism: Feminist Intersections with Other Animals and the Earth*, 191–207. London: Bloomsbury Academic Press.

Twine, R. (2014b). Vegan killjoys at the table: contesting happiness and negotiating relationships with food practices. *Societies* 4(4): 623–39. https://bit.ly/47xz9kP.

Twine, R. (2013a). Is biotechnology deconstructing animal domestication? Movements toward liberation. *Configurations* 21(2): 135–58. https://bit.ly/3RVoXfR.

Twine, R. (2013b). Animals on drugs: understanding the role of pharmaceutical companies in the animal-industrial complex. *Journal of Bioethical Inquiry* 10(4): 505–14. https://doi.org/10.1007/s11673-013-9476-1.

Twine, R. (2012). Revealing the "animal-industrial complex": a concept and method for critical animal studies? *Journal for Critical Animal Studies* 10(1): 12–39.

Twine, R. (2010a). *Animals as Biotechnology: Ethics, Sustainability, and Critical Animal Studies*. London: Routledge.

Twine, R. (2010b). Genomic natures read through posthumanisms. *Sociological Review* 58(1_Suppl): 175–95. https://tinyurl.com/4wwdw48x.

Twine, R. (2010c). Intersectional disgust? Animals and (eco)feminism. *Feminism and Psychology* 20(3): 397–406.

Twine, R. (2007). Thinking across species: a critical bioethics approach to enhancement. *Theoretical Medicine and Bioethics* 28(6): 509–23.

Twine, R. (2005). Constructing a critical bioethics by deconstructing culture/ nature dualism. *Medicine, Health Care and Philosophy* 8(3): 285–95.

Twine, R. (2002). Ecofeminism and the "new" sociologies: a collaboration against dualism. Unpublished doctoral dissertation. Manchester Metropolitan University.

Twine, R. (2001). Ma(r)king essence: ecofeminism and embodiment. *Ethics and the Environment* 6(2): 31–58.

Twine, R. (1997). Masculinity, nature, ecofeminism. Personal website. http://richardtwine.com/ecofem/masc.pdf.

UK Government (2021). *Net Zero Strategy: Build Back Greener*. UK Government. https://www.gov.uk/government/publications/net-zero-strategy.

UKHACC (UK Health Alliance on Climate Change) (2020). *All-consuming: Building a Healthier Food System for People and Planet*. London: UK Health Alliance on Climate Change.

UN Nutrition (2021). *Livestock-derived Foods and Sustainable Healthy Diets*. Rome: United Nations Nutrition Secretariat.

UNCED (United Nations Conference on Environment and Development) (1993). Agenda 21. In *Report of the United Nations Conference on Environment and Development*, Rio de Janeiro, 3–14 June 1992, vol. 1, Resolutions Adopted by the Conference, annex 2. New York: United Nations.

UNEP (United Nations Environment Programme) (2021a) Methane emissions are driving climate change. Here's how to reduce them. 20 August. https://tinyurl.com/n7cefjk6.

UNEP (United Nations Environment Programme) (2021b). *Making Peace with Nature: A Scientific Blueprint to Tackle the Climate, Biodiversity, and Pollution*

Emergencies. [Baste, I.A., R.T. Watson, K.I. Brauman, C. Samper and C. Walzer]. Rome: UNEP.

UNEP (United Nations Environment Programme) (2020). *Emissions Gap Report 2020.* Rome: UNEP.

UNEP (United Nations Environment Programme) (2013). *Our Nutrient World: The Challenge to Produce More Food and Energy with Less Pollution.* [Sutton, M.A., A. Bleeker, C.M. Howard, M. Bekunda, B. Grizzetti, W. de Vries et al.]. Edinburgh: Centre for Ecology and Hydrology and United Nations Environment Programme.

UNEP (United Nations Environment Programme) (2012). *The Critical Role of Global Food Consumption Patterns in Achieving Sustainable Food Systems and Food for All.* Rome: UNEP.

UNEP (United Nations Environment Programme) (2010). *Assessing the Environmental Impacts of Consumption and Production: Priority Products and Materials.* Report of the Working Group on the Environmental Impacts of Products and Materials to the International Panel for Sustainable Resource Management. Rome: UNEP.

UNEP (United Nations Environment Programme) (2009). *The Environmental Food Crisis: The Environment's Role in Averting Future Food Crises: A UNEP Rapid Response Assessment.* United Nations Environment Programme and Earthprint. Rome: UNEP.

Unigwe, C. (2019). It's not just Greta Thunberg: Why are we ignoring the developing world's inspiring activists? *Guardian,* 5 October. https://tinyurl.com/2fcvnp4z.

United Nations General Assembly (2010). *Report Submitted by the Special Rapporteur on the Right to Food: Human Rights Council Sixteenth Session; Agenda Item 3, Promotion and Protection of All Human Rights, Civil, Political, Economic, Social and Cultural Rights, Including the Right to Develop.* Rome: United Nations General Assembly.

Urry, J. (2011). *Climate Change and Society.* Cambridge, UK: Polity Press.

Van Dooren, C., M. Marinussen, H. Blonk, H. Aiking and P. Vellinga (2014). Exploring dietary guidelines based on ecological and nutritional values: a comparison of six dietary patterns. *Food Policy* 44: 36–46. https://bit.ly/3vs5Ked.

Van Gils, J.A., S. Lisovski, T. Lok, W. Meissner, A. Ożarowska, J. de Fouw et al. (2016). Body shrinkage due to Arctic warming reduces red knot fitness in tropical wintering range. *Science* 352(6287): 819–21. https://bit.ly/47tNxdO.

VanDerWal, J., H.T. Murphy, A.S. Kutt, G.C. Perkins, B.L. Bateman, J.J. Perry et al. (2013). Focus on poleward shifts in species' distribution underestimates the fingerprint of climate change. *Nature Climate Change* 3(3): 239–43. https://doi.org/10.1038/nclimate1688.

Vargas, S.P., P.J. Castro-Carrasco, N.A. Rust and J.L. Riveros F. (2021). Climate change contributing to conflicts between livestock farming and guanaco conservation in central Chile: a subjective theories approach. *Oryx* 55(2): 275–83. https://doi.org/10.1017/S0030605319000838.

Vaughan, A. (2015). One in six of world's species faces extinction due to climate change. *Guardian*, 30 April. https://tinyurl.com/2p9ctsn8.

Vegconomist (2021). Landmark political agreement in Denmark: $90 million will go to farmers who produce plant-based foods. *Vegconomist*, 5 October. https://tinyurl.com/4fu5hsbf.

Vergès, F. (2017). Racial Capitalocene. In T. Johnson and A. Lubin, eds, *Futures of Black Radicalism*, 72–82. London: Verso.

Vinnari, M. and E. Vinnari (2014). A framework for sustainability transition: the case of plant-based diets. *Journal of Agricultural and Environmental Ethics* 27(3): 369–96. https://doi.org/10.1007/s10806-013-9468-5.

Wadiwel, D. (2015). *The War against Animals*. Leiden: Brill.

Wall, J. (2022). From childhood studies to childism: reconstructing the scholarly and social imaginations. *Children's Geographies* 20(3): 257–70. https://bit.ly/47vdkC7.

Walsh, D.F. (1998). Structure/agency. In C. Jenks, ed., *Core Sociological Dichotomies*, 8–33. London: Sage.

Wang, X. and K. Lo (2021). Just transition: a conceptual review. *Energy Research and Social Science* 82: 102291. https://doi.org/10.1016/j.erss.2021.102291.

Warde, A. (2016). *The Practice of Eating*. New York: John Wiley & Sons.

Watson, M. (2017). Placing power in practice theory. In A. Hui, T. Schatzki and E. Shove, eds. *The Nexus of Practices: Connections, Constellations, Practitioners*, 169–82. London: Routledge.

Watson, M. (2012). How theories of practice can inform transition to a decarbonised transport system. *Journal of Transport Geography* 24: 488–96. https://doi.org/10.1016/j.jtrangeo.2012.04.002.

Watts, J., G. Blight, L. McMullan and P. Gutiérrez (2019). Half a century of dither and denial – a climate crisis timeline. *Guardian*, 9 October. https://tinyurl.com/y9dzwpme.

Watts, N., M. Amann, N. Arnell, S. Ayeb-Karlsson, J. Beagley, K. Belesova et al. (2021). The 2020 report of the *Lancet* countdown on health and climate change: responding to converging crises. *Lancet* 397(10269): 129–70. https://doi.org/10.1016/S0140-6736(20)32290-X.

WCED (World Commission on Environment and Development) (1987). *Our Common Future*. Oxford: Oxford University Press.

Weis, T. (2007). *The Global Food Economy: The Battle for the Future of Farming*. London: Zed Books.

Welch, D. and L. Yates (2018). The practices of collective action: practice theory, sustainability transitions and social change. *Journal for the Theory of Social Behaviour* 48(3): 288–305. https://doi.org/10.1111/jtsb.12168.

Weldemariam, K. (2017). Challenging and expanding the notion of sustainability within early childhood education: perspectives from post-humanism and/or new materialism. In O. Franck and C. Osbeck, eds, *Ethical Literacies and Education for Sustainable Development: Young People, Subjectivity and Democratic Participation*, 105–26. Heidelberg: Springer.

Weldemariam, K., D. Boyd, N. Hirst, B.M. Sageidet, J.K. Browder, L. Grogan et al. (2017). A critical analysis of concepts associated with sustainability in early childhood curriculum frameworks across five national contexts. *International Journal of Early Childhood* 49(3): 333–51. https://bit.ly/4aOZ4ap.

Welsh, T. (2021). Despite pressure, WHO review keeps status quo malnutrition treatment. *Devex*, 19 March. https://tinyurl.com/kmtmaj6d.

Weston, S. (2020). "The losses could be profound": how floods are wreaking havoc on wildlife. *Guardian*, 1 April. https://tinyurl.com/2e99umcz.

Whatmore, S., P. Stassart and H. Renting (2003). What's alternative about alternative food networks? *Environment and Planning A: Economy and Space* 35: 389–91.

Whitbourn, M. (2022). Animal welfare activists lose High Court challenge over "hidden camera" laws. *Sydney Morning Herald*, 10 August. https://tinyurl.com/ycejkaf2.

White, R. (2018). Looking backward/moving forward: articulating a "Yes, BUT ...!" response to lifestyle veganism, and outlining post-capitalist futures in critical veganic agriculture. *EuropeNow*, 5 September. https://tinyurl.com/2nfusmw9.

White, R.R. and M.B. Hall (2017). Nutritional and greenhouse gas impacts of removing animals from US agriculture. *Proceedings of the National Academy of Sciences* 114(48): E10301–E10308. https://tinyurl.com/2suzu6nh.

Whitley, C.T. and L. Kalof (2014). Animal imagery in the discourse of climate change. *International Journal of Sociology* 44(1): 10–33. https://bit.ly/4aUSOOy.

Wilkie, R.M. (2018). "Minilivestock" farming: who is farming edible insects in Europe and North America? *Journal of Sociology* 54(4): 520–37. https://doi.org/10.1177/1440783318815304.

Wilks, M., L. Caviola, G. Kahane and P. Bloom (2021). Children prioritize humans over animals less than adults do. *Psychological Science* 32(1): 27–38. https://doi.org/10.1177/0956797620960398.

References

Willett, W., J. Rockström, B. Loken, M. Springmann, T. Lang, S. Vermeulen et al. (2019). Summary Report of the EAT-*Lancet* Commission on healthy diets from sustainable food systems. https://bit.ly/3Se288G.

Williston, B. (2018). *The Ethics of Climate Change: An Introduction*. London: Routledge.

Winograd, K., ed. (2016). *Education in Times of Environmental Crises: Teaching Children to be Agents of Change*. London: Routledge.

Wirsenius, S., C. Azar and G. Berndes (2010). How much land is needed for global food production under scenarios of dietary changes and livestock productivity increases in 2030? *Agricultural Systems* 103(9): 621–38. https://doi.org/10.1016/j.agsy.2010.07.005.

Wirsenius, S., F. Hedenus and K. Mohlin (2011). Greenhouse gas taxes on animal food products: rationale, tax scheme and climate mitigation effects. *Climatic Change* 108(1): 159–84. https://doi.org/10.1007/s10584-010-9971-x.

Wolfe, C. (2010). *What Is Posthumanism?* Minneapolis: University of Minnesota Press.

Wolfe, C. (2003). *Animal Rites: American Culture, the Discourse of Species, and Posthumanist Theory*. Chicago: University of Chicago Press.

World Resources Institute (2019). *Creating a Sustainable Food Future: A Menu of Solutions to Feed Nearly 10 Billion People by 2050*. [Searchinger, T., R. Waite, C. Hanson, J. Ranganathan, P. Dumas, E. Matthews et al.]. Washington D.C.: World Resources Institute.

Wright, P.J., J.K. Pinnegar and C. Fox (2020). Impacts of climate change on fish, relevant to the coastal and marine environment around the UK. *MCCIP Science Review 2020*, 354–81. https://doi.org/10.14465/2020.arc16.fsh.

WSPA (World Society for the Protection of Animals) (2008). *Eating Our Future: The Environmental Impact of Industrial Animal Agriculture*. London: WSPA.

WWF (World Wildlife Fund) (2022a). *Land of Plenty: A Nature-Positive Pathway to Decarbonise UK Agriculture and Land Use*. WWF-UK.

WWF (World Wildlife Fund) (2022b). *Living Planet Report 2022: Building a Nature-Positive Society*. R.E.A. Almond, M. Grooten and T. Petersen, eds, Gland: WWF.

WWF (World Wildlife Fund) (2020). *Bending the Curve: The Restorative Power of Planet-based Diets*. Gland: WWF.

WWF (World Wildlife Fund) (2018). *Living Planet Report 2018: Aiming Higher*. M. Grooten and R.E.A. Almond, eds, Gland: WWF.

WWF (World Wildlife Fund) (2017). *Appetite for Destruction*. Gland: WWF.

Xu, X., P. Sharma, S. Shu, T.-S. Lin, P. Ciais, F.N. Tubiello et al. (2021). Global greenhouse gas emissions from animal-based foods are twice those of

plant-based foods. *Nature Food* 2(9): 724–32.
https://bit.ly/41QTTCP.

Yates, L. (2015). Rethinking prefiguration: alternatives, micropolitics and goals in social movements. *Social Movement Studies* 14(1): 1–21.
https://doi.org/10.1080/14742837.2013.870883.

Yong, E. (2018). Humans are destroying animals' ancestral knowledge. *Atlantic*, 6 September. https://tinyurl.com/2rca793e.

York, R. (2012). Do alternative energy sources displace fossil fuels? *Nature Climate Change* 2(6): 441–43. https://doi.org/10.1038/nclimate1451.

Young, T. and J. Bone (2020). Troubling intersections of childhood/animals/education: narratives of love, life, and death. In A. Cutter-Mackenzie-Knowles, K. Malone and E.B. Hacking, eds. *Research Handbook on Childhoodnature: Assemblages of Childhood and Nature Research*, 1379–97. Heidelberg: Springer.

Yusoff, K. (2018). *A Billion Black Anthropocenes or None*. Minneapolis: University of Minnesota Press.

Yusoff, K. (2010). Biopolitical economies and the political aesthetics of climate change. *Theory, Culture and Society* 27(2–3): 73–99. https://bit.ly/3Sejd2f.

Zeller, U., N. Starik and T. Göttert (2017). Biodiversity, land use and ecosystem services: an organismic and comparative approach to different geographical regions. *Global Ecology and Conservation* 10: 114–25. https://bit.ly/3NXzYfv.

Zylinska, J. (2018). *The End of Man: A Feminist Counterapocalypse*. Minneapolis: University of Minnesota Press.

Index

www.ingramcontent.com/pod-product-compliance
Lightning Source LLC
Chambersburg PA
CBHW040828300326
41914CB00059B/1257